FROM ATOMIC TO MESOSCALE

The Role of Quantum Coherence in Systems of Various Complexities

FROM ATOMIC TO MESOSCALE

The Role of Quantum Coherence in Systems of Various Complexities

Edited by

Svetlana A. Malinovskaya

Stevens Institute of Technology, USA

Irina Novikova

College of William and Mary, USA

World Scientific

NEW JERSEY · LONDON · SINGAPORE · BEIJING · SHANGHAI · HONG KONG · TAIPEI · CHENNAI

Published by

World Scientific Publishing Co. Pte. Ltd.
5 Toh Tuck Link, Singapore 596224
USA office: 27 Warren Street, Suite 401-402, Hackensack, NJ 07601
UK office: 57 Shelton Street, Covent Garden, London WC2H 9HE

Library of Congress Cataloging-in-Publication Data
From atomic to mesoscale : the role of quantum coherence in systems of various complexities / edited by
Svetlana A. Malinovskaya (Stevens Institute of Technology, USA), Irina Novikova (College of William and
Mary, USA).
    pages cm
  Includes bibliographical references and index.
  ISBN 978-9814678698 (hard cover : alk. paper)
  1. Coherence (Nuclear physics). 2. Quantum systems. I. Malinovskaya, Svetlana A., editor.
II. Novikova, Irina, 1975– editor.
  QC794.6.C58F76 2015
  530.12--dc23

                       2015005911

British Library Cataloguing-in-Publication Data
A catalogue record for this book is available from the British Library.

Foreword

The latest advancements and future directions of atomic, molecular and optical (AMO) physics and their vital role in modern sciences and technologies form the content of this volume. The chapters are devoted to studies of a wide range of quantum systems, with an emphasis on understanding of quantum coherence and other quantum phenomena originated from light-matter interactions. The book intends to survey the current research landscape and to highlight major scientific trends in AMO physics as well as those interfacing with interdisciplinary sciences. The volume may be particularly useful for young researchers working on establishing their scientific interests and goals.

The content of the book was inspired by presentations at the workshop from Atomic to Mesoscale, which was held at the Institute for Theoretical Atomic, Molecular and Optical Physics (ITAMP) in March, 2014. All talks are available on the ITAMP website.

The book is organized in six sections; most of them are a synergy of theory and experiment. Section on "Collective phenomena and long-range interactions in ultracold atoms and molecules" describes the realization of quantum magnetism with ultracold molecules in an optical lattice and reviews the experimental and theoretical progress in this field. This section provides an overview of quantum optics in high-density, cold atomic matter in which disorder-driven electromagnetic interactions can develop strongly correlated character. This section also discusses spin dynamics that occur within spinor Bose-Einstein gases, composed of bosonic atoms. It highlights the advantages of spin-sensitive imaging for understanding and utilizing such dynamics.

The section on "Atom-like coherent solid state systems" introduces the nitrogen vacancy (NV) center, being a quantum defect in diamonds and belonging to cutting edge scientific and technological objects. The imaging of biological cells, cellular temperature gradient measurement and nanoscale electric field sensing are discussed as applications of NV-diamond quantum sensors. This section also presents the studies on self-assembled quantum dots interacting with light to investigate coherence properties and confirm entanglement between a spontaneously emitted photon and an electron spin qubit confined to the quantum dot. The combination of the solid-state nature of quantum dots and their close approximation to atomic systems makes them an attractive platform for quantum information-based technologies.

The section on "Coherent nanophotonics and plasmonics" emphasizes the central role of quantum photonics in the development of quantum computation and communication technologies owing to the high transmission capacity and outstanding low-noise properties of photonic information channels. Theoretical and experimental studies of single-photon sources are presented focusing on the NV center in a diamond placed near a hyperbolic metamaterial. This section also presents the studies of the optical properties of hybrid systems in the linear regime accompanied by strong coupling between the molecules and the surface plasmon polaritons.

The section on "Fundamental Physics" reports on a boson-sampling computational model as the key to a non-universal quantum computer and discusses the feasibility of building a boson-sampling device using existing technology. Moreover, this section introduces a new approach to quantum amplification by superradiant emission of radiation (QASER).

The section on "Ultrafast dynamics in strong laser fields" describes an approach to create attosecond magnetic fields by making use of few cycle circularly polarized attosecond UV pulses produced by HHG. Those pulses generate electronic currents in molecular media on an attosecond time scale, which are the sources of attosecond magnetic pulses. Another chapter of this section reviews theoretical methods to study and predict the response of fullerenes with interior atoms to UV and soft x-ray radiation. Various effects of coherence are discussed here as they are manifested in plasmon resonances, many-body correlations and Auger-intercoulombic hybrid multicenter decays of inner-shell holes.

The section "Ultracold chemistry" presents theoretical investigations of ultracold chemical reactions at a single partial wave level such as barrierless chemical reactions. Studies of highly efficient energy transfer processes are introduced involving rotational and vibrational energy exchange in ultracold molecular collisions.

We hope this overview will be useful and inspirational to many scientists. We would like to acknowledge ITAMP and Dr. Hossein Sadeghpour for support in carrying out the workshop, which became so important in the preparation of the volume. We are grateful to Dr. Christian Buth for technical help at the initial stage of the project. We would also like to thank Dr. Duncan Steel for insightful discussions regarding the volume proposal.

Hoboken, NJ
Williamsburg, VA

Svetlana A. Malinovskaya
Irina Novikova

Contents

I

Collective Phenomena and Long-Range Interactions in Ultracold Atoms and Molecules

Chapter 1

Quantum Magnetism with Ultracold Molecules

M. L. Wall, K. R. A. Hazzard*, and A. M. Rey

*JILA, NIST, Department of Physics, University of Colorado,
440 UCB, Boulder, Colorado 80309, USA*

This article gives an introduction to the realization of effective quantum magnetism with ultracold molecules in an optical lattice, reviews experimental and theoretical progress, and highlights future opportunities opened up by ongoing experiments. Ultracold molecules offer capabilities that are otherwise difficult or impossible to achieve in other effective spin systems, such as long-ranged spin-spin interactions with controllable spatial and spin anisotropy and favorable energy scales. Realizing quantum magnetism with ultracold molecules provides access to rich many-body behaviors, including many exotic phases of matter and interesting excitations and dynamics. Far-from-equilibrium dynamics plays a key role in our exposition, just as it did in recent ultracold molecule experiments realizing effective quantum magnetism. In particular, we show that dynamical probes allow the observation of correlated many-body spin physics, even in polar molecule gases that are not quantum degenerate. After describing how quantum magnetism arises in ultracold molecules and discussing recent observations of quantum magnetism with polar molecules, we survey prospects for the future, ranging from immediate goals to long-term visions.

1. Introduction

The realization of a Bose-Einstein condensate (BEC) in an ultracold, dilute gas of alkali atoms[1–3] was a landmark achievement in several respects. For one, the production of an atomic BEC required significant technical advances in cooling and trapping atoms with electromagnetic radiation, as well as evaporative cooling. In addition, atomic BECs provided the most direct evidence for a many-body phenomenon predicted more than 80 years prior. The fact that nearly all aspects of the atomic system are amenable to experimental control not only enabled the realization of this new state of matter, but also facilitated the study of its properties out of equilibrium, such as the dynamics of vortices[4] and solitons.[5] As we shall see, ultracold molecules parallel this, providing entirely new phases of matter and non-equilibrium behaviors that are otherwise unrealized in ultracold matter.

---

*Current address: Department of Physics, Rice University, Houston, Texas 77005, USA.

Following the realization of a BEC the field of many-body physics with ultracold atoms has grown steadily,[6] including degenerate fermionic gases[7] as well as many non-alkali species.[8–11] A burgeoning subfield of research involves optical lattices standing wave arrangements of laser light that form a periodic potential for atoms or molecules.[12] Such a periodic potential mimics the crystal potential felt by electrons in a solid, enabling the atoms to "simulate" the behavior of interacting electrons in a crystal lattice. The power of the atom-electron analogy comes from the fact that characteristics of the atomic system, such as lattice geometry, degree of disorder, and strength of interactions, are all highly tunable. This enables the atoms to behave as a *quantum simulator*, a quantum system that behaves analogously to another system (which may be much harder to microscopically control or measure),[13] has already led to many spectacular observations, such as the transition from a weakly-interacting gas to a Mott insulator for both bosonic[14] and fermionic[15,16] atoms.

In spite of the successes of ultracold atom experiments, some phenomena remain difficult to manifest and probe in cold atoms. One such phenomenon is quantum magnetic interactions between effective spins. Quantum magnetism, which studies the many-body physics of coupled spins, is of key importance in condensed matter physics (see Sec. 2). In atomic realizations of quantum magnetism, usually the "spin" is formed from some discrete set of internal states, for example hyperfine sublevels. The reason why realizing effective magnetic interactions between such spins is difficult, as Sec. 2.2 describes in more detail, is that the dominant interactions between neutral atoms are short ranged; this requires any effective non-local spin-spin interactions between internal states to be mediated by motion. Hence, magnetic correlations become visible only when the motional temperature is less than the effective spin-spin coupling energy. To date, such temperatures are extraordinarily difficult to reach.

Ultracold ground state molecules are a newly realized platform for quantum magnetism in which the magnetism arises in a qualitatively distinct way from atoms, and this underlies the molecules' many favorable qualities. As opposed to atoms, polar molecules have strong, long-range electric dipole-dipole interactions[a]. The basic idea is to encode effective degrees of freedom in long lived, low-lying, and easily accessible internal degrees of freedom such as rotational and vibrational modes. Dipole-dipole interactions can mediate coupling between these effective spins even when molecules are pinned in a lattice (i.e. when their tunneling is completely suppressed). One consequence is that polar molecules can be used to study far-from-equilibrium quantum magnetism even in non-degenerate quantum gases. Such far-from-equilibrium dynamics has been observed in ultracold polar molecule experiments and will play a central role in our exposition. Additionally, taking a broad view of molecular diversity and experimental constraints such as temperature and lattice-scale probe resolution, we provide an overview of the "quantum

---

[a]Homonuclear molecules are not polar, and so are not amenable for simulating quantum magnetism in the fashion discussed in this work.

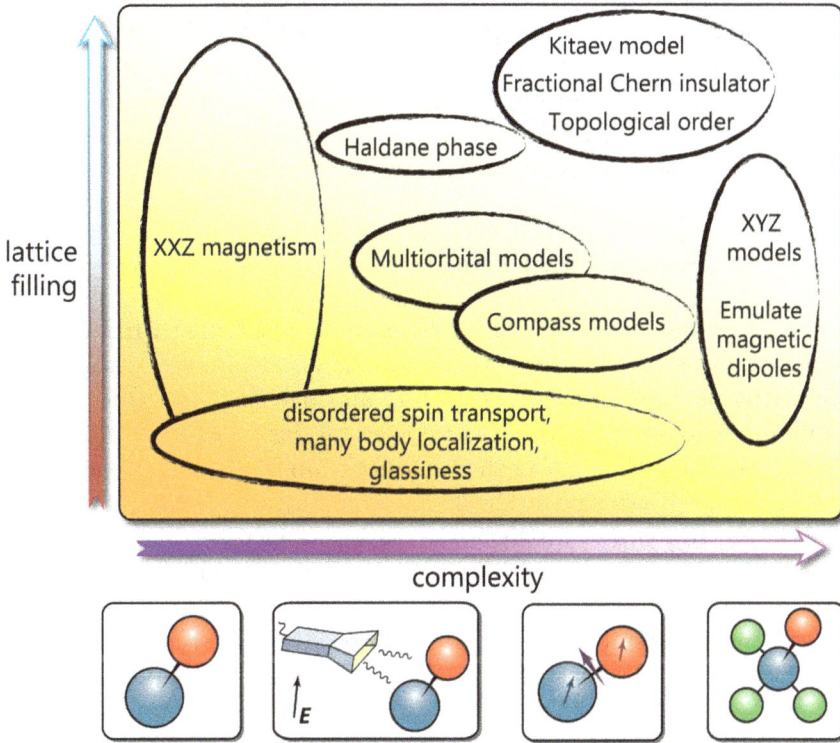

Fig. 1.   Regimes of quantum magnetism in ultracold ground state polar molecules in optical lattices, encompassing ongoing experiments (lower left) and directions being pursued (upper right). The vertical "lattice filling" axis represents the fraction of occupied lattice sites, that is the molecule density relative to the lattice spacing. Higher fillings generally correspond to lower motional entropies and temperatures. Current experiments reach filling fractions in the range of 10%-20%. Interesting physics exists in this regime, but exciting prospects also occur as one increases the filling fraction towards unity. The horizontal "complexity" axis represents complexity in two senses: the molecules' degrees of freedom and the experimental requirements to harness them. Diverse, rich, and novel phenomena at the forefront of modern quantum many-body physics occur in all of the indicated regimes.

simulation landscape" of quantum magnetic phenomena achievable with ultracold molecules, and summarize it in Fig. 1. This figure shows the new regimes of quantum magnetism that become available both as the motional entropy and temperature decrease (vertical axis) and as the complexity of the internal molecular structure increases (horizontal axis). This figure will be discussed in more detail in Sec. 5.

Our paper is organized as follows. Sec. 2 derives effective spin Hamiltonians describing the internal state dynamics of polar molecules in optical lattices (Sec. 2.1) and neutral atoms in optical lattices for comparison (Sec. 2.2), focusing on the simplest scenarios. Sec. 3 describes recent experiments in which effective quantum magnetism has been experimentally probed via far-from-equilibrium dynamics, as well as new theoretical tools that were developed to verify and understand the experimental observations. In Sec. 4 we explore molecules with complex internal structure, review methods of producing ultracold molecules, and go over basic molecular structure. In Sec. 5, we identify future directions for quantum magnetism with ultracold molecules, considering both advances in experimental technology and the structural complexity of molecules on the experimental horizon. Finally, in Sec. 6, we conclude.

2. Quantum Magnetism with Ultracold Molecules and Atoms

Exploring quantum magnetism with ultracold matter is a particularly fruitful direction of research, in part because quantum magnetic phenomena lie at the core of condensed matter physics. Moreover, despite the apparent simplicity of many models of quantum magnetism, these models are in general extraordinarily hard to solve with classical resources. This makes them excellent candidates for "quantum simulation" with ultracold systems. Although quantum magnetism is a vast field that is well beyond the capacity of this review to cover, we mention here some of the broad ideas that motivate its study. More complete reviews and introductions can be found, e.g., in Refs. 17–19.

One reason for the intense study of quantum magnetism is its relevance to materials and experimental phenomena – for example, antiferromagnets, multiferroic materials, spin glasses, and spin nematics – and the frequent proximity of quantum magnetism to unconventional superconductivity. Another motivation is the numerous *exotic* phenomena that have been theoretically predicted, including topologically ordered phases and (algebraic) spin liquids. These harbor physics which cannot be described within the "Landau paradigm" of symmetry breaking, as is also the case with the fractional quantum Hall effect. Observing a broader range of such phenomena, which lie outside of conventional classification, would clearly deepen our knowledge of quantum many-body physics. Finally, an understanding of quantum magnetism can have a great impact in advancing current technology including better and faster hard drives, computers, and spintronic devices.

Quantum magnetism in the solid state usually refers to interactions between electron spins localized in a crystal lattice. As the Coulomb interaction, which provides the microscopic interaction between electrons, is spin-independent, interactions between spins should be interpreted as *effective* interactions which arise from Coulomb interactions in conjunction with Fermi statistics, i.e. the required antisymmetry of electrons under exchange. In Sec. 2.2, we show how such effective magnetic interactions arise from particles with short-range interactions when they can tunnel in a lattice. This spin interaction mechanism, known as superexchange, is the most common mechanism by which effective magnetic phenomena arise in cold *atomic* gases loaded in optical lattices. Before we discuss the superexchange mechanism, however, Sec. 2.1 describes the simplest example of effective quantum magnetism mediated by dipole-dipole interactions in polar molecules, the main focus of this review. As we shall see, the resulting models for dipole-mediated quantum magnetism and the superexchange mechanism are very similar, even though the physical mechanism is very different. Finally, Sec. 2.3 discusses control and experimental consequences of the terms appearing in the effective spin models.

2.1. *Effective Magnetism with Polar Molecules*

Let us now consider how effective quantum magnetism arises for polar molecules.[20–23] For clarity, we discuss the simplest manifestation in this section before discussing how more complex magnetic interactions may be engineered in Secs. 4.2 and 5. Our starting point is shown schematically in the top left panel of Fig. 2. Here, molecules are pinned in a deep optical lattice with exactly one molecule per lattice site. By "pinned" we mean that molecules do not move between lattice sites on the timescales of an experiment. We now wish to encode an effective spin-$1/2$ in the internal degrees of freedom of the molecule. Considering $^1\Sigma$ molecules, in which there are no unpaired spins or orbital angular momentum[b], and neglecting hyperfine structure, the lowest-lying degrees of freedom to encode spin in are the rotational degrees of freedom[c]. The rotational states are described by a rigid rotor Hamiltonian[24] and can be labeled by $|N M_N\rangle$, where N is the rotational angular momentum quantum number and $-N \leq M_N \leq N$ is the projection of the rotational angular momentum along a space-fixed quantization axis. In the absence of external fields, the rotational energy spectrum is $E_{N M_N} = B_N N (N + 1)$, where B_N is called the rotational constant and is inversely proportional to the moment of inertia of the molecule. Typical rotational constants are a few GHz, which is much larger than the dipolar interaction energies of molecules at typical (~ 500 nm) optical lattice spacings, and also much larger than ultracold temperatures. These facts, together with the anharmonic spectrum and very long (> 10s) lifetimes of rotational excitations, imply that the number of molecules in each of the excited rotational states is conserved over the timescale of an experiment[d].

The $(2N+1)-$fold degeneracy of rotational excited states makes isolating a pair of rotational levels in which to encode a spin-$1/2$ challenging, and so we would like to split this degeneracy. One way to do so is to introduce a DC electric field $\mathbf{E}_{\mathrm{DC}}$; the resulting splitting is illustrated in Fig. 2. The key feature that we need in order to understand the emergence of quantum magnetism is that states such as $|0,0\rangle$ and $|1,0\rangle$ in Fig. 2, are now well-isolated from all other states, and so we can use them to encode a spin-$1/2$. The details of the coupling and notation will be explained below.

---

[b]A review of molecular structure and terminology is given in Sec. 4.2.
[c]In fact, these non-rotational degrees of freedom can sometimes also be neglected in molecules with more complex molecular structures under appropriate circumstances.
[d]Provided, of course, that rotational excitations are not generated by external means, e.g. by a microwave field.

Now that we have isolated an effective spin-1/2, we investigate the effect of dipole-dipole interactions within this subspace of states. The dipole-dipole interaction between molecules i and j is

$$\hat{H}_{\mathrm{DDI}} = \frac{\hat{\mathbf{d}}_i \cdot \hat{\mathbf{d}}_j - 3\left(\hat{\mathbf{d}}_i \cdot \mathbf{e}_r\right)\left(\hat{\mathbf{d}}_j \cdot \mathbf{e}_r\right)}{r^3}, \tag{1}$$

where $\mathbf{e}_r$ is a unit vector connecting molecules i and j, r is the distance between these molecules, and $\hat{\mathbf{d}}_i$ is the dipole operator of molecule i. Within the subspace of states $\{|\downarrow\rangle, |\uparrow\rangle\} \equiv \{|0,0\rangle, |1,0\rangle\}$ forming our spin-1/2, and in the limit of the interaction being much smaller than the rotational splitting, the interaction Eq. (1) is simple; for example

$$\hat{H}_{\mathrm{DDI}} |\uparrow\downarrow\rangle = a\,|\uparrow\downarrow\rangle + b\,|\downarrow\uparrow\rangle \tag{2}$$

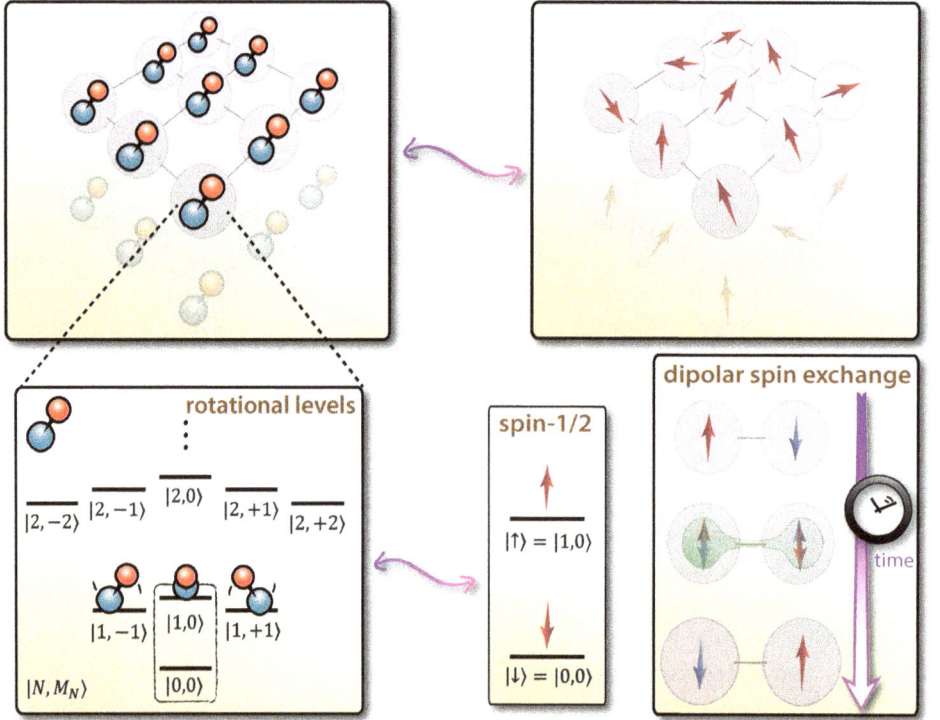

Fig. 2. Quantum magnetism of ultracold molecules in a lattice. Molecules in a deep lattice (top left) realize lattice spin models (top right), when the lattice is deep enough to *suppress* tunneling. The spin degree of freedom is encoded in rotational states of the molecule (bottom left). Two types of interactions occur naturally: a "spin exchange" interaction that exchanges pairs of spin states (illustrated, bottom right) and, in the presence of a dc electric field, an "Ising" interaction that splits the energies of aligned and anti-aligned pairs of spins [see Eq. (17)]. Both processes are capable of correlating and entangling spins.

with $a = \langle \uparrow \downarrow | \hat{H}_{\text{DDI}} | \uparrow \downarrow \rangle$ and $b = \langle \downarrow \uparrow | \hat{H}_{\text{DDI}} | \uparrow \downarrow \rangle$. Processes that change the total magnetization, such as $| \uparrow \downarrow \rangle \to | \uparrow \uparrow \rangle$, are energetically far off-resonant and therefore negligible. This implies that the spin-spin interaction of the molecules is

$$\hat{H}_{ij} = \left[\frac{J_\perp(\hat{\mathbf{d}}_i, \hat{\mathbf{d}}_j, \mathbf{r}_i - \mathbf{r}_j)}{2} \left(\hat{S}_i^+ \hat{S}_j^- + \text{h.c.} \right) + J_z(\hat{\mathbf{d}}_i, \hat{\mathbf{d}}_j, \mathbf{r}_i - \mathbf{r}_j) \hat{S}_i^z \hat{S}_j^z \right] \quad (3)$$

with $S_i^{\pm,z}$ the usual spin-1/2 operators acting on molecule i. We will derive below the forms of $J_\perp(\hat{\mathbf{d}}_i, \hat{\mathbf{d}}_j, \mathbf{r}_i - \mathbf{r}_j)$ and $J_z(\hat{\mathbf{d}}_i, \hat{\mathbf{d}}_j, \mathbf{r}_i - \mathbf{r}_j)$ and their dependence on electric field and choice of rotational states, as well as determine additional single spin terms that are omitted in Eq. (3) [see Eq. (11) for the final result].

However, before giving a more complete derivation of Eq. (3) and a determination of the coefficients in it, we first will describe how an electric field may be used to achieve the level splitting required to energetically isolate the spin-1/2 degree of freedom. This also provides useful background for how the interactions in Eq. (3) may be manipulated with electric fields.

The coupling Hamiltonian of the molecule's dipole operator, $\hat{\mathbf{d}}$, with the external field is $-\hat{\mathbf{d}} \cdot \mathbf{E}_{\text{DC}}$. We take $\mathbf{E}_{\text{DC}}$ to set the space-fixed z axis, $\mathbf{E}_{\text{DC}} = E_{\text{DC}} \mathbf{e}_z$ and thus the coupling Hamiltonian has matrix elements

$$\langle N'M_{N'} | - \hat{d}_z E_{\text{DC}} | N M_N \rangle = -E_{\text{DC}} \delta_{M_{N'}, M_N} \langle N'M_{N'} | \hat{d}_0 | N M_N \rangle \quad (4)$$

where

$$\langle N'M_{N'} | \hat{d}_p | N M_N \rangle = d (-1)^{M_{N'}} \sqrt{(2N'+1)(2N+1)} \begin{pmatrix} N' & 1 & N \\ -M_{N'} & p & M_N \end{pmatrix} \begin{pmatrix} N' & 1 & N \\ 0 & 0 & 0 \end{pmatrix} \quad (5)$$

are the matrix elements of the dipole operator in the basis of rotational states. In Eq. (5), $(\dots)$ is a $3j$-symbol and $\hat{d}_{\pm 1} = \mp(\hat{d}_x \pm i\hat{d}_y)/\sqrt{2}$ and $\hat{d}_0 = \hat{d}_z$ are spherical components of the dipole operator. Only the $p = 0$ component is used above, but the $p = \pm 1$ components will be useful to us later. The electric field mixes rotational states while preserving their projection M_N on the field axis. Because the electric field does not cause level crossings, we can still label the eigenstates of rotation in the presence of a DC field with $|NM_N\rangle$, where N is now interpreted as a label corresponding to the number of rotational quanta if the field were to be ramped adiabatically to zero. The energies of these states in weak fields, calculated to lowest order in the small parameter $\beta_{\text{DC}} = dE_{\text{DC}}/B_N$, are

$$E_{NM_N}/B_N = N(N+1) + \frac{\beta_{\text{DC}}^2}{2} \frac{N(N+1) - 3M_N^2}{(2N-1)(2N+3)N(N+1)}. \quad (6)$$

As shown in Fig. 2, all states with the same value of $|M_N|$ remain degenerate, and are separated from all other states with the same value of N by an energy $\sim \beta_{\text{DC}}^2 B_N$.

Now turning to the full derivation of how the dipole-dipole interaction, Eq. (1), projects onto the spin-1/2 degree of freedom we note that the dipolar interaction is the contraction of two rank-two tensors, one acting on the internal state of the

molecules through the dipole operators and one acting on the orbital motion of the
two molecules through the angular dependence. Hence, the dipole-dipole interaction can be written as a sum of terms which transfer q units of rotational angular
momentum projection to orbital angular momentum projection with $-2 \leq q \leq 2$.
Since our spin-1/2 is comprised of states with $M_N = 0$, there is no way to change
the rotational projection quantum number within this set of states (as is implicit
in Eq. (2)). Hence, only the $q = 0$ terms are important. Explicitly, these terms are

$$\hat{H}_{\mathrm{DDI};q=0} = \frac{1 - 3\cos^2\theta}{r^3}\left[\hat{d}_0\hat{d}_0 + \frac{\hat{d}_1\hat{d}_{-1} + \hat{d}_{-1}\hat{d}_1}{2}\right] \tag{7}$$

with θ the angle between the quantization axis and the vector connecting the two
molecules. Within the subspace of states $\{|\downarrow\rangle, |\uparrow\rangle\} = \{|0,0\rangle, |1,0\rangle\}$ the expectations
of $\hat{d}_{\pm 1}$ all vanish due to selection rules. The matrix elements of the $\hat{d}_0$ operator,
computed using Eq. (5) to lowest order in β_{DC}, are

$$\langle\downarrow|\hat{d}_0|\downarrow\rangle = \frac{\beta_{\mathrm{DC}}}{3} \equiv d_\downarrow, \quad \langle\uparrow|\hat{d}_0|\uparrow\rangle = -\frac{\beta_{\mathrm{DC}}}{5} \equiv d_\uparrow, \tag{8}$$

$$\langle\uparrow|\hat{d}_0|\downarrow\rangle = \langle\downarrow|\hat{d}_0|\uparrow\rangle = \frac{1}{\sqrt{3}}\left(1 - \frac{43\beta_{\mathrm{DC}}^2}{360}\right) \equiv d_{\downarrow\uparrow}. \tag{9}$$

The expected dipole moments in Eq. (8) describe the degree of orientation of the
states $|\downarrow\rangle$ and $|\uparrow\rangle$ in the presence of the external field. Importantly, these dipole
moments vanish in zero field, which says that the average dipole moment of a pure
rotational state along any direction fixed in space is zero. In contrast, the *transition*
dipole moments in Eq. (9) are nonzero even in zero field.[e] These encapsulate the
strength of a dipole-allowed transition from $|\downarrow\rangle$ to $|\uparrow\rangle$ and vice-versa. For example,
these transition dipole matrix elements determine the magnitude of the coupling
between $|\downarrow\rangle$ to $|\uparrow\rangle$ when a molecule is illuminated with near-resonant microwave
radiation. Fig. 3 shows the behavior of these dipole moments, going beyond the
perturbative $\beta_{\mathrm{DC}} \ll 1$ limit.

Let us now consider two molecules at lattice sites i and j, and write the dipole-
dipole interaction, Eq. (7), in the basis $\{|\uparrow_i\uparrow_j\rangle, |\uparrow_i\downarrow_j\rangle, |\downarrow_i\uparrow_j\rangle, |\downarrow_i\downarrow_j\rangle\}$. We find

$$\hat{H}_{ij} = \frac{1 - 3\cos^2\theta_{ij}}{|\mathbf{r}_i - \mathbf{r}_j|^3}\begin{pmatrix} d_\uparrow^2 & 0 & 0 & 0 \\ 0 & d_\downarrow d_\uparrow & d_{\downarrow\uparrow}^2 & 0 \\ 0 & d_{\downarrow\uparrow}^2 & d_\downarrow d_\uparrow & 0 \\ 0 & 0 & 0 & d_\downarrow^2 \end{pmatrix}, \tag{10}$$

with $\mathbf{r}_i$ the location of molecule i. If we define the spin-1/2 operators $\hat{S}_i^z = \frac{1}{2}(|\uparrow_i\rangle\langle\uparrow_i| - |\downarrow_i\rangle\langle\downarrow_i|)$, $\hat{S}_i^+ = |\uparrow_i\rangle\langle\downarrow_i|$, $\hat{S}_i^- = (\hat{S}_i^+)^\dagger$ obeying the commutation relations[f]

---

[e]Physically, the dipole operator is a vector operator, and so is odd under parity. The parity of the
state $|NM_N\rangle$ is $(-1)^N$, and therefore in the absence of an electric field a nonzero matrix element
can exist only between states N and $N \pm 1$.

[f]Note that $\hat{S}^\pm = (\hat{S}^x \pm i\hat{S}^y)$ is the "raising operator" form of the spin operators, but the dipole
operators $\hat{d}^{\pm 1} = \mp(\hat{d}^x \pm i\hat{d}^y)/\sqrt{2}$ are written in the standard form for spherical tensor operators
(with the overall minus sign on $\hat{d}^1$).

$\left[\hat{S}_i^z, \hat{S}_j^{\pm}\right] = \pm\delta_{ij}\hat{S}_i^{\pm}$, we can re-write Eq. (10) as a spin Hamiltonian:

$$\hat{H}_{ij} = \frac{1 - 3\cos^2\theta_{ij}}{|\mathbf{r}_i - \mathbf{r}_j|^3} \left[\frac{J_{\perp}}{2}\left(\hat{S}_i^+\hat{S}_j^- + \text{h.c.}\right) + J_z\hat{S}_i^z\hat{S}_j^z + W\left(\mathbb{I}_i\hat{S}_j^z + \hat{S}_i^z\mathbb{I}_j\right) + V\mathbb{I}_i\mathbb{I}_j\right],$$

$$(11)$$

where

$$J_{\perp} \equiv 2d_{\downarrow\uparrow}^2, \tag{12}$$
$$J_z \equiv (d_{\uparrow} - d_{\downarrow})^2, \tag{13}$$
$$W \equiv (d_{\uparrow}^2 - d_{\downarrow}^2)/2, \tag{14}$$
$$V \equiv (d_{\uparrow} + d_{\downarrow})^2/4, \tag{15}$$

and $\mathbb{I}$ is the identity operator in spin space. This result gives the full determination of the Hamiltonian motivated in Eq. (3). Eq. (11) readily generalizes to a many-site lattice in which each site can be filled with $n_i = 0$ or 1 molecules: one sums over all pairs i and j, replaces the identity operator in spin space with the total molecular density operator $\mathbb{I}_i \to \hat{n}_i \equiv \sum_{\sigma=\uparrow,\downarrow} \hat{a}_{i\sigma}^{\dagger}\hat{a}_{i\sigma}$, and replaces the spin operators by their second-quantized counterparts

$$\hat{S}_i^a = \frac{1}{2}\sum_{\mu\nu} \hat{a}_{i\mu}^{\dagger}\,[\sigma^a]_{\mu\nu}\,\hat{a}_{i\nu}. \tag{16}$$

Here, $\hat{a}_{i\nu}$ is a fermionic or bosonic operator in second quantization destroying a particle at site i in spin state ν and σ^a is a Pauli matrix. The term proportional to V represents coupling of the molecule density to itself, and so is a constant for pinned molecules and may be ignored. Similarly, in a unit-filled lattice the term proportional to W is a constant of the motion (to the extent the density is homogeneous) and may be ignored. The resulting effective spin model description for pinned molecules in a unit-filled lattice reads

$$\hat{H} = \frac{1}{2}\sum_{i \neq j} \frac{1 - 3\cos^2\theta_{ij}}{|\mathbf{r}_i - \mathbf{r}_j|^3} \left[\frac{J_{\perp}}{2}\left(\hat{S}_i^+\hat{S}_j^- + \text{h.c.}\right) + J_z\hat{S}_i^z\hat{S}_j^z\right]. \tag{17}$$

The Hamiltonian $\hat{H}$ is called an XXZ spin model, for reasons that we will clarify in Sec. 2.3. The effective magnetic interactions are long ranged, decaying as $1/r^3$ with the distance between lattice sites, and inherit the $(1 - 3\cos^2\theta)$ anisotropy characteristic of the dipole-dipole interaction, Eq. (7). The relative importance of the J_z and $J_{\perp}$ terms can be tuned by the strength of the external field, in accord with the β_{DC}-dependence of the dipole matrix elements shown in Fig. 3. In particular, the J_z term (as well as the W and V terms, which can be neglected under some circumstances) vanishes in zero electric field. We will investigate more complex means to tune effective magnetic interactions, even beyond the form in Eq. (17), in Sec. 5.

The spin-spin interactions can also be modified by varying the choice of the two rotational levels used to encode the spin-1/2. To demonstrate this, let us perform the same analysis as above using the states $\{|\downarrow\rangle, |\tilde{\uparrow}\rangle\} \equiv \{|0,0\rangle, |1,1\rangle\}$. The relevant dipole matrix elements, analogous to Eqs. (8)-(9), are

$$\langle\downarrow|\hat{d}_0|\downarrow\rangle = \frac{\beta_{\mathrm{DC}}}{3} \equiv d_\downarrow, \quad \langle\tilde{\uparrow}|\hat{d}_0|\tilde{\uparrow}\rangle = \frac{\beta_{\mathrm{DC}}}{10} \equiv d_{\tilde{\uparrow}}, \tag{18}$$

$$\langle\tilde{\uparrow}|\hat{d}_1|\downarrow\rangle = -\langle\downarrow|\hat{d}_{-1}|\tilde{\uparrow}\rangle = \frac{1}{\sqrt{3}}\left(1 - \frac{49\beta_{\mathrm{DC}}^2}{1440}\right) \equiv d_{\downarrow\tilde{\uparrow}}, \tag{19}$$

where we have again used Eq. (5). Two differences are apparent when comparing to Eqs. (8)-(9). Namely, $d_{\tilde{\uparrow}}$ is positive while $d_\uparrow$ is negative (See Eq. (8)) and the transition dipole matrix elements $d_{\downarrow\tilde{\uparrow}}$ came from the operators $\hat{d}_{\pm 1}$ instead of $\hat{d}_0$. Another consequence of this is that when we project the $q = 0$ part of the dipole-dipole interaction, Eq. (7), into the space of our effective spin-1/2, both the $\propto \hat{d}_0\hat{d}_0$ and $\propto (\hat{d}_1\hat{d}_{-1} + \hat{d}_{-1}\hat{d}_1)$ terms contribute[g]. Projecting into the basis $\{|\tilde{\uparrow}_i\tilde{\uparrow}_j\rangle, |\tilde{\uparrow}_i\downarrow_j\rangle, |\downarrow_i\tilde{\uparrow}_j\rangle, |\downarrow_i\downarrow_j\rangle\}$, we find a matrix of the same form as Eq. (10), only with $d_\uparrow$ replaced by $d_{\tilde{\uparrow}}$ and $d_{\downarrow\uparrow}^2$ replaced by $-d_{\downarrow\tilde{\uparrow}}^2/2$. The factor of 1/2 comes from the second term in the brackets of Eq. (7), and the minus sign from Eq. (19) (see also footnote f). The $-1/2$ can alternatively be semiclassically understood as coming

---

[g]For molecules with no nuclear spin structure, the $|1,1\rangle$ and $|1,-1\rangle$ states are degenerate and hence will be resonantly coupled by the $q = \pm 2$ terms of the dipole-dipole interaction that were ignored when simplifying Eq. (1) to Eq. (7). However, in the alkali dimers, e.g. KRb, hyperfine structure splits this degeneracy by an amount large compared to typical interaction energies. Hence, we do not consider the $q = \pm 2$ terms here.

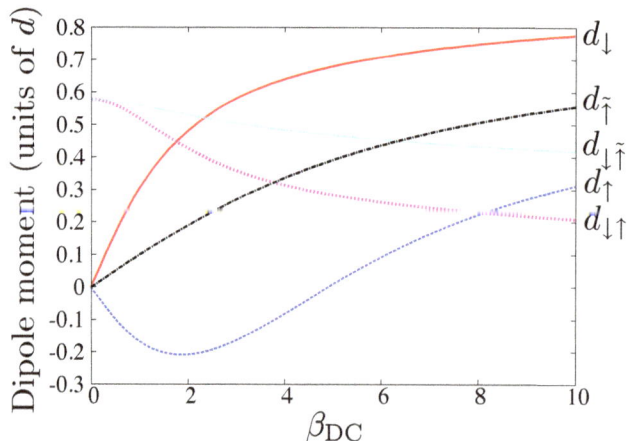

Fig. 3. Tuning spin-spin interactions with an electric field. An applied electric field with dimensionless strength $\beta_{\mathrm{DC}} = dE_{\mathrm{DC}}/B_N$ alters the expected dipole moments of the states $|\downarrow\rangle$, $|\uparrow\rangle$, and $|\tilde{\uparrow}\rangle$; see Eqs. (8)-(9) and (18)-(19). In particular, the expected dipole moments increase linearly for small fields and saturate asymptotically to the permanent dipole moment. The transition dipole moments are nonzero in zero field and monotonically decrease as the field is increased. The dipole moments determine the spin-spin couplings through, e.g., Eqs. (12)–(15).

from the time average of two rotating dipoles. Thus, the effective spin model description is again an XXZ model of precisely the same form as Eq. (17), the only difference being that $d_\uparrow \to d_{\bar\uparrow}$ and $J_\perp = -d_{\downarrow\bar\uparrow}^2$ rather than $2d_{\downarrow\uparrow}^2$. When a magnetic coupling is negative, spins prefer to align to minimize the energy, resulting in a *ferromagnetic* tendency. Likewise, a positive value represents *anti-ferromagnetic* coupling. The two simple examples given above demonstrate that the form of the effective Hamiltonian, Eq. (17), namely the XXZ coupling and the $(1 - 3\cos^2\theta)/r^3$ dependence of interaction matrix elements, is a universal characteristic of the effective magnetic interactions between diatomic molecules prepared in two dipole-coupled rotational levels. The tunable, non-universal aspects are the actual coupling constants $J_\perp$, J_z, etc.; these can be controlled and even made to change sign by external fields and the choice of rotational states forming the effective spin-1/2 (see for example Fig. 3).

2.2. *Effective Magnetism with Two-component Atoms*

As an example of how quantum magnetism arises due to short-range, spin-independent interactions (in contrast to molecules), we consider the simplest model of interacting spin-1/2 fermionic particles on a lattice, the single-band Hubbard model

$$\hat{H}_{\text{Hubbard}} = -t \sum_{\langle i,j \rangle} \sum_{\sigma \in \{\uparrow,\downarrow\}} \left[\hat{a}_{i\sigma}^\dagger a_{j\sigma} + \text{h.c.} \right] + U \sum_i \hat{n}_{i\uparrow}\hat{n}_{i\downarrow}. \tag{20}$$

Here, t is the tunneling energy, U the interaction energy, $\langle i,j \rangle$ represents a sum over all nearest-neighbor pairs of lattice sites i and j, $\hat{a}_{i\sigma}$ is a fermionic operator in second quantization destroying a particle at site i in spin state σ, and $\hat{n}_{i\sigma} = \hat{a}_{i\sigma}^\dagger a_{j\sigma}$. In a solid state system, U represents an approximation to the screened Coulomb potential, and the Hubbard model is rarely expected to provide anything more than a crude guide to qualitative features. However, ultracold neutral atoms dominantly interact via a short-ranged ($\sim 1/r^6$) van der Waals potential, which is well-modeled with a contact pseudopotential at low energies. Consequently, the Hubbard model provides an excellent microscopic description of ultracold two-component fermionic atoms in an optical lattice.[25] In the ultracold atomic realization of the Hubbard model, two magnetic sublevels of the hyperfine manifold play the role of the electronic spin-1/2.

When $U \gg t$, an effective magnetic interaction arises between singly-occupied neighboring sites due to virtual occupation of sites by two fermions,[17] a mechanism known as *superexchange*. To see how this arises, let us consider two sites i and j which are nearest neighbors and are spanned by the basis $\{|\uparrow,\uparrow\rangle, |\uparrow\downarrow,0\rangle, |\uparrow,\downarrow\rangle, |\downarrow,\uparrow\rangle,$ $|0,\uparrow\downarrow\rangle, |\downarrow,\downarrow\rangle\}$, where the left and right sides denote the occupations of sites i and j, respectively, and the states $|\uparrow\downarrow,0\rangle$ and $|0,\uparrow\downarrow\rangle$ denote spin singlet states. The

Hubbard Hamiltonian in this basis is

$$\hat{H}_{\text{Hubbard}} = \begin{pmatrix} 0 & 0 & 0 & 0 & 0 & 0 \\ 0 & U & -t & -t & 0 & 0 \\ 0 & -t & 0 & 0 & -t & 0 \\ 0 & -t & 0 & 0 & -t & 0 \\ 0 & 0 & -t & -t & U & 0 \\ 0 & 0 & 0 & 0 & 0 & 0 \end{pmatrix}. \tag{21}$$

In the case $U \gg t$, tunneling to the energetically high-lying states $|\uparrow\downarrow, 0\rangle$ and $|0, \uparrow\downarrow\rangle$ happens only "virtually," and so we can consider the effects of these states in lowest-order perturbation theory. The effective Hamiltonian in the degenerate subspace spanned by $\{|\uparrow, \uparrow\rangle, |\uparrow, \downarrow\rangle, |\downarrow, \uparrow\rangle, |\downarrow, \downarrow\rangle\}$, to lowest order in t/U, is

$$\hat{H}_{\text{eff}} = \begin{pmatrix} 0 & 0 & 0 & 0 \\ 0 & -2t^2/U & -2t^2/U & 0 \\ 0 & -2t^2/U & -2t^2/U & 0 \\ 0 & 0 & 0 & 0 \end{pmatrix}. \tag{22}$$

This effective Hamiltonian has a form similar to the matrix appearing in our derivation of effective magnetic interactions for polar molecules, Eq. (1). Accordingly, we can write this effective Hamiltonian as a spin model

$$\hat{H}_{\text{eff}} = \frac{J_\perp}{2} \left(\hat{S}_i^+ \hat{S}_j^- + \text{h.c.} \right) + J_z \hat{S}_i^z \hat{S}_j^z + V \mathbb{I}_i \mathbb{I}_j, \tag{23}$$

where $J_z = J_\perp = 4t^2/U$ and $V = -t^2/U$. Physically, the singlet state $(|\uparrow, \downarrow\rangle - |\downarrow, \uparrow\rangle)/\sqrt{2}$ can lower its energy by virtual tunneling to a doubly occupied site. In contrast, the triplet states $|\uparrow, \uparrow\rangle$, $|\downarrow, \downarrow\rangle$, and $(|\uparrow, \downarrow\rangle + |\downarrow, \uparrow\rangle)/\sqrt{2}$ are not connected to any physical states by tunneling, as occupancy of a site by two particles of the same spin is forbidden by the Pauli principle. Hence, quantum statistics plays a key role in the generation of effective magnetic couplings for spin-independent interactions.

Just as was the case for the polar molecules above, the two-site Hamiltonian, Eq. (23), can be immediately generalized to the many-site case. Simplifications occur if we consider the case of half filling (one fermion per lattice site), since in this case tunneling processes involving only singly occupied sites do not occur, and the term proportional to V becomes a constant of the motion which we neglect (again ignoring density inhomogeneities). The resulting effective spin dynamics is then governed by the celebrated *Heisenberg model*

$$\hat{H}_{\text{Heisenberg}} = J \sum_{\langle i,j \rangle} \hat{\mathbf{S}}_i \cdot \hat{\mathbf{S}}_j, \tag{24}$$

where $J = 4t^2/U$ and $\hat{\mathbf{S}}_i = (\hat{S}_i^x, \hat{S}_i^y, \hat{S}_i^z)$. The Heisenberg model, realized through the superexchange mechanism with a short-range interaction U, represents the minimal realization of quantum magnetism seen in ultracold gases in optical lattices. The spin coupling in the Heisenberg model is a special case of XXZ coupling with

$J_\perp = J_z$. An XXZ coupling can be realized in cold atom systems by superexchange with two-component bosons due to differences in interaction strengths between the two components.[26] An important aspect of magnetism generated through superexchange is that it is mediated through motion via the tunnel-coupling t. Hence, magnetic correlations are only visible when the temperature of the motional degrees of freedom is of order $J \sim t^2/U \sim 1\,\text{nK}$. This is an extremely cold temperature, even by ultracold standards!

2.3. *Interpretation of Processes in the XXZ Model*

For the spin-1/2 case, the $J_\perp$ term in Eq. (17) flips a spin up at site j to spin down while simultaneously flipping a down spin at a neighboring site i to up and vice versa. Schematically, we can draw this process as $|\downarrow_i\uparrow_j\rangle \to |\uparrow_i\downarrow_j\rangle$, as in Fig. 2. As this effectively swaps the spin labels of the two sites, we will refer to this process as *spin exchange*[h]. One proposal to study this mechanism with polar molecules was presented in Ref. 27. Note that configurations of identical spin projections $|\uparrow_i\uparrow_j\rangle$ and $|\downarrow_i\downarrow_j\rangle$ do not participate in spin exchange. The model Eq. (17) with only spin exchange, $J_z = 0$, can be written as

$$\hat{H}_{\text{XX}} = \frac{J_\perp}{4} \sum_{i \neq j} \frac{1 - 3\cos^2\theta_{ij}}{|\mathbf{r}_i - \mathbf{r}_j|^3} \left(\hat{S}_i^+ \hat{S}_j^- + \text{h.c.} \right), \tag{25}$$

$$= \frac{J_\perp}{2} \sum_{i \neq j} \frac{1 - 3\cos^2\theta_{ij}}{|\mathbf{r}_i - \mathbf{r}_j|^3} \left(\hat{S}_i^x \hat{S}_j^x + \hat{S}_i^y \hat{S}_j^y \right). \tag{26}$$

A spin Hamiltonian with such coupling is called an *XX model*. The nomenclature for an XXZ model, Eq. (17), should now be clear, as it represents equal coupling (XX) for the x and y components of the spins, with a different coupling (Z) for the z component of the spins. In this same notation, Heisenberg coupling as in Eq. (24) would be called XXX coupling.

In the opposite limit $J_\perp \to 0$ of the XXZ model, we recover an *Ising model*

$$\hat{H}_{\text{Ising}} = \frac{J_z}{2} \sum_{i \neq j} \frac{1 - 3\cos^2\theta_{ij}}{|\mathbf{r}_i - \mathbf{r}_j|^3} \hat{S}_i^z \hat{S}_j^z. \tag{27}$$

As opposed to the XX coupling, which causes transitions between pairs of spins which are anti-aligned in the z basis and does not affect aligned spins, the Ising model generically has non-zero diagonal elements for all configurations in the z basis. In particular, aligned configurations such as $|\uparrow_i\uparrow_j\rangle$ or $|\downarrow_i\downarrow_j\rangle$ have an Ising interaction energy proportional to J_z, while anti-aligned configurations $|\uparrow_i\downarrow_j\rangle$ or $|\downarrow_i\uparrow_j\rangle$ have an energy proportional $-J_z$. Because all operators appearing in the Ising model commute with one another, the static properties of the Ising model are

---

[h]We remark that for higher-dimensional spin representations this term does not necessarily exchange the two spin components. For example, two spin-particles can undergo the transition $|1,0\rangle|1,0\rangle \to |1,-1\rangle|1,1\rangle$ under the $J_\perp$ term.

effectively classical, and it displays no quantum phase transitions. Nevertheless, the dynamics is highly *nonclassical*, and can generate strong correlations and maximally entangled states;[28] however, because of its special commuting property, the dynamics of correlation functions in Ising models can be determined analytically.[29–31]

3. Dynamical Investigations of Quantum Magnetism: Experiment and Theory

In this section, we describe the experimental realization and observation of the dipolar XXZ model described by Eq. (17) in ultracold molecules and the theoretical techniques that were developed in order to understand the resulting behavior. Subsection 3.1 describes the experimental evidence that the ultracold polar molecule experiments are operating in the regime quantitatively described by Eq. (17). This evidence is based on extensive measurements of far-from-equilibrium dynamics and comparisons with theory.

When the experiments were performed, their dynamics could not be treated quantitatively by existing theoretical methods. As a result, they stimulated the development of new theoretical ideas and methods. Subsection 3.2 describes techniques that have been developed with the aid of experiments and applied to them, and it gives a short survey of theoretical methods for which ultracold molecule experiments are expected to provide a testing ground.

3.1. *Observation of Spin-exchange and Verification of Eq. (17)*

Reference 32 theoretically argued that the dipolar spin-spin interactions in Eq. (17) could be observed even in systems with lattice filling (number of particles divided by number of lattice sites) well below unity, extending to densities so low that more sites are empty than are occupied. The proposal was to observe dynamics using Ramsey spectroscopy, a standard dynamic protocol in atomic and molecular experiments (described below). We will describe how three goals of this proposal have now been successfully achieved: observation of spin-spin interactions in ultracold molecules, benchmarking of the accuracy of Eq. (17) to describe these interactions, and the exploration of physics beyond the regime accessible to previously existing exact methods.

Yan *et al.* employed Ramsey spectroscopy, illustrated in Fig. 4, to observe spin-exchange interactions in ultracold KRb.[33] Ramsey spectroscopy may be described as follows: all of the spins are aligned along the same direction at time $t = 0$, then they are allowed to evolve dynamically under no external influence – only Eq. (17) – for a time t, and finally one reads out the spin component along a chosen direction $\hat{n}$ summed over all particles, $\langle \sum_i \mathbf{S_i} \cdot \hat{n} \rangle$.[i]

---

[i]In the lab, the Ramsey protocol is implemented as a pair of microwave pulses, separated in time, that couple the rotational states. The first aligns the spins in the desired direction, and the latter rotates the spins so that the various spin components can be measured by doing standard number

An example measurement of such dynamics, similar to Refs. 33 and 35, is shown in the graph at the bottom left of Fig. 4 for $N = 11,000$ molecules in the lattice and with the spin-1/2 degree of freedom realized in the $|\uparrow\rangle = |1,0\rangle$ and $|\downarrow\rangle = |0,0\rangle$ rotational states (see Fig. 2). This panel shows the measured spin component along the x direction as a function of the evolution time. Experiments conclusively identify dipolar interactions from a number of these measurements; there are four main pieces of evidence, which we describe now.

Two features are apparent in Fig. 4 (bottom left): a relatively small oscillation superposed on a roughly exponential decay to zero. Both of these features are due to dipolar spin-exchange interactions. A first indication is that the observed oscillation frequency ν coincides with the expected oscillation frequency of two KRb molecules in these rotational states occupying nearest neighbor sites separated along the $\hat{z}$ lattice direction (see Fig. 5) namely $\nu = J_\perp/(2h) \approx 100$Hz. The direction is relevant because of the $1 - 3\cos^2\theta$ anisotropy of the dipolar interaction, illustrated in Fig. 5. That this single coupling is so important is not an accident: the current experiments

---

measurements of each rotational state. Furthermore, experiments so far have used a related, but more robust measurement, for spin readout. Instead, the collective spin vector is swept around a circle of constant z component and the resulting "Ramsey contrast" – the oscillation amplitude as a function of this angle – is the measured variable at each time. Moreover, the data we will show is taken with a spin-echo sequence, which adds an extra pulse to (partially) remove the effects of an inhomogeneous field $\sum_i h_i S_i^z$ that is present due to "non-magic" trapping conditions,[34] i.e. the fact that trapping frequencies for the two rotational states differ slightly in this experiment.

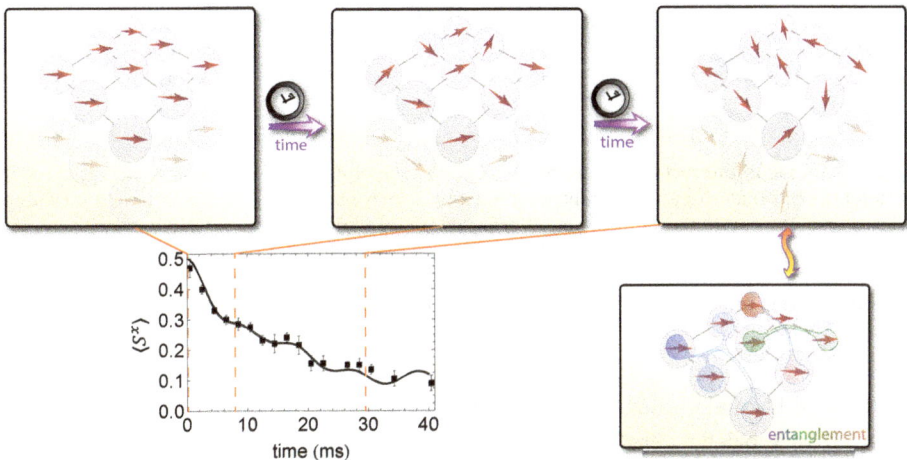

Fig. 4.   Non-equilibrium spin dynamics in a lattice. Bottom left plot and top dynamic sequence: initially aligned spins (along the $\hat{x}$ direction) evolve in time due to interaction. Both inhomogeneous precession and the growth of entanglement can occur. Such dynamics, implemented using Ramsey spectroscopy, has been used in ultracold KRb experiments to verify that the system accurately realizes Eq. (17).

estimate about a $f = 0.1$ lattice filling, and in this regime, the oscillations come from configurations of molecules having a single nearest neighbor. These configurations have clear oscillatory dynamics, and the fastest frequency is the most visible.[32,33,35] Further data analysis was, in fact, able to identify in the data other oscillation frequencies $\nu/\sqrt{2}$ and $\nu/2$, which corresponded to the next strongest interaction strengths.[35]

A second signature is that the time-constant for decay of $\langle S^x \rangle$ decreases with increasing density. In contrast, any single particle effect would remain unaffected by a simple change of density. A further indication of interactions effects is that the coherence time was observed to decreased as $1/N$, with N the total number of molecules. This scaling is consistent with the power law decay of the dipolar interactions: Increasing the lattice filling fraction $f \propto N$ decreases the mean distance between molecules ($\bar{R} \propto f^{-1/3}$), leading to an average dipolar interaction that scales as $1/\bar{R}^3 \propto f \propto N$.

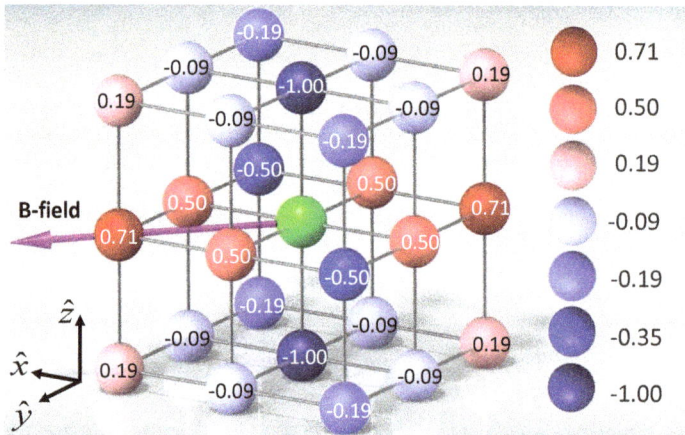

Fig. 5. Spin-spin interaction strengths. Labels on each site indicate the magnitude of the interaction between that site and the central green site, relative to the maximum interaction strength. This is given by the function $1 - 3\cos^2 \theta$, where θ is the angle between the intermolecular separation vector and the quantization axis. We take the quantization axis to be along the axis halfway between the $\hat{x}$ and $\hat{y}$ directions, corresponding to the experimental set up of Ref. 33. [Here the *magnetic* field **B**, rather than an electric field, sets the quantization axis through nuclear magnetic effects.] Figure after Ref. 33.

Two additional pieces of evidence were presented in Refs. 33 and 35 that we describe very briefly here. First, the experiments in Ref. 33 showed that some of the effects of interactions could be suppressed by using a more involved sequence of microwave pulses than the simple Ramsey spin-echo sequence, specifically the "WAHUHA" sequence.[36] This suppression was expected because this pulse sequence removes interactions completely in the case of two isolated molecules and is expected to partially remove the interactions in general. Second, as discussed above, changing the rotational state pair used for the spin-1/2 system from $\{|\!\uparrow\rangle = |1,0\rangle\,, |\!\downarrow\rangle = |0,0\rangle\}$ to $\{|\!\uparrow\rangle = |1,-1\rangle\,, |\!\downarrow\rangle = |0,0\rangle\}$ decreases the interaction strength by a factor of two, but otherwise leaves the interaction unchanged. The experiments in Ref. 35 performed the measurements with both pairs of rotational states and showed that only the timescale, not the shape, of the $\langle S^x(t)\rangle$ dynamics changed when the interaction strength was varied. This showed that *only* the expected dipolar interactions contributed to the evolution over the measurement time.

These four pieces of evidence – oscillation frequency, density-dependent contrast decay rate, response to the WAHUHA pulse sequence, and dependence of the dynamics on the rotational state used – clearly show that the JILA experiments have observed dipolar "spin-exchange" interactions between molecules in different rotational states, but on their own provide no direct evidence that the spin model describing the system is Eq. (17) and do not tell what role various aspects of this equation play in the dynamics. For example, the evidence described above does not constrain whether the long-range interactions are playing a crucial role, and whether the spin interactions are of the XXZ form claimed. The next subsection describes how these questions were addressed in Ref. 35. Doing this required comparing experiments to calculations using newly developed theoretical methods. This subsection also discusses other theoretical methods that are expected to be useful in the future.

We raise one broader point before turning to the theoretical comparison. All of the observations discussed were possible despite the non-degenerate temperature of the molecules in the lattice and the consequent low filling. The key capability that enabled complex quantum many-body behavior to manifest in these measurements was the ability to perform precise quantum control of *internal* degrees of freedom, namely the rotational states, i.e. the ability to prepare almost pure states in the spin degree of freedom. Another aspect of these experiments is the decoupling of the spin and motional degrees of freedom: the motion is completely frozen in the lattice, and only affects the spin degrees of freedom by determining their (static) coupling strengths. It is interesting to note that the ability to control the internal spin degrees of freedom has been utilized in other non-degenerate systems – hydrogen gases,[37] optical lattice alkaline earth atomic clocks,[38,39] and warm (even room temperature) alkali vapors in highly excited Rydberg states[40] – as well as degenerate systems, including interacting spin-1/2 Fermi gases,[41–43] and high-spin Fermi gases[44] (a highly non-comprehensive list). An intriguing question, therefore, is to what extent the quantum correlated physics survives at elevated temperatures in the presence of spin-motion decoupling. In many cases these prior experiments were not thought about as strongly interacting many-body systems and it would be interesting to revisit them from this perspective with this question in mind. One area that is particularly relevant to this review is the study of molecules at cold, but non-degenerate or possibly even not-ultracold temperatures. There are several experiments aiming to produce or that have produced molecules in this regime (see Sec. 4.2).

3.2. Quantum Simulation of Strongly Correlated Quantum Magnetism: Theoretical Developments

In order to quantitatively validate the molecules' realization of Eq. (17), it is necessary to compare their behavior to theoretical predictions. However, this required a new theoretical method, as all theories that existed prior to these experiments were extremely inaccurate for this problem. Here we focus on the method introduced in Ref. 35, termed the MACE (moving average cluster expansion).

To understand how rich the physics of these systems is, as well as the complexity of calculating their dynamics, note that these systems combine several features that would seem to render the calculations intractable: they are three dimensional, long-ranged interacting, disordered, far-from-equilibrium quantum many-body systems. Moreover, the dynamics is *completely* beyond mean-field theory: mean-field theory predicts *no* dynamic evolution of the contrast for the initial conditions with all spins along the $\hat{x}$ direction, which is the situation studied experimentally in Refs. 33 and 35. The absence of mean-field dynamics is most easily seen for the Ising term: the mean field Hamiltonian contains terms of the form $S_i^z \langle S_j^z \rangle$, and these quantities vanish if the spins point along $\hat{x}$. A similar type of analysis can be done for the XX term, and this analysis shows that there is no mean field dynamical evolution due to this term, either.

Even in one dimension, where special theoretical techniques can be applied to solve for the dynamics, one finds that substantial entanglement emerges. The amount of entanglement generally increases even beyond that exhibited by the ground state.[32] Because of all of these features, the dynamics might appear to be nearly impossible to solve exactly, either analytically or numerically. Indeed, the methods that existed prior to the experiments fail to capture it.[35]

In spite of these challenges, Ref. 35 developed a new theoretical method – the MACE – to accurately calculate the $\langle S^x(t) \rangle$ dynamics of Eq. (17) and applied it to the experiments. The idea of MACE is to break the system into independent clusters, but calculate observables (e.g. $\langle S_i^x(t) \rangle$) only for the spin i at the center of the cluster. This avoids effects of the cluster boundary, at least until times where correlations have had time to propagate inward to the cluster center. Specifically, for each spin i one creates a cluster of g spins by enumerating the $(g-1)$ largest coupling constants connecting i to other spins $\{j\}$. One exactly solves the dynamics for each cluster, and then calculates $\langle S_i^x(t) \rangle$ for each i using the cluster optimized for spin i. It turns out that the MACE converges orders-of-magnitude faster than previously existing state-of-the-art methods. The MACE calculations quantitatively agree with experiment with no fitting parameters, for a wide range of molecule lattice fillings and for different choices for rotational state pairs to realize the spin-$1/2$ degrees of freedom. (The filling fraction was determined from one dataset, and then this parameter was used for comparison with all of the numerous other datasets.) This agreement confirmed that Eq. (17) accurately describes the molecules. Moreover alternative models are ruled out, suggesting that potential experimental imperfections do not alter the interaction dynamics. For example, one clearly observes the long-range nature of the interactions in this manner, as truncating the theoretical calculation to nearest neighbor interactions fails to reproduce the dynamics even qualitatively.

Even for the MACE, it is challenging to rigorously and quantitatively assess its convergence, and consequently the observed quantitative agreement with experimental measurements of $\langle S^x(t) \rangle$ also provides independent evidence that the MACE is converging. In this sense, the experiments are serving as quantum simulators: systems whose dynamics are beyond the ability of existing methods to calculate in a controlled way. These experiments and newly developed theory are all the more interesting because the XXZ spin model is realized in numerous important systems, even outside of ultracold atomic and molecular gases, such as excitons in solids and large molecules, nitrogen-vacancy (NV) centers in diamond, and magnetic defects in other solid state systems.

We emphasize that the capability of the MACE to model the particular measurements carried out so far, however, does not mean that it will be able to capture other measurements – even those accessible to current experiments. Examples of experiments likely beyond the capacity of MACE are measurements of spin transport and correlation functions, which involve connections between two or more regions that are several lattice sites apart. In those cases, the experiments themselves are expected to shed light on the dynamics dictated by Eq. (17) and will help to benchmark theories used to model it.

Although at present it is unclear what the most promising theoretical methods will be for calculating correlations, transport, and other future experimental scenarios, we mention techniques that are likely to provide insight:

- **Mean-field theory.** Although entirely inapplicable to the $\langle S^x(t) \rangle$ dynamics initiated from the state polarized along the $\hat{x}$ axis that we have concentrated on here, mean-field theory can be a useful tool in general.
- **MACE.** The MACE and further development of this method may capture much behavior qualitatively, even in cases where it ceases to be quantitatively accurate.
- **DMRG.** In one dimensional systems, density matrix renormalization group (DMRG) methods[45–49] can solve for a variety of behaviors in these spin models . It is possible to extend calculations to quasi-1D "ladders" of width ~ 5, to obtain insight into two-dimensional systems.[50]
- **Many-body perturbation theory.** In special cases, perturbation theory allows one to accurately calculate response properties. For example if a

simple spin state is driven slightly out of equilibrium – for example, the all-down spin state $|\cdots \downarrow\downarrow \cdots\rangle$ has a few spins flipped – the spin excitations created are dilute and their behavior may be tractable with many-body Green's function techniques (see e.g., Refs. 51,52).

- **Truncated Wigner approximation.** The truncated Wigner approximation (TWA)[53,146] introduces the quantum noise in the initial state and then propagates the mean-field equations of motion. This is a promising route since in many of the proposed current experiments, the initial state has a product state structure and thus relatively small quantum noise. Consequently, the TWA may provide a way to capture many features of time dynamics, out to some time that, while not too large, may go well beyond a naive short-time perturbation theory, and, in some cases, even the measurement capabilities of current experiments.

- **Exact diagonalization.** For a system with a finite number of spins, in principle one can numerically construct the full Hamiltonian as a matrix in the finite Hilbert space, and then one can diagonalize this matrix using standard algorithms.[54] However, the accessible system sizes are severely limited, because the Hilbert space dimension grows exponentially, like 2^N for an N particle spin-$1/2$ system. In particular, studying more than ~ 30-35 spins is generally impossible. Nevertheless, understanding the behavior for these relatively small systems can lend much insight into the general behavior.

It will be fascinating to test the zoo of many-body approximations with experiments, discovering when various techniques are accurate, when they are not, and stimulated by this, discovering new concepts and methods.

3.3. *Future Directions for Molecules out of Equilibrium*

As suggested above, there are many interesting facets that can be explored fairly immediately with the ongoing experiments. One general possibility is measurement of spin conductivities in strongly interacting dipolar spin models. This includes the study of many-body localization,[55–61] see Sec. 5. Another possibility is to measure correlation growth[28] either by looking at global fluctuations in the Ramsey signal[62,63] (analogous to noise correlation spectroscopy[62]) or by employing local spin flips and probes.[64] Another possibility is to explore intrinsically dynamic phenomena and look for dynamical phase transitions or universality out of equilibrium.[28,65–79]

4. Increasing Molecular Complexity

The simple derivation of an effective spin Hamiltonian for polar molecules in Sec. 2.1 assumed that the molecules had no internal structure besides the rotational degrees of freedom. In Sec. 5 we will describe what new types of magnetic phenomena occur for molecules with more complex structure. The present section discusses progress towards and challenges for the production of ultracold molecules, focusing on molecules which are under active experimental investigation. An excellent overview of all ultracold molecule experiments and their stage of progress circa 2012 may be found in Ref. 80. Sec. 4.2 provides a brief review of molecular structure and terminology in order to keep our discussion self-contained and accessible to researchers outside the field of molecular physics.

4.1. *Cooling and Trapping of Molecular Gases*

Laser cooling is a workhorse of ultracold atomic physics. It removes entropy from a selectively excited atomic sample by utilizing spontaneous emission to return the system to a fiducial state. Naturally, one wishes to extend this technology to molecules, and indeed the first proposal to do so appeared one year after the realization of atomic BEC.[81] However, certain difficulties become immediately evident when attempting to laser cool molecules. In particular, a molecule in an excited electronic state can spontaneously radiate ("branch") to one – or many – states that are in the ground electronic manifold, but are rotationally or vibrationally excited. Hence, rather than returning to the ground state and lowering entropy, a molecule can distribute the photon energy among its various internal modes. The vibrational modes are the most troublesome in this respect, as transitions between them do not obey strict selection rules, but are only governed by state overlaps known as Franck-Condon factors. Simply adding additional lasers to pump molecular population out of unwanted levels back into the cooling cycle, as is done for hyperfine branching in atoms, adds additional complexity which quickly becomes unsustainable as molecular complexity itself increases. Hence, the application of laser cooling to molecules has instead focused on special classes of molecules in which the vibrational spectrum is almost decoupled from the electronic structure, resulting in nearly closed vibrational cycling transitions.[82]

Since direct laser cooling of molecules using the standard techniques developed for atoms is generally difficult, the most successful ultracold polar molecule experiments to date instead create ultracold molecules by "assembling" atoms which have themselves been brought to quantum degeneracy. For example, in the KRb experiment at JILA, this is performed by magneto-association of ultracold gases of K and Rb across a Feshbach resonance[83,j]. Many other such intra- and inter-species resonances have been identified. The resulting molecules are very highly excited, but may be transferred to the ground rotational-vibrational-electronic state using stimulated Raman adiabatic passage (STIRAP),[84] a feat which is enabled by modern developments in highly-stable laser technology. Once in the rotational-vibrational-electronic ground state, molecules can be transferred to their hyperfine

---

jIt is amusing to note that the first near-degenerate ground state molecule was formed by assembling the first bosonic (Rb) and fermionic (K) atoms to reach quantum degeneracy.

ground state by using further manipulation with microwaves.[85] More details on the production of ultracold molecules can be found in a number of other recent reviews.[86–90]

Polar molecules interact through electric dipole-dipole interactions which can be controlled by an external electric field. These interactions are repulsive when two molecules collide in a plane perpendicular to their oriented dipoles, but become attractive when the two molecules collide along their orientation axis [see e.g. Eq. (7)]. In addition, many of the experimentally relevant molecular species, including half of the alkali dimers,[91] are chemically reactive. Hence, in order to avoid rapid losses from chemical reactions, these molecules must be confined in a reduced-dimensional geometry that energetically suppresses the reactive collisions by geometrically allowing only repulsive interactions. In Ref. 92 it was shown that trapping molecules whose dipoles were aligned by an external electric field in a quasi-2D geometry generated by a 1D optical lattice reduced the chemical reaction rate by two orders of magnitude. Chemical reactions between molecules can be further suppressed by loading the molecules into a 3D optical lattice. In this case, the suppression either comes from freezing out the motion of the molecules altogether or through the quantum Zeno mechanism.[33,93] For the discussion of effective quantum magnetism in Sec. 2.1 we considered the case of molecules pinned in a deep optical lattice where only the internal degrees of freedom are relevant for the dynamics. Many experiments, including the JILA KRb experiment, form molecules directly in a 3D lattice by assembling lattice-confined atoms, and this scenario is particularly relevant for them.

4.2. *Molecular Structure and Progress Cooling Complex Molecules*

Because the Coulomb interaction is isotropic, the electronic state of an atom can be expressed in terms of its conserved angular momenta. One way to label atoms' electronic states is through term symbols $^{2S+1}L_J$, in which, e.g., $\hat{\mathbf{J}}^2|JM\rangle = J(J+1)|JM\rangle$, with $|JM\rangle$ a state with J quanta of angular momentum and a projection M of $\mathbf{J}$ along a space-fixed axis. Unlike atoms, which have spherical symmetry, (diatomic) molecules have only cylindrical symmetry with respect to the internuclear axis. This means that the relevant quantities for classifying molecules are not the "length" of the angular momentum vectors, e.g. J, but rather the projections of these vectors onto the internuclear axis. For example, the projection of the total electronic angular momentum along this axis, Ω, plays a similar role for molecules as the total angular momentum J for atoms. Table 1 overviews the spectroscopic term notation used for molecules and compares it with the notation for atoms. The appearance of an angular momentum is not meant to imply that this angular momentum is in fact a good quantum number. In atoms, relativistic effects couple the electronic orbital angular momentum $\mathbf{L}$ to the electron spin $\mathbf{S}$, and so L and S are no longer good quantum numbers. Similarly, the angular momenta of

Table 1. Labeling of electronic angular momenta for atoms and molecules

	Atomic species		Molecular species	
Angular momentum	Notation	Values	Notation	Values
Orbital angular momentum	L	S,P,D,F,...	Λ	Σ, Π, Δ,Φ, ...
Spin angular momentum	S	0, 1/2, 1,...	Σ	0, 1/2, 1,...
Total angular momentum	J	0,1/2, 1, ...	Ω	0,$\pm$1/2, $\pm$1, ...
Term symbol	$^{2S+1}L_J$		$^{2S+1}\|\Lambda\|_{\|\Omega\|}$	

a molecule can be coupled in a variety of ways by the intramolecular interactions and so the projections, e.g. Ω, may not be good quantum numbers. The various coupling schemes of the internal angular momenta of a molecule are classified by Hund's cases – for details, see e.g. Ref. 94.

It is an experimental fact that the majority of diatomic molecules have $^1\Sigma$ ground states, which is to say that the electronic wave function is invariant with respect to all symmetry transformations (e.g. rotations and reflections) of the molecule and the total spin is zero. In particular, the alkali metal dimers, including the species KRb,[83,95,96] LiCs,[97,98] RbCs,[99–102] NaK,[103,104] LiNa,[105] and LiRb[106] which are under active experimental investigation, all have $^1\Sigma$ ground states. Molecules that are formed from one open-shell atom, such as an alkali metal atom, and one closed-shell atom, such as an alkaline earth atom, have $^2\Sigma$ ground states due to the additional unpaired electron spin. Examples of such molecules being pursued in experiment include RbSr,[107] RbYb,[108] and LiYb.[109,110] Additionally, many molecules which have highly diagonal Franck-Condon factors and so are amenable to direct laser cooling, such as SrF,[111] YO,[112] and the alkaline-earth monohydrides[82] have $^2\Sigma$ ground states[k]. Finally, free radicals such as ClO, BrO, CH, NO, and OH all have $^2\Pi$ ground states which arise due to significant non-adiabatic effects in the electronic structure. The hydroxyl radical OH, which was the first radical to be studied with microwave spectroscopy,[114] has been shown to be amenable to evaporative cooling,[115] and experiments are underway to cool it to quantum degeneracy. While comparatively little work has been focused on cooling molecules with more than two atoms to degeneracy, a notable exception features polyatomic molecules which have both cylindrical symmetry and a small inversion splitting, known as symmetric top molecules. Such molecules, for example methyl fluoride, CH_3F, display a very strong coupling to external electric fields which allows them to be cooled in a modified Sisyphus scheme known as opto-electrical cooling[116,117] combined with guided deceleration.[118]

---

[k]Laser cooling does not require a molecule with a $^2\Sigma$ ground state; $^1\Sigma$ ground states with highly diagonal Franck-Condon factors would be preferred in principle due to their simpler internal structure. However, all $^1\Sigma$ species identified to have highly diagonal Franck-Condon factors have transition frequencies which are challenging to accommodate with currently available laser technology.[113]

5. Future Prospects

In this section we look to the future of quantum magnetic phenomena manifested in ultracold molecule experiments and identify both immediately available avenues as well as longer term prospects. The vista can be visualized in the quantum simulation landscape shown in Fig. 1. One immediate direction is to add an electric field to the Ramsey-type experiments described in Sec. 3. An electric field induces nonzero $d_\uparrow$ and $d_\downarrow$ dipole moments (see Fig. 3) and therefore enables the study of tunable XXZ magnetism rather than the spin-exchange (i.e. XX) magnetism observed in present experiments. In addition to tuning the $J_\perp$ and J_z terms, which can drive quantum phase transitions between magnetically ordered phases, the electric field also introduces density-density interactions [the V terms in Eq. (11)] and, more interestingly, density-spin couplings [W terms in Eq. (11)] which do not appear for magnetism generated by superexchange, see Eq. (2.2). The electric field can also aid the experiment in practical ways. For example, at a certain "magic" value of the electric field, the polarizabilities of the $|0,0\rangle$ and $|1,0\rangle$ state coincide[119] and the two states feel identical optical lattice potentials. This assists in reducing decoherence similar to "magic" wavelength conditions in atomic clocks.[120] Finally, as an electric field mixes rotational levels of different parity, it introduces small ($\mathcal{O}\left(\beta_{DC}\right)$ and higher) transition dipole matrix elements between states whose rotational quantum numbers differ by more than one. An electric field hence allows for higher rotational states to be populated with direct, single-photon microwave transitions, a key step towards harnessing the full richness of the molecule's internal structure.

Another near-term experimental enhancement which enables exciting new physics is the introduction of local state preparation and probes. As an example, one can imagine that only a specific, localized region of molecules are transferred to the rotational excited state. This localized excitation will propagate throughout the system via the dipolar spin-exchange process described by Eq. (17), and high-resolution imaging could be used to track this spin transport as well as the spread of correlations. The propagation of correlations in a long-range interacting system can display fundamentally different behavior from short-range interacting systems, such as the breakdown of locality.[30,121–125,146] In addition, such long-range spin transport has strong analogies to the transport of energy in light-harvesting complexes and exciton transport, leading to diverse quantum simulation prospects.[126] In the study of spin transport phenomena, the disordered filling of the lattice due to the relatively hot temperature of the molecules can be advantageous, as disordered, long-range interacting spin systems have been identified as prime candidates for observing many-body localization (MBL).[55,56,59–61] A system displaying MBL transports neither heat nor charge, even when the amount of energy is macroscopic. MBL is currently of great interest, in part because a system displaying MBL can break continuous symmetries or display topological order in situations where such order would be forbidden in equilibrium.[56,127] Hence, introducing local probes to present ultracold molecule experiments immediately provides access to a wealth of new phenomena.

A third direction for future experiments is to decrease the temperature of the molecules loaded into an optical lattice, which will in turn increase the lattice filling and allow for the study of quantum magnetism in equilibrium. Here, we mention only a few examples of novel physics, and refer the interested reader to recent review articles which focus on the equilibrium many-body physics of polar molecules.[128] An interesting feature of long-range interacting spin models is that they can display features normally associated with gapless, quantum critical phases, such as algebraic decay of correlations and large (area-law violating[129]) entanglement entropy, even in gapped phases.[130–134] Additionally, the interplay between coherent tunneling, spin interactions and chemical reactions[33,93] will allow us to study itinerant quantum magnetism such as generalized $t - J$ models.[22,23]

A final, longer-term direction for ultracold molecule realizations of quantum magnetism is to increase molecular diversity, both in the form of many different species with the same essential molecular structure as well as molecules with more complex structure. As an example of the former point, all of the alkali metal dimer molecules have $^1\Sigma$ ground states and hence essentially the same molecular structure, but differ in terms of energy scales: the dipole moments of the alkali dimers vary over an order of magnitude, as does the degree over which they can be polarized by an electric field of a given strength.[135] In addition, half of the alkali dimers are chemically reactive, while the other half are not.[91] An increase in the complexity of the internal structure beyond two rotational levels of a rigid rotor can be realized either by considering molecules with more complex structure or by accessing additional levels with more complex microwave/optical dressing schemes; we display both possibilities along the complexity axis of Fig. 1. These possibilities are expounded upon further in Fig. 6. In the far left panel, we display that even Σ-state molecules whose rotational structure is that of a simple rigid rotor often have other internal degrees of freedom with energies much smaller than rotational scales. In $^1\Sigma$ molecules such as the alkali dimers, the internal structure consists of hyperfine degrees of freedom which can be quite numerous; ^{40}K^{87}Rb has a hyperfine degeneracy of 36. The hyperfine structure is coupled to the rotational structure for nuclear spins $I \geq 1$ by nuclear quadrupole couplings. Such couplings allow for controllable population of nuclear spin states through microwave transitions.[85] Additional internal structure in Σ state molecules can also arise from unpaired electron spin, such as $^2\Sigma$ states for alkali metal-alkaline earth diatomics, long-lived $^3\Sigma$ excited states of many $^1\Sigma$ ground state molecules, and even exotic high-spin states such as the $^6\Sigma$ state of CrRb. Such additional degrees of freedom may be useful in simulating multi-orbital models with ultracold molecules, and can also be used to design complex spin models.[136,137]

Dressing of molecules with microwave or optical fields[22,23,138] leads to a very rich landscape of possibilities by accessing the full tensorial structure and anisotropy of the dipole-dipole interaction, Eq. (1), opening up components beyond the $q = 0$ one in Eq. (7). This enables the study of compass-type models in which the coupling strength depends on the spatial direction of the coupling, which has been proposed to lead to symmetry protected topological phases[139] and models with true topological order such as the Kitaev honeycomb model.[140] In addition, the full tensorial structure of the dipole-dipole interaction can exchange internal rotational angular momentum with orbital angular momentum, and hence lead to spin-orbit coupled rotational excitations which are chiral and feature a non-trivial Berry phase.[141] The essential idea of the microwave dressing procedure is to both (1) isolate d-level systems comprised of rotational levels linked by near-resonant radiation and (2) tune the interactions between these levels by choosing the polarizations, field strengths, and frequencies of the dressing radiation. Microwave dressing enables more complex interactions because many rotational levels can be coherently superposed, see Fig. 6, meaning that many dipole matrix elements are involved. Additionally, provided that such superpositions can be efficiently prepared, many such "dressed states" can be considered, leading to effectively high-dimensional spin representations, e.g. the spin-1 realized by the black, green, and red dressed levels in Fig. 6. Dressing with optical radiation which has spatial variation on the lattice

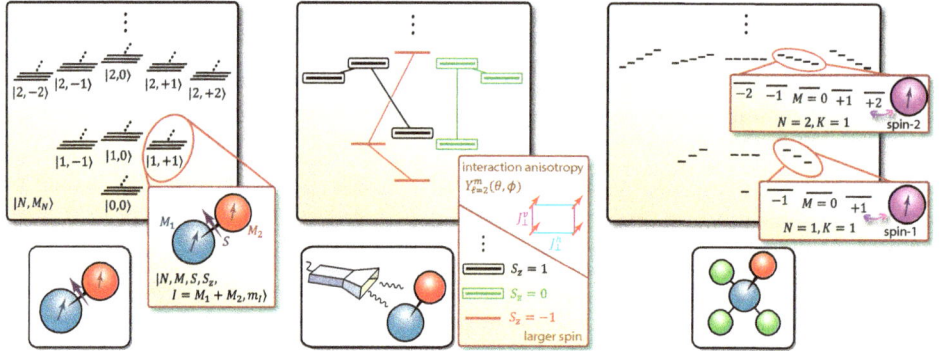

Fig. 6.   More complex physics from more complex molecules: beyond the rigid rotor. Left: most molecules have many hyperfine states, determined by the hyperfine degrees of freedom of the constituent atoms. (^{40}K^{87}Rb, for example, has 36 hyperfine states.) These degrees of freedom may serve to realize, for example, multi-orbital spin models. Middle: microwaves used to generate dressed states as the effective spin degrees of freedom allow the manipulation of the dipolar interaction in ways that do not occur for bare rotational states. This allows much greater control of both the spatial and spin anisotropy in the interaction. Right: molecules with more than two atoms give rise to new rotational states due to their ability to rotate around more than one axis. (The energy splittings are not to scale, and the magnitude of the "small" grouped splittings may even be larger than the large separations in some molecules.) These allow polar molecules to effectively behave as magnetic dipoles. Some possible physics arising from these more complex situations is depicted in Fig. 1.

scale enables even more diverse phenomena, such as topological flat bands[142] and fractional Chern insulators.[143]

Finally, new quantum magnetic phenomena arise in molecules with more complex internal structure than diatomic molecules with Σ ground states. As one example, we mention *symmetric top molecules* (STMs), molecules with (1) cylindrical symmetry and (2) nonzero rotational or orbital angular momentum projection along the body frame of the molecule. While diatomic Σ state molecules are cylindrically symmetric, they rotate about an axis perpendicular to their body frame, and hence do not satisfy (2). An example of an STM is the polyatomic molecule methyl fluoride, CH_3F, shown in Fig. 6. In methyl fluoride the rotational angular momentum can have some projection K onto the body frame of the molecule, and this leads to the rotational spectrum shown in the right panel of Fig. 6. In contrast to the Stark spectrum of Σ-state molecules shown in the left and center panels of Fig. 6, STMs display a linear coupling to an external electric field. This allows STMs to behave like elemental magnetic dipoles, leading to a large number of quantum simulation prospects.[144,147] For example, STMs in static fields can realize models of arbitrarily large integer spins interacting through anisotropic, long-range interactions.[144] We note that STMs do not have to contain more than two atoms; diatomic molecules with orbital angular momentum projection $|\Lambda| > 0$ such as the $^2\Pi$ ground state of OH[145] also behave as STMs in modest electric fields.

6. Conclusions

In conclusion, we have shown how to realize effective quantum magnetism with polar molecules pinned in optical lattices. The basic idea is to encode an effective spin 1/2 degree of freedom in two isolated rotational levels and to generate spin interactions via dipolar interactions. Even in the simplest scenario, the effective spin-spin couplings can be tuned by an external electric field as well as by the choice of rotational states forming the effective spin-1/2. This way of realizing quantum magnetism was contrasted with the superexchange mechanism. In the latter, the spin-spin couplings are induced via short-range interactions through virtual tunneling processes and hence the motional temperature plays an essential role. In polar molecule quantum magnetism, motional and spin temperatures are decoupled, and so quantum magnetism can be seen even in hot gases as long as a well characterized initial spin state is prepared. We reviewed the experimental confirmation of dipolar quantum magnetism in optical lattice experiments with KRb molecules at JILA, and discussed the new theoretical tools that were developed to verify and explain the experimental results. We expect that these tools, as well as the general theme that far from equilibrium dynamics can produce correlated quantum many-body physics in long-range interacting gases without motional degeneracy, will be of use in the cold gases community. Finally, we suggest some directions for future realizations of quantum magnetism with ultracold molecules, ranging from near-term experimental upgrades to long-term realizations of molecules with greater complexity.

Acknowledgments

We would like to acknowledge conversations with Lincoln Carr, Michael Foss-Feig, Alexey V. Gorshkov, Kenji Maeda, Salvatore Manmana, Alex Pikovski, Johannes Schachenmayer, Bihui Zhu, and especially the countless discussions with the JILA KRb experimental group over the last few years: Debbie Jin, Jun Ye, Bryce Gadway, Bo Yan, Jacob Covey, Brian Neyenhuis, and Steven Moses. This work was supported by the NSF (PIF-1211914 and PFC-1125844), AFOSR and ARO individual investigator awards, and the ARO with funding from the DARPA-OLE program. The authors also thank the the Aspen Center for Physics and KITP (NSF-PHY11-25915). KRAH and MLW thank the NRC for support.

References

1. M. H. Anderson, J. R. Ensher, M. R. Matthews, C. E. Wieman, and E. A. Cornell, Observation of Bose-Einstein condensation in a dilute atomic vapor, *Science.* **269**, 198–201, (1995).
2. K. B. Davis, M.-O. Mewes, M. R. Andrews, N. J. van Druten, D. S. Durfee, D. M. Kurn, and W. Ketterle, Bose-Einstein condensation in a gas of sodium atoms, *Phys. Rev. Lett.* **75**, 3969–3972, (1995).
3. C. C. Bradley, C. A. Sackett, J. J. Tollett, and R. G. Hulet, Evidence of Bose-Einstein condensation in an atomic gas with attractive interactions, *Phys. Rev. Lett.* **75**, 1687, (1995).
4. M. R. Matthews, B. P. Anderson, P. C. Haljan, D. S. Hall, C. E. Wieman, and E. A. Cornell, Vortices in a Bose-Einstein condensate, *Phys. Rev. Lett.* **83**, 2498–2501 (1999).
5. B. P. Anderson, P. C. Haljan, C. A. Regal, D. L. Feder, L. A. Collins, C. W. Clark, and E. A. Cornell, Watching dark solitons decay into vortex rings in a Bose-Einstein condensate, *Phys. Rev. Lett.* **86**, 2926–2929 (2001).
6. I. Bloch, J. Dalibard, and W. Zwerger, Many-body physics with ultracold gases, *Rev. Mod. Phys.* **80**, 885, (2008).
7. B. DeMarco and D. S. Jin, Onset of Fermi degeneracy in a trapped atomic gas, *Science.* **285**, 1703–1706, (1999).
8. M. A. Cazalilla and A. M. Rey, Ultracold Fermi gases with emergent SU(N) symmetry, *Rep. Prog. Phys.* **77**, 124401 (2014).
9. M. Lu, N. Q. Burdick, and B. L. Lev, Quantum degenerate dipolar fermi gas, *Phys. Rev. Lett.* **108**, 215301 (2012).
10. M. Lu, N. Q. Burdick, S. H. Youn, and B. L. Lev, Strongly dipolar Bose-Einstein condensate of Dysprosium, *Phys. Rev. Lett.* **107**, 190401 (2011).
11. K. Aikawa, A. Frisch, M. Mark, S. Baier, A. Rietzler, R. Grimm, and F. Ferlaino, Bose-einstein condensation of Erbium, *Phys. Rev. Lett.* **108**, 210401 (2012).
12. I. Bloch, Ultracold quantum gases in optical lattices, *Nature Physics.* **1**, 23–30, (2005).

13. R. P. Feynman, Simulating physics with computers, *Int. J. Theor. Phys.* **21**, 467, (1982).
14. M. Greiner, O. Mandel, T. Esslinger, T. W. Hansch, and I. Bloch, Quantum phase transition from a superfluid to a Mott insulator in a gas of ultracold atoms, *Nature.* **415**, 39–44, (2002).
15. R. Jördens, N. Strohmaier, K. Günter, H. Moritz, and T. Esslinger, A Mott insulator of fermionic atoms in an optical lattice, *Nature.* **455**, 204–207, (2008).
16. U. Schneider, L. Hackermüller, S. Will, T. Best, I. Bloch, T. A. Costi, R. W. Helmes, D. Rasch, and A. Rosch, Metallic and insulating phases of repulsively interacting fermions in a 3D optical lattice, *Science.* **322**(5907), 1520–1525, (2008).
17. A. Auerbach, *Interacting Electrons and Quantum Magnetism.* (Springer, Berlin, 1994).
18. S. Sachdev, Quantum magnetism and criticality, *Nature Physics.* **4**, 173, (2008).
19. C. Lacroix, P. Mendels, and F. Mila, Eds., *Introduction to frustrated magnetism.* (Springer, Heidelberg, 2011).
20. R. Barnett, D. Petrov, M. Lukin, and E. Demler, Quantum magnetism with multicomponent dipolar molecules in an optical lattice, *Phys. Rev. Lett.* **96**, 190401 (2006).
21. M. L. Wall and L. D. Carr, Hyperfine molecular Hubbard Hamiltonian, *Phys. Rev. A.* **82**, 013611 (2010).
22. A. V. Gorshkov, S. R. Manmana, G. Chen, J. Ye, E. Demler, M. D. Lukin, and A. M. Rey, Tunable superfluidity and quantum magnetism with ultracold polar molecules, *Phys. Rev. Lett.* **107**, 115301 (2011).
23. A. V. Gorshkov, S. R. Manmana, G. Chen, E. Demler, M. D. Lukin, and A. M. Rey, Quantum magnetism with polar alkali-metal dimers, *Phys. Rev. A.* **84**, 033619 (2011).
24. R. Zare, *Angular Momentum: Understanding Spatial Aspects in Chemistry and Physics.* (Wiley, New York, 1988).
25. D. Jaksch, C. Bruder, J. I. Cirac, C. W. Gardiner, and P. Zoller, Cold Bosonic Atoms in Optical Lattices, *Phys. Rev. Lett.* **81**, 3108–3111 (1998).
26. L.-M. Duan, E. Demler, and M. D. Lukin, Controlling spin exchange interactions of ultracold atoms in optical lattices, *Phys. Rev. Lett.* **91**, 090402 (2003).
27. A. Pikovski, M. Klawunn, A. Recati, and L. Santos, Nonlocal state swapping of polar molecules in bilayers, *Phys. Rev. A.* **84**, 061605 (2011).
28. K. R. A. Hazzard, M. van den Worm, M. Foss-Feig, S. R. Manmana, E. D. Torre, T. Pfau, M. Kastner, and A. M. Rey, Quantum correlations and entanglement in far-from-equilibrium spin systems, *Phys. Rev. A.* **90**, 063622 (2014).
29. M. Foss-Feig, K. R. A. Hazzard, J. J. Bollinger, and A. M. Rey, Nonequilibrium dynamics of arbitrary-range Ising models with decoherence: An exact analytic solution, *Phys. Rev. A.* **87**, 042101 (2013).
30. M. van den Worm, B. C. Sawyer, J. J. Bollinger, and M. Kastner, Relaxation timescales and decay of correlations in a long-range interacting quantum simulator, *New J. Phys.* **15**, 083007, (2013).
31. M. Foss-Feig, K. R. A. Hazzard, J. J. Bollinger, A. M. Rey, and C. W. Clark, Dynamical quantum correlations of Ising models on an arbitrary lattice and their resilience to decoherence, *New J. Phys.* **15**, 113008, (2013).
32. K. R. A. Hazzard, S. R. Manmana, M. Foss-Feig, and A. M. Rey, Far-from-equilibrium quantum magnetism with ultracold polar molecules, *Phys. Rev. Lett.* **110**, 075301 (Feb., 2013). ISSN 0031-9007.

33. B. Yan, S. A. Moses, B. Gadway, J. P. Covey, K. R. A. Hazzard, A. M. Rey, D. S. Jin, and J. Ye, Observation of dipolar spin-exchange interactions with lattice-confined polar molecules, *Nature.* **501**, 521, (2013).

34. B. Neyenhuis, B. Yan, S. A. Moses, J. P. Covey, A. Chotia, A. Petrov, S. Kotochigova, J. Ye, and D. S. Jin, Anisotropic polarizability of ultracold polar $^{40}K^{87}Rb$ molecules, *Phys. Rev. Lett.* **109**, 230403 (2012).

35. K. R. A. Hazzard, B. Gadway, M. Foss-Feig, B. Yan, S. A. Moses, J. P. Covey, N. Y. Yao, M. D. Lukin, J. Ye, D. S. Jin, and A. M. Rey, Many-body dynamics of dipolar molecules in an optical lattice, *Phys. Rev. Lett.* **113**, 195302 (2014).

36. J. S. Waugh, L. M. Huber, and U. Haeberlen, Approach to high-resolution nmr in solids, *Phys. Rev. Lett.* **20**, 180–182 (1968).

37. B. R. Johnson, J. S. Denker, N. Bigelow, L. P. Lévy, J. H. Freed, and D. M. Lee, Observation of nuclear spin waves in spin-polarized atomic hydrogen gas, *Phys. Rev. Lett.* **52**, 1508–1511 (1984).

38. N. D. Lemke, J. von Stecher, J. A. Sherman, A. M. Rey, C. W. Oates, and A. D. Ludlow, p-wave cold collisions in an optical lattice clock, *Phys. Rev. Lett.* **107**, 103902 (2011).

39. M. J. Martin, M. Bishof, M. D. Swallows, X. Zhang, C. Benko, J. von Stecher, A. V. Gorshkov, A. M. Rey, and J. Ye, A quantum many-body spin system in an optical lattice clock, *Science.* **341**, 632–636, (2013).

40. R. C. Stoneman, M. D. Adams, and T. F. Gallagher, Resonant-collision spectroscopy of Rydberg atoms, *Phys. Rev. Lett.* **58**, 1324–1327 (1987).

41. B. DeMarco and D. S. Jin, Spin excitations in a Fermi gas of atoms, *Phys. Rev. Lett.* **88**, 040405 (2002).

42. M. Koschorreck, D. Pertot, E. Vogt, and M. Kohl, Universal spin dynamics in two-dimensional Fermi gases, *Nature Physics.* **9**, 405–409 (July, 2013). ISSN 1745-2473.

43. A. B. Bardon, S. Beattie, C. Luciuk, W. Cairncross, D. Fine, N. S. Cheng, G. J. A. Edge, E. Taylor, S. Zhang, S. Trotzky, and J. H. Thywissen, Transverse demagnetization dynamics of a unitary Fermi gas, *Science.* **344**(6185), 722–724, (2014).

44. J. S. Krauser, U. Ebling, N. Fläschner, J. Heinze, K. Sengstock, M. Lewenstein, A. Eckardt, and C. Becker, Giant spin oscillations in an ultracold Fermi sea, *Science.* **343**, 157–160, (2014).

45. S. R. White, Density matrix formulation for quantum renormalization groups, *Phys. Rev. Lett.* **69**, 2863–2866 (1992).

46. G. Vidal, Efficient simulation of one-dimensional quantum many-body systems, *Phys. Rev. Lett.* **93**, 040502 (2004).

47. A. J. Daley, C. Kollath, U. Schollwöck, and G. Vidal, Time-dependent density-matrix renormalization-group using adaptive effective Hilbert spaces, *J. Stat. Mech.* p. P04005, (2004).

48. U. Schollwöck, The density-matrix renormalization group, *Rev. Mod. Phys.* **77**, 259–315 (2005).

49. U. Schollwöck, The density-matrix renormalization group in the age of matrix product states, *Annals of Physics.* **326**, 96–192, (2011).

50. E. Stoudenmire and S. R. White, Studying two-dimensional systems with the density matrix renormalization group, *Annual Review of Condensed Matter Physics.* **3**(1), 111–128, (2012).

51. A. A. Abrikosov, L. P. Gorkovn and I. E. Dzyaleshinki, *Methods of Quantum Field Theory in Statistical Physics.* (Dover Publications, New York, USA, Oct. 1975), revised edition. ISBN 0486632288.

52. J. W. Negele and H. Orland, *Quantum Many-Particle Systems.* (Westview Press, USA, 1998).

53. A. Polkovnikov, Phase space representation of quantum dynamics, *Annals of Physics.* **325**, 1790–1852, (2010).
54. R. M. Noack and S. R. Manmana, Diagonalization- and numerical renormalization-group-based methods for interacting quantum systems, *AIP Conference Proceedings.* **789**, 93–163, (2005).
55. D. M. Basko, I. L. Aleiner, and B. L. Altshuler, *Annals of Physics.* **321**, 1126, (2006).
56. A. Pal and D. A. Huse, Many-body localization phase transition, *Phys. Rev. B.* **82**, 174411 (2010).
57. L. Mathey, K. J. Günter, J. Dalibard, and A. Polkovnikov, Dynamic Kosterlitz-Thouless transition in 2D Bose mixtures of ultra-cold atoms, *arXiv:1112.1204*.
58. L. Mathey and A. Polkovnikov, Light cone dynamics and reverse Kibble-Zurek mechanism in two-dimensional superfluids following a quantum quench, *Phys. Rev. A.* **81**, 033605 (2010).
59. N. Y. Yao, C. R. Laumann, S. Gopalakrishnan, M. Knap, M. Mueller, E. A. Demler, and M. D. Lukin, Many-body localization with dipoles, *Phys. Rev. Lett.* **113**, 24302 (2014).
60. M. P. Kwasigroch and N. R. Cooper, Bose-Einstein condensation and many-body localization of rotational excitations of polar molecules, *Phys. Rev. A.* **90**, 021605(R) (2014).
61. M. Serbyn, M. Knap, S. Gopalakrishnan, Z. Papić, N. Y. Yao, C. R. Laumann, D. A. Abanin, M. D. Lukin, and E. A. Demler, Interferometric probes of many-body localization, *Phys. Rev. Lett.* **113**, 147204 (2014).
62. S. Fölling, Quantum noise correlation experiments with ultracold atoms, in Chapter 8 in "Quantum gas experiments — exploring many-body states," Päivi Törmä and K. Songetook, eds. (Imperial College Press, Singapore, 2014).
63. A. Rey, A. Gorshkov, C. Kraus, M. Martin, M. Bishof, M. Swallows, X. Zhang, C. Benko, J. Ye, N. Lemke, and A. Ludlow, Probing many-body interactions in an optical lattice clock, *Annals of Physics.* **340**, 311–351, (2014).
64. M. Knap, A. Kantian, T. Giamarchi, I. Bloch, M. D. Lukin, and E. Demler, Probing real-space and time-resolved correlation functions with many-body ramsey interferometry, *Phys. Rev. Lett.* **111**, 147205 (2013).
65. R. Bistritzer and E. Altman, Intrinsic dephasing in one-dimensional ultracold atom interferometers, *Proc. Natl. Acad. Sci.* **104**, 9955–9959, (2007).
66. C. De Grandi, V. Gritsev, and A. Polkovnikov, Quench dynamics near a quantum critical point, *Phys. Rev. B.* **81**, 012303 (Jan, 2010).
67. C. De Grandi, V. Gritsev, and A. Polkovnikov, Quench dynamics near a quantum critical point: Application to the sine-Gordon model, *Phys. Rev. B.* **81**, 224301 (2010).
68. V. Gritsev and A. Polkovnikov. Universal dynamics near quantum critical points. In ed. L. Carr, *Understanding in Quantum Phase Transitions*. Taylor & Francis, 2010, (2010).
69. A. Mitra, Time evolution and dynamical phase transitions at a critical time in a system of one-dimensional bosons after a quantum quench, *Phys. Rev. Lett.* **109**, 260601 (2012).
70. A. Mitra and T. Giamarchi, Thermalization and dissipation in out-of-equilibrium quantum systems: A perturbative renormalization group approach, *Phys. Rev. B.* **85**, 075117 (2012).
71. A. Mitra and T. Giamarchi, Mode-coupling-induced dissipative and thermal effects at long times after a quantum quench, *Phys. Rev. Lett.* **107**, 150602 (2011).

72. A. Polkovnikov, K. Sengupta, A. Silva, and M. Vengalattore, Colloquium: Nonequilibrium dynamics of closed interacting quantum systems, *Rev. Mod. Phys.* **83**, 863–883 (2011).

73. C. Karrasch, J. Rentrop, D. Schuricht, and V. Meden, Luttinger-liquid universality in the time evolution after an interaction quench, *Phys. Rev. Lett.* **109**, 126406 (2012).

74. R. Vosk and E. Altman, Many-body localization in one dimension as a dynamical renormalization group fixed point, *Phys. Rev. Lett.* **110**, 067204 (2013).

75. E. G. Dalla Torre, E. Demler, T. Giamarchi, and E. Altman, Dynamics and universality in noise-driven dissipative systems, *Phys. Rev. B.* **85**, 184302 (2012).

76. E. G. Dalla Torre, E. Demler, and A. Polkovnikov, Universal rephasing dynamics after a quantum quench via sudden coupling of two initially independent condensates, *Phys. Rev. Lett.* **110**, 090404 (2013).

77. S. De Sarkar, R. Sensarma, and K. Sengupta, A perturbative renormalization group approach to driven quantum systems, *J. Phys.: Condens. Matter* **26**, 325602 (2014).

78. A. Mitra, Correlation functions in the prethermalized regime after a quantum quench of a spin-chain, *Phys. Rev. B.* **87**, 205109, (2013).

79. O. Acevedo, L. Quiroga, F. J. Rodríguez, and N. F. Johnson, New dynamical scaling universality for quantum networks across adiabatic quantum phase transitions, *Phys. Rev. Lett.* **112**, 030403 (2014).

80. G. Quéméner and P. S. Julienne, Ultracold molecules under control!, *Chemical Reviews.* **112**(9), 4949–5011, (2012).

81. J. T. Bahns, W. C. Stwalley, and P. L. Gould, Laser cooling of molecules: A sequential scheme for rotation, translation, and vibration, *The Journal of Chemical Physics.* **104**, 9689–9697, (1996).

82. M. Rosa, Laser-cooling molecules, *The European Physical Journal D - Atomic, Molecular, Optical and Plasma Physics.* **31**, 395–402, (2004).

83. K.-K. Ni, S. Ospelkaus, M. H. G. de Miranda, A. Pe'er, B. Neyenhuis, J. J. Zirbel, S. Kotochigova, P. S. Julienne, D. S. Jin, and J. Ye, A high phase-space-density gas of polar molecules, *Science.* **322**, 231–235, (2008).

84. E. A. Shapiro, A. Pe'er, J. Ye, and M. Shapiro, Piecewise adiabatic population transfer in a molecule via a wave packet, *Phys. Rev. Lett.* **101**, 023601 (2008).

85. S. Ospelkaus, K.-K. Ni, G. Quéméner, B. Neyenhuis, D. Wang, M. H. G. de Miranda, J. L. Bohn, J. Ye, and D. S. Jin, Controlling the hyperfine state of rovibronic groundstate polar molecules, *Phys. Rev. Lett.* **104**, 030402 (2010).

86. S. Y. T. van de Meerakker, H. L. Bethlem, N. Vanhaecke, and G. Meijer, Manipulation and control of molecular beams, *Chemical Reviews.* **112**, 4828–4878, (2012).

87. N. R. Hutzler, H.-I. Lu, and J. M. Doyle, The buffer gas beam: An intense, cold, and slow source for atoms and molecules, *Chemical Reviews.* **112**, 4803–4827, (2012).

88. L. D. Carr, D. Demille, R. V. Krems, and J. Ye, Cold and ultracold molecules: Science, technology, and applications, *New J. Phys.* **11**, 055049, (2009).

89. C. P. Koch and M. Shapiro, Coherent control of ultracold photoassociation, *Chemical Reviews.* **112**, 4928–4948, (2012).

90. T. Köhler, K. Góral, and P. S. Julienne, Production of cold molecules via magnetically tunable feshbach resonances, *Rev. Mod. Phys.* **78**, 1311–1361 (2006).

91. P. S. Żuchowski and J. M. Hutson, Reactions of ultracold alkali-metal dimers, *Phys. Rev. A.* **81**, 060703 (2010).

92. M. H. G. de Miranda, A. Chotia, B. Neyenhuis, D. Wang, G. Quéméner, S. Ospelkaus, J. L. Bohn, J. Ye, and D. S. Jin, Controlling the quantum stereodynamics of ultracold bimolecular reactions, *Nature Physics.* **7**, 502–507 (2011).

93. B. Zhu, B. Gadway, M. Foss-Feig, J. Schachenmayer, M. L. Wall, R. A. Hazzard, K. B. Yan, A. Moses, S. P. Covey, J. S. Jin, D. J. Ye, M. Holland, and M. Rey, A. Suppressing the loss of ultracold molecules via the continuous quantum Zeno effect, *Phys. Rev. Lett.* **112**, 070404 (2014).

94. J. Brown and A. Carrington, *Rotational Spectroscopy of Diatomic Molecules.* (Cambridge University Press, Cambridge, 2003).

95. S. Ospelkaus, A. Pe'er, K. K. Ni, J. J. Zirbel, B. Neyenhuis, S. Kotochigova, P. S. Julienne, J. Ye, and D. S. Jin. Ultracold dense gas of deeply bound heteronuclear molecules, *Nature Phsics*, **4**, 622–626 (2008).

96. S. Ospelkaus, K. K. Ni, M. H. G. de Miranda, A. Pe'er, B. Neyenhuis, D. Wang, S. Kotochigova, P. S. Julienne, D. S. Jin, and J. Ye, Ultracold polar molecules near quantum degeneracy, *Faraday Discuss.* **142**, 351–359, (2009).

97. M. Repp, R. Pires, J. Ulmanis, R. Heck, E. D. Kuhnle, M. Weidemüller, and E. Tiemann, Observation of interspecies[6]Li-[133]Cs Feshbach resonances, *Phys. Rev. A.* **87**, 010701 (2013).

98. S.-K. Tung, C. Parker, J. Johansen, C. Chin, Y. Wang, and P. S. Julienne, Ultracold mixtures of atomic [6]Li and [133]Cs with tunable interactions, *Phys. Rev. A.* **87**, 010702 (2013).

99. H. Cho, D. McCarron, D. L. Jenkin, M. P. Köppinger, and S. L. Cornish, A high phase-space density mixture of [87]Rb and [133]Cs: towards ultracold heteronuclear molecules, *European Physical Journal D.* pp. 1–7, (2011).

100. M. Debatin, T. Takekoshi, R. Rameshan, L. Reichsöllner, F. Ferlaino, R. Grimm, R. Vexiau, N. Bouloufa, O. Dulieu, and H.-C. Nagerl, Molecular spectroscopy for ground-state transfer of ultracold RbCs molecules, *Phys. Chem. Chem. Phys.* **13**, 18926–18935, (2011).

101. T. Takekoshi, M. Debatin, R. Rameshan, F. Ferlaino, R. Grimm, H.-C. Nägerl, C. R. Le Sueur, J. M. Hutson, P. S. Julienne, S. Kotochigova, and E. Tiemann, Towards the production of ultracold ground-state RbCs molecules: Feshbach resonances, weakly bound states, and the coupled-channel model, *Phys. Rev. A.* **85**, 032506 (2012).

102. T. Takekoshi, L. Reichsöllner, A. Schindewolf, J. M. Hutson, C. R. L. Sueur, O. Dulieu, F. Ferlaino, R. Grimm, and H.-C. Nägerl, Ultracold dense samples of dipolar RbCs molecules in the rovibrational and hyperfine ground state, *Phys. Rev. Lett.* **113**, 205301 (2014).

103. J. W. Park, C.-H. Wu, I. Santiago, T. G. Tiecke, S. Will, P. Ahmadi, and M. W. Zwierlein, Quantum degenerate Bose-Fermi mixture of chemically different atomic species with widely tunable interactions, *Phys. Rev. A.* **85**, 051602 (2012).

104. C.-H. Wu, J. W. Park, P. Ahmadi, S. Will, and M. W. Zwierlein, Ultracold fermionic Feshbach molecules of [23]Na[40]K, *Phys. Rev. Lett.* **109**, 085301 (2012).

105. M.-S. Heo, T. T. Wang, C. A. Christensen, T. M. Rvachov, D. A. Cotta, J.-H. Choi, Y.-R. Lee, and W. Ketterle, Formation of ultracold fermionic NaLi Feshbach molecules, *Phys. Rev. A.* **86**, 021602 (2012).

106. S. Dutta, J. Lorenz, A. Altaf, D. S. Elliott, and Y. P. Chen, Photoassociation of ultracold LiRb* molecules: Observation of high efficiency and unitarity-limited rate saturation, *Phys. Rev. A.* **89**, 020702 (2014).

107. B. Pasquiou, A. Bayerle, S. M. Tzanova, S. Stellmer, J. Szczepkowski, M. Parigger, R. Grimm, and F. Schreck, Quantum degenerate mixtures of strontium and rubidium atoms, *Phys. Rev. A.* **88**, 023601 (2013).

108. M. Borkowski, P. S. Żuchowski, R. Ciuryło, P. S. Julienne, D. Kędziera, L. Mentel, P. Tecmer, F. Münchow, C. Bruni, and A. Görlitz, Scattering lengths in isotopologues of the RbYb system, *Phys. Rev. A.* **88**, 052708 (2013).

109. H. Hara, Y. Takasu, Y. Yamaoka, J. M. Doyle, and Y. Takahashi, Quantum degenerate mixtures of alkali and alkaline-earth-like atoms, *Phys. Rev. Lett.* **106**, 205304 (2011).

110. A. H. Hansen, A. Y. Khramov, W. H. Dowd, A. O. Jamison, B. Plotkin-Swing, R. J. Roy, and S. Gupta, Production of quantum-degenerate mixtures of ytterbium and lithium with controllable interspecies overlap, *Phys. Rev. A.* **87**, 013615 (2013).

111. E. S. Shuman, J. F. Barry, and D. DeMille, Laser cooling of a diatomic molecule, *Nature.* **467**, 820–823 (2010).

112. M. T. Hummon, M. Yeo, B. K. Stuhl, A. L. Collopy, Y. Xia, and J. Ye, 2D magneto-optical trapping of diatomic molecules, *Phys. Rev. Lett.* **110**, 143001 (2013).

113. J. Barry. *Laser cooling and slowing of a diatomic molecule.* PhD thesis, Yale University, (2013).

114. G. C. Dousmanis, T. M. Sanders, and C. H. Townes, Microwave spectra of the free radicals OH and OD, *Phys. Rev.* **100**, 1735–1754 (1955).

115. B. K. Stuhl, M. T. Hummon, M. Yeo, G. Quéméner, J. L. Bohn, and J. Ye, Evaporative cooling of the dipolar hydroxyl radical, *Nature.* **492**, 396–400, (2012).

116. B. G. U. Englert, M. Mielenz, C. Sommer, J. Bayerl, M. Motsch, P. W. H. Pinkse, G. Rempe, and M. Zeppenfeld, Storage and adiabatic cooling of polar molecules in a microstructured trap, *Phys. Rev. Lett.* **107**, 263003 (2011).

117. M. Zeppenfeld, B. G. Englert, R. Glöckner, A. Prehn, M. Mielenz, C. Sommer, L. D. van Buuren, M. Motsch, and G. Rempe, Sisyphus cooling of electrically trapped polyatomic molecules, *Nature.* **491**, 570–573 (2012).

118. S. Chervenkov, X. Wu, J. Bayerl, A. Rohlfes, T. Gantner, M. Zeppenfeld, and G. Rempe, Continuous centrifuge decelerator for polar molecules, *Phys. Rev. Lett.* **112**, 013001 (2014).

119. S. Kotochigova and D. DeMille, Electric-field-dependent dynamic polarizability and state-insensitive conditions for optical trapping of diatomic polar molecules, *Phys. Rev. A.* **82**, 063421 (2010).

120. J. Ye, H. J. Kimble, and H. Katori, Quantum state engineering and precision metrology using state-insensitive light traps, *Science.* **320**, 1734–1738, (2008).

121. J. Schachenmayer, B. P. Lanyon, C. F. Roos, and A. J. Daley, Entanglement growth in quench dynamics with variable range interactions, *Phys. Rev. X.* **3**, 031015 (2013).

122. P. Hauke and L. Tagliacozzo, Spread of correlations in long-range interacting quantum systems, *Phys. Rev. Lett.* **111**, 207202 (2013).

123. J. Eisert, M. van den Worm, S. R. Manmana, and M. Kastner, Breakdown of quasilocality in long-range quantum lattice models, *Phys. Rev. Lett.* **111**, 260401 (2013).

124. P. Richerme, Z.-X. Gong, A. Lee, C. Senko, J. Smith, M. Foss-Feig, S. Michalakis, A. V. Gorshkov, and C. Monroe, Non-local propagation of correlations in long-range interacting quantum systems, *Nature* **511**, 198 (2014).

125. P. Jurcevic, B. P. Lanyon, P. Hauke, C. Hempel, P. Zoller, R. Blatt, and C. F. Roos, Observation of entanglement propagation in a quantum many-body system, *Nature* **511**, 202 (2014).

126. G. Günter, H. Schempp, M. Robert-de Saint-Vincent, V. Gavryusev, S. Helmrich, C. S. Hofmann, S. Whitlock, and M. Weidemüller, Observing the dynamics of dipole-mediated energy transport by interaction-enhanced imaging, *Science.* **342**, 954–956, (2013).

127. A. Chandran, V. Khemani, C. R. Laumann, and S. L. Sondhi, Many-body localization and symmetry-protected topological order, *Phys. Rev. B.* **89**, 144201 (2014).

128. M. A. Baranov, M. Dalmonte, G. Pupillo, and P. Zoller, Condensed matter theory of dipolar quantum gases, *Chemical Reviews.* **112**, 5012–5061, (2012).

129. J. Eisert, M. Cramer, and M. B. Plenio, Colloquium, *Rev. Mod. Phys.* **82**, 277–306 (2010).

130. X.-L. Deng, D. Porras, and J. I. Cirac, Effective spin quantum phases in systems of trapped ions, *Phys. Rev. A.* **72**, 063407 (2005).

131. J. Eisert and T. J. Osborne, General entanglement scaling laws from time evolution, *Phys. Rev. Lett.* **97**, 150404 (2006).

132. P. Hauke, F. M. Cucchietti, A. Müller-Hermes, M.-C. Bañuls, J. I. Cirac, and M. Lewenstein, Complete devil's staircase and crystal-superfluid transitions in a dipolar XXZ spin chain: a trapped ion quantum simulation, *New J. Phys.* **12**, 113037, (2010).

133. T. Koffel, M. Lewenstein, and L. Tagliacozzo, Entanglement entropy for the long-range Ising chain in a transverse field, *Phys. Rev. Lett.* **109**, 267203 (2012).

134. D. Vodola, L. Lepori, E. Ercolessi, A. V. Gorshkov, and G. Pupillo, Kitaev chains with long-range pairing, *Phys. Rev. Lett.* **113**, 156402 (2014).

135. M. L. Wall, E. Bekaroglu, and L. D. Carr, Molecular Hubbard hamiltonian: Field regimes and molecular species, *Phys. Rev. A.* **88**, 023605 (2013).

136. A. Micheli, G. K. Brennen, and P. Zoller, A toolbox for lattice-spin models with polar molecules, *Nature Physics.* **2**, 341, (2006).

137. G. K. Brennen, A. Micheli, and P. Zoller, Designing spin-1 lattice models using polar molecules, *New J. Phys.* **9**, 138, (2007).

138. N. R. Cooper and G. V. Shlyapnikov, Stable topological superfluid phase of ultracold polar fermionic molecules, *Phys. Rev. Lett.* **103**, 155302 (2009).

139. S. R. Manmana, E. M. Stoudenmire, K. R. A. Hazzard, A. M. Rey, and A. V. Gorshkov, Topological phases in ultracold polar-molecule quantum magnets, *Phys. Rev. B.* **87**, 081106 (2013).

140. A. V. Gorshkov, K. R. A. Hazzard, and A. M. Rey, Kitaev honeycomb and other exotic spin models with polar molecules, *Molecular Physics.* **111**, 1908–1916, (2013).

141. S. V. Syzranov, M. L. Wall, V. Gurarie, and A. M. Rey, Spin-orbital dynamics in a system of polar molecules, *Nature Communications* **5**, 5391, (2014).

142. N. Y. Yao, C. R. Laumann, A. V. Gorshkov, S. D. Bennett, E. Demler, P. Zoller, and M. D. Lukin, Topological flat bands from dipolar spin systems, *Phys. Rev. Lett.* **109**, 266804 (2012).

143. N. Y. Yao, A. V. Gorshkov, C. R. Laumann, A. M. Läuchli, J. Ye, and M. D. Lukin, Realizing fractional Chern insulators in dipolar spin systems, *Phys. Rev. Lett.* **110**, 185302 (2013).

144. M. L. Wall, K. Maeda, and L. D. Carr, Simulating quantum magnets with symmetric top molecules, *Annalen der Physik.* **525**, 845–865, (2013).

145. B. K. Stuhl, M. T. Hummon, M. Yeo, G. Quéméner, J. L. Bohn, and J. Ye, Evaporative cooling of the dipolar hydroxyl radical, *Nature.* **492**, 396–400, (2012).

146. J. Schachenmayer, A. Pikovski, and A. M. Rey, Many-body quantum spin dynamics with Monte Carlo trajectories on a discrete phase space, *Phys. Rev. X.* **5**, 011022 (2015).

147. M. L. Wall, K. Maeda, and L. D. Carr, Realizing unconventional quantum magnetism wih symmetric top molecules, *New J. Phys.* **17**, 025001 (2015).

Chapter 2

Optical Manipulation of Light Scattering in Cold Atomic Rubidium

R. G. Olave*, A. L. Win, K. Kemp, S. J. Roof, S. Balik, and M. D. Havey

Old Dominion University, Department of Physics, Norfolk, Virginia 23529, USA

I. M. Sokolov[1,2] and D. V. Kupriyanov[1]

[1]*Department of Theoretical Physics, State Polytechnic University, 195251, St.-Petersburg, Russia*

[2]*Institute for Analytical Instrumentation, RAS, 198103, St.-Petersburg, Russia*

A brief perspective on light scattering in dense and cold atomic rubidium is presented. We particularly focus on the influence of auxiliary applied fields on the system response to a weak and nearly resonant probe field. Auxiliary fields can strongly disturb light propagation; in addition to the steady state case, dynamically interesting effects appear clearly in both the time domain, and in the optical polarization dependence of the processes. Following a general introduction, two examples of features found in such studies are presented. These include nonlinear optical effects in (a) comparative studies of forward- and fluorescence-configuration scattering under combined excitation of a control and probe field, and (b) manipulation of the spatial structure of the optical response due to a light shifting strong applied field.

1. Introduction

The research presented in this chapter forms one part of the broad interface between atomic and condensed matter physics. A particular focus is an overview of quantum optics in high-density and cold atomic matter, in which disorder-driven electromagnetic interactions can develop strongly correlated character.[1] These emergent effects are typified by Anderson localization of light, the onset of random lasing and collective interatomic interactions such as super and sub radiance involving the system particles as a whole. The overriding theme is

---

*rolave@odu.edu

then investigation of the physics of strongly correlated atomic-radiative systems, including the collective properties and phase transitions in ultracold atomic vapor. A closely related secondary theme appears when we consider the resonant interparticle interactions among the system constituents. Then the macroscopic optical response is modified from the microscopic single atom Lorentzian response to include so-called local field shifts, including the collective Lamb shift[2,3]. It is important to note that much of the phenomenology appears most strongly in high density gases at low temperatures. By this we mean that the temperatures on the order of 100 μK or less, and atomic densities are characteristically greater than 10^{13} atoms/cm^3. For such densities there are typically several atoms in a volume λ^3, where λ is the wavelength of light in the atomic medium.

We generally emphasize that the processes of emission and absorption of radiation in dense gases are quite representative of near-resonant many body interactions among the fields and the atoms in the ensemble.[2,3] In the approximation where the atoms have a pairwise interaction of the dipole type, there contributes both a long range radiative interaction decreasing as $1/r$ with increasing atom-atom separation r, and short range interactions of the form of the Coulomb electrostatic dipole-dipole interaction. Among other effects which appear when virtual photons are exchanged among the different atoms in the system there emerges a collective Lamb shift which is essential to description of the global response of the system. In general, the collective aspects yield a modification of the well-known local field shift[4,5] of the resonance frequency,[6] and (with proper state preparation) an associated superradiant emission from the system.[7,8]

In this chapter we focus on the interaction of a weak and near resonance probe beam with an ensemble of cold atoms. Even more specifically, we consider the influence of a second auxiliary field on the optical response of the probe. Overall, we study influence of auxiliary fields in both the time domain and the dependence on the polarization of both detected light and the relative optical polarization of the auxiliary and probe fields. There are two cases we consider. In the first, the spectral structure of the atomic susceptibility is modified by the auxiliary field. This is the case for the reported experimental results, and associated theoretical calculations, on polarized fluorescence excited in the so called lambda configuration characteristically employed in studies of electromagnetically induced transparency. In the second, we consider modification of the spatial properties of an ultracold ensemble by an auxiliary field which produces a spatially varying light shift of the atomic levels. This approach allows simulation of a quasi one dimensional spatial configuration of atoms which, for instance, can have applications to studies of superradiance,[9] Anderson localization[10] of light and random lasing[11–14] in reduced spatial geometry.

2. Illustrative Topics

2.1. *Polarized Fluorescence Studies under Conditions of Electromagnetically induced Transparency in Cold Atomic Rubidium*

2.1.1. *Introduction*

In this portion of the chapter we report on a primarily experimental study directed towards optical control of weak localization[15–19] in ultracold atomic vapor. Generally, control is achieved by application of a dressing field to the atomic medium in the presence of the weaker probe incident radiation.[20,21] Such a configuration has a wide range of applications[20,22–26,30,31] in nonlinear optics,[26–29] quantum information,[20,22,23] single photon manipulation,[20,22,23] and precision magnetometry.[30,31] Here we consider manipulation of the weak probe beam and light scattered from it, thus undergoing multiple scattering in the weak localization regime. The particular scheme used is not optimum for large orders of multiple scattering, as the amount of multiple scattering is limited to at most several orders by the low optical depth on the relevant probe transitions, and by spontaneous emission into dark states which are only weakly coupled by either the control or probe field. In the experiments reported here we compare measured and theoretical polarization dependent forward and sideways scattered light. Finally we note that in these studies we do not use relative phase locked pump and control fields, thus realizing quite a bit shorter, and more readily measurable, time scales for the time evolution of both the probe and control field polarized fluorescence, and the forward scattered light.

In the remaining sections, we first describe the experimental approach. This is followed by a presentation of our investigation of the forward scattered probe light and the light polarization dependence of the scattered light in a ninety degree fluorescence geometry. Comparison and discussion is made through theoretical results[32–34] obtained for a model of the experimental configurations.

2.1.2. *Experimental Configuration*

The sample of cold ^{87}Rb gas is obtained by cooling and loading the atomic vapor into a magneto optical trap (MOT),[35] which operates near the closed hyperfine transition $F = 2 \rightarrow F' = 3$. The trapping laser is derived from an external-cavity diode laser (ECDL) system in a master-slave configuration, and is split into three pairs of retro reflected beams, delivering an estimated trapping laser intensity of $\simeq 40$ mW/cm^2. To prevent optical pumping of the atomic vapor into the lower energy $F = 1$ hyperfine level, a repumper laser is set to the $F = 1 \rightarrow F' = 2$ hyperfine transition. The trap and repumper lasers can be frequency tuned and switched on or off through acousto optical modulators (AOM). Fluorescence imaging measurements show that the sample has an effective Gaussian atom distribution, which can be described by $\rho(r) = \rho_o e^{-r^2/2r_o^2}$; for this sample $r_o \approx 0.3$ mm. The temperature of the ^{87}Rb atomic gas was determined via ballistic expansion measurements, and is estimated to be $\leq 100\ \mu$K. Finally, a maximum optical depth of $b \sim 10$ on the trapping transition was determined through absorption imaging measurements, resulting in a peak atomic density of $\rho_o \sim 9 \cdot 10^{10}$ atoms/cm^3.

In this experiment, several overlapping Zeeman Λ schemes operating within a three energy-level configuration are shown in Fig. 1. There the resonance transition frequencies are given by f_{11} for the $F = 1 \rightarrow F' = 1$ transition, and f_{21} for the $F = 2 \rightarrow F' = 1$ transition. Before any measurement, the repumper laser is turned off for 1 ms, and in this period approximately 86% of the atoms are optically pumped to the $F = 1$ hyperfine ground state. Initially, an intense linearly polarized control laser beam with Rabi frequency Ω_c couples on bare atomic resonance the $F = 2$ ground state to the $F' = 1$ excited state ($\Delta_c = 0$, where $\Delta_c = f_c - f_{21}$ is the control field detuning from resonance). The medium is then probed by a weak pulse with orthogonal linear polarization and Rabi frequency Ω_p, with a frequency detuning from the bare atomic resonance given by Δ_p. Here $\Delta_p = f_p - f_{11}$.

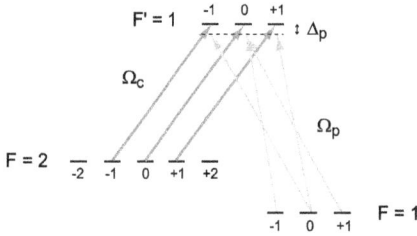

Fig. 1. Relevant hyperfine lambda scheme. The near-resonant strong coupling laser has Rabi frequency Ω_c. The probe laser has Rabi frequency Ω_p, and detuning from the bare atomic resonance of Δ_p.

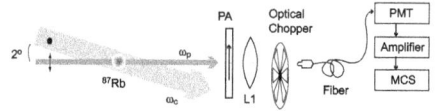

Fig. 2. Experimental setup for detection of transmission of a probe pulse through the atomic gas. The coupling laser has a frequency f_c, while the probe laser has a frequency f_p. A lens L1 collects the probe pulse transmitted through the Rb cloud and launches the beam into a fiber. PA is a polarization analyzer, PMT is a photomultiplier tube, and MCS is a multichannel scaler.

The control and probe lasers come from separate ECDL systems, with a combined root-mean-square bandwidth of 1.3 MHz, determined via optical heterodyning measurements. The control beam has a Gaussian spatial intensity profile, with a radius of $r_{wc} = 0.9$ mm, and is closely uniform in intensity over the volume of the MOT. The probe beam also has a Gaussian spatial intensity profile, and a radius of $r_{wp} = 0.2$ mm. The probe is directed through the central region of the atomic sample.

2.1.3. *Results and Discussion*

In this section we first present our results for forward scattering geometry, where we study the time-dependent optical behavior of a near resonance probe pulse transmitted through a ^{87}Rb vapor cloud dressed by an optical control field. This is followed by time-resolved observation of the polarized probe and control field induced sample fluorescence in a 90° geometry.

2.1.4. *Forward Scattering Geometry*

A schematic setup for the detection of the transmitted probe pulse is shown in Fig. 2. A coupling laser with frequency f_c together with a probe laser of frequency f_p forms a coupled set of near-resonance lambda configurations, and the forward scattered probe light is detected. To minimize background and to avoid saturation of the detector, the coupling laser has been offset by about 2° from the probe pulse propagation axis. A polarization analyzer (PA) selects light with the same polarization as the incident probe pulse, thus strongly suppressing the control field background at the detector. A lens L1 focuses the transmitted light into a multimode fiber which further directs the light to an infrared-sensitive photomultiplier tube (PMT). The time-resolved signals are then amplified using a fast preamplifier and recorded by a multichannel scaler. The system time resolution is about 5 ns.

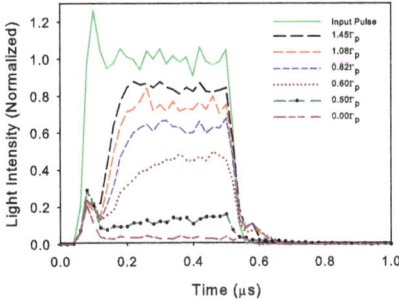

Fig. 3. Measurement of the time evolution of the transmitted probe pulse for varying Ω_c. The nearly rectangular probe pulse is 500 ns long. Data is taken at $\Delta_p = \Delta_c = 0$, for an atomic sample with optical depth of $b \simeq 3$. The slow light components can be observed as well. The probe pulse in the absence of a sample, and the probe pulse when there is no coupling laser ($\Omega_c = 0.0\ \Gamma_p$) are included for comparison. Here, $\Gamma_p = 3.8 \cdot 10^7\,s^{-1}$.

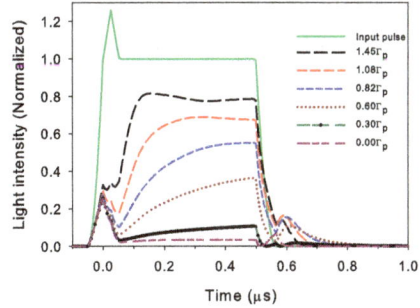

Fig. 4. Theoretical calculations of the time evolution of the transmitted probe pulse for varying Ω_c, as indicated in the caption. The probe pulse is approximately 500 ns long, at $\Delta_p = \Delta_c = 0$, for an atomic sample with optical depth of $b = 3.4$. The calculations assume a ground states decoherence rate of $0.07\ \Gamma_p$. Here, $\Gamma_p = 3.8 \cdot 10^7\,s^{-1}$.

For a resonant optical depth of $b \simeq 3$ on the probe transition, measurements were taken to compare the time evolution and intensity of the transparency of a 500 ns long probe pulse as a function of the control field intensity Ω_c; the experimental results, for a resonant probe $\Delta_p = \Delta_c = 0$, can be seen in Fig. 3. We point out that, strictly speaking this detuning relation does not precisely define the point of maximal transparency because of the additional light shift induced by the coupling with other upper-state hyperfine sublevels.[36] However in the discussed experimental conditions, when the Rabi frequency has the same order as the natural decay rate $\Gamma_p = 3.8 \cdot 10^7 s^{-1}$ this correction is very small. The measurements are time averaged so as to have a 20 ns time resolution; this improves the signal to noise ratio, while at the same time giving a temporal resolution shorter than the natural lifetime of 26 ns. Clearly, the transparency develops for all employed values of the coupling field strength, but it reaches a steady state faster for greater values of Ω_c. In these data the delayed pulses associated with slow light can be seen as well. It should be noted that the incident probe pulse is not exactly temporally square; it has an estimated rise and fall time of 40 ns. The pulse initially overshoots (about 10% above the main signal), this is due in part to initial overshooting of the acousto-optic modulator (AOM) that controls the switching of the pulse. This overshoot does not take place when the AOM switches off.

Theoretical calculations for the transmission of the probe pulse under similar conditions are presented in Fig. 4. The calculational approach we have used is presented in detail in several of our earlier papers,[32–34] and is not reiterated here. We do point out that the probe pulse shape, including the mentioned overshoot, and the measured rise and fall times, is built into the calculation. In comparing Figs. 3 and 4, we see overall qualitative agreement between the experimental and theoretical time responses at the various control field Rabi frequencies.

From results similar to those of Fig. 3, the steady state response of the configuration as a function of the control field Rabi frequency may be extracted. Figure 5 shows such results for Rabi frequencies Ω_c ranging from well smaller than the natural width up to about 8 MHz. For this figure, the signals have been normalized such that the steady-state intensity of the incident probe pulse corresponds to 1 (100% transparency). The uncertainty in the data comes primarily from signal counting statistics and from run-to-run variations in experimental conditions.

Figure 6(a) presents a set of measurements taken, for a control field Rabi frequency $\Omega_c = 1.45\ \Gamma_p$, to investigate storage and retrieval of light pulses. For these measurements the control and probe fields are simultaneously turned off (at $t = 0.6\ \mu s$), and the control field is switched back on after some variable time delay. As can be seen, when the control field is turned back on, some portion of the stored light pulse is retrieved. Figure 6(b) is a plot of the integrated transmission of the probe pulse versus storage time. A fit to the data yields a $1/e$ decay time of the atomic coherence of 196 ns, or a decoherence rate on the order of 0.9 MHz. In order to physically understand this relatively large decoherence rate, we consider that when the pump and control fields are turned off, a dark state polariton is formed with some degree of efficiency. Once the probe and control fields are off,

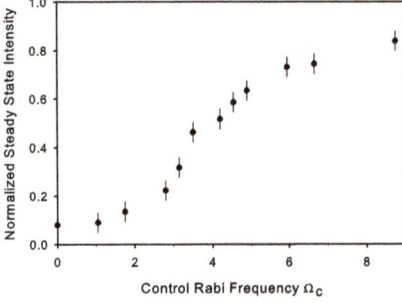

Fig. 5. Intensity of the steady-state transmitted probe signal as a function of Ω_c. The probe and control fields are on resonance ($\Delta_p = \Delta_c = 0$). The signals are normalized to the intensity of the undisturbed probe pulse.

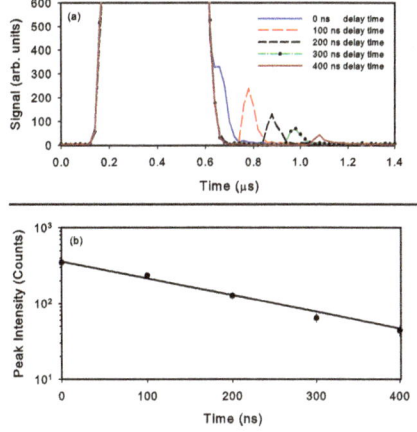

Fig. 6. Probe pulse light storage and retrieval with variable control field delays for $\Omega_c = 1.45\ \Gamma_p$. (a) The control field is switched off at $t = 0.6\ \mu s$, and turned on after different time intervals. (b) Comparison of the retrieved probe signal for variable storage times. A fit to the data yields a $1/e$ decay rate of the atomic coherence of 195 ns. Here, $\Gamma_p = 3.8 \cdot 10^7 s^{-1}$.

the quasiparticle decays relatively slowly in the dark with decoherence time limited mainly by residual magnetic fields over the volume of the atomic sample. However, when the control field is turned back on, in order to regenerate the probe optical field, the control field phase has evolved from when it was turned off. For each realization of the experiment, this phase in relation to that of the created polariton will be different. This phase evolution leads to a decay of the coherence that is proportional to the linewidth of the control laser field (~ 1 MHz), which becomes relevant when it is switched back on to regenerate the forward scattered light.

2.1.5. *Fluorescence Detection Geometry*

A schematic setup for the detection of fluorescence is shown in Fig. 7. In these measurements, the linearly polarized control laser with frequency f_c is initially switched on, and 50 μs later a weak laser probe pulse with frequency f_p and orthogonal linear polarization probes the system. A linear polarization analyzer PA selects the polarization state of the light to be detected. A lens L1 collects and collimates fluorescence coming from the location of the MOT, and a lens L2 focuses the light into a multimode optical fiber which transfers the light to the PMT. The signal is then amplified and time-resolved using a multichannel scaler. Note that for these measurements, in contrast with the transmission experiments, there is no 2° offset between the control and the probe beams. The relative response of the optical and electronic detection system is calibrated for two orthogonal linear polarizations, corresponding to directions collinear with, or orthogonal to, the probe or control field. This allows meaningful measurements of the linear polarization degree of the time-resolved fluorescence to be made.

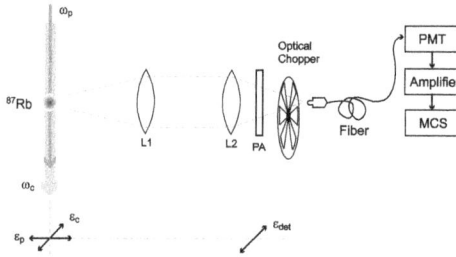

Fig. 7. Experimental setup for detection of fluorescence from a probe pulse propagating through the atomic gas. The coupling laser has a frequency t_c. The probe laser has a frequency t_L. A lens L1 collects fluorescence emitted from the atomic cloud, while a lens L2 focuses this light into a fiber. PA is a polarization analyzer, PMT is a photomultiplier tube, MCS is a multichannel scaler. The probe and control lasers have linear and mutually orthogonal polarization. The detected polarization is variable, and is shown in the figure to be collinear with the control field polarization. ϵ_p, ϵ_c, and ϵ_{det} are the probe, control, and detected light polarization vectors respectively. Drawing not to scale.

The fluorescence generated by optical excitation of the system by the probe and control fields, as a function of the control laser Rabi frequency Ω_c, is presented in Fig. 8. In these measurements the control field is linearly polarized and perpendicular to the detector plane. The detector linear polarization analyzer is set to maximally pass light polarized in the same direction as the control field. For a control field with a low intensity ($\Omega_c < \Gamma_p$), a slower build up of fluorescence is initially observed, but the transparency is small and it has a narrow frequency width, so the main contribution to the fluorescence signal comes from scattered light having frequency components outside the transparency window. On the other hand, for large coupling field intensities ($\Omega_c > \Gamma_p$) the medium develops a high degree of transparency for a wider frequency range, which results in a suppression of fluorescence clearly observed in the data.

For weak control field Rabi frequencies, there is a slowly decaying portion of the fluorescence signal, as revealed by Fig. 9. To explore this slow decay, we employed a protocol like that used in storage and retrieval type of measurements of slow fluorescence light, and similar to our transmission measurements (Fig. 6). The results, for a control field Rabi frequency of $\Omega_c = 0.69 \; \Gamma_p$, with 5 ns resolution, are presented in Fig. 9. In these measurements, the control field was turned off at $t = 0.5 \; \mu s$, and switched back on with variable time delays. We see that the obtained fluorescence signals, upon switching back on the control field, with no apparent amplitude decay, for periods of time much longer than the decay time

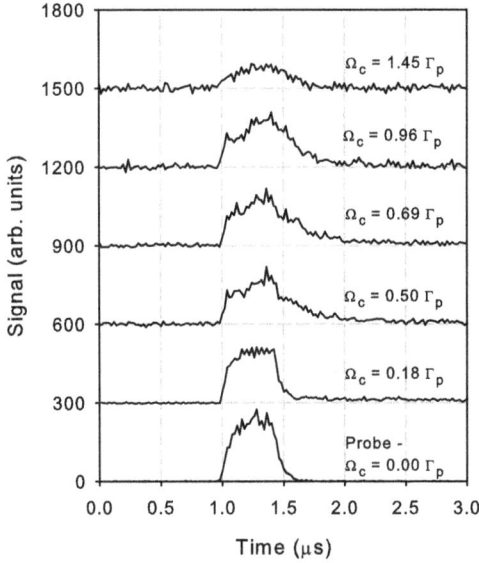

Fig. 8.   Probe fluorescence from an optically dressed ^{87}Rb cloud, as a function of the control laser Rabi frequency Ω_c. The probe pulse is 500 ns long, on resonance ($\Delta_p = \Delta_c = 0$). When $\Omega_c = 0$ a characteristic probe-only fluorescence signal is detected, shown here for comparison. The control field is turned on 50 μs before the probe pulse, and stays on for 60 μs.

associated with the coherence time in the vapor. As we saw earlier in this section, the $1/e$ decay rate for the transmission measurements is about 200 ns. Although there is a qualitative change in the shape of the generated fluorescence pulse on the time scale of 200 ns, the persistence of the main features of this signal, which may be generated for delay times greater than 10 μs, has a different origin. We attribute these fluorescence signals to optical pumping by the control field (initiated by spontaneous Raman scattering from the $F = 1$ level), which populates the $F = 2$, $m = \pm 2$ Zeeman levels (see Fig. 1). The control beam has ~ 0.1 % of light linearly polarized orthogonal to the main beam, which is enough to slowly optically pump the atoms into the $F = 2$ level, and to also contribute to the fluorescence signal in the process. For the highest control field Rabi frequencies, optical pumping occurs more quickly, and this parasitic signal, although the mechanism remains active, is not apparent in the data taken here. Such effects may occur in other circumstances, and can occur even under the circumstances when there is a high degree of phase coherence between the control field and the probe field.

We now turn to measurements of the polarization dependence of the time-resolved fluorescence, using the geometry of Fig. 7. For comparison purposes, we point out that we have previously reported linear polarization measurements for the case of a weak probe beam only.[37] In these measurements, as in the others reported

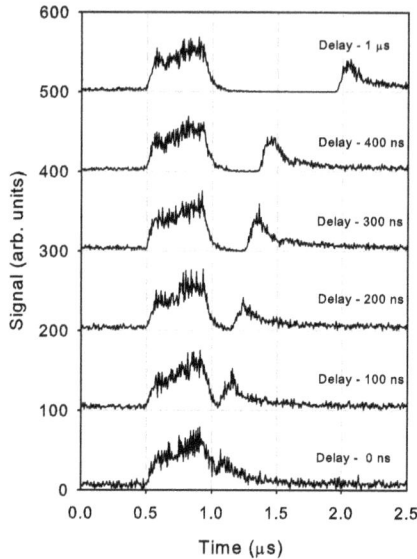

Fig. 9.   Probe fluorescence as the control field, with a Rabi frequency of $\Omega_c = 0.69\ \Gamma_p$, is turned off/on for various time delays. The probe pulse frequency is on resonance ($\Delta_p = 0$). The slowly decaying signal obtained when the control field is switched back on could be the result of optical pumping of atoms into the $F = 1$ level by the control field. Here, $\Gamma_p = 3.8 \cdot 10^7 s^{-1}$.

in this paper, the probe and control field have mutually orthogonal linear polariza-tion directions. In the fluorescence detection arm there is a linear polarization analyzer which may be rotated so as to change the detected state of polarization. The detection plane, formed by the wave vectors of the probe and control fields, and the detected fluorescence direction, is horizontal. Four different measurements are made. In two of them, the probe field is vertical, the control field polarization vector is in the detection plane, and measurements are made of the fluorescence signals with the detector polarization analyzer either collinear with the probe field direction or perpendicular to it. In the other two measurements, the control field polarization is vertical, and the probe field vector lies in the detection plane. Again, measure-ments are made of the fluorescence signals alternately with the detector polarization analyzer collinear with the control field direction and perpendicular to it. These polarization configurations are illustrated by inserts to Figs. 10–13.

The measured time-resolved fluorescence signals with the *probe field* vertically polarized, and for two orthogonal states of detected light linear polarization, are shown in Fig. 10. In these measurements the data, taken in 5 ns bins, are regrouped into 20 ns bins in order to improve the signal to noise ratio. In the figure, the over-all evolution of the intensity is a rise (with an overshoot, as discussed previously) to an approximate steady state followed by decay of the signal after the probe is extinguished. We also see that the fluorescence has a nonzero degree of linear polar-ization, with an average linear polarization degree in the range 30−35%, while the

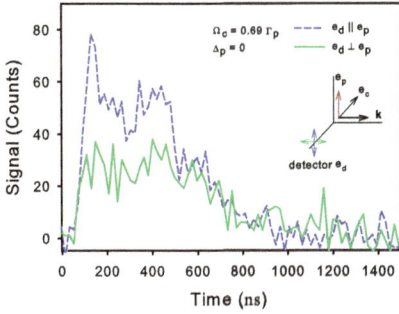

Fig. 10. Experimental results for the time-dependent polarized fluorescence under similar conditions. For these results the probe linear polarization is vertical while the control field linear polarization is horizontal. As indicated, two results are shown; in one case the linear polarization of the detected light is in the same direction as the probe linear polarization, while in the other they are orthogonal.

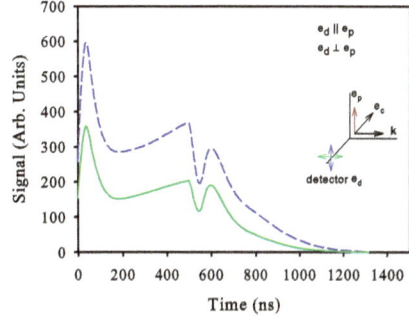

Fig. 11. Theoretical calculations of the time-dependent polarized fluorescence under similar conditions. For these results the probe linear polarization is vertical while the control field linear polarization is horizontal. As indicated, two results are shown; in one case the linear polarization of the detected light is in the same direction as the probe linear polarization, while in the other they are orthogonal.

probe and control field are both on. Here, as is customary, the linear polarization degree is defined as the difference in the measured intensities for the two orthogonal polarization states, normalized to the sum of these intensities. We first note that if this excitation were purely from the probe $F = 1 \to F' = 1 \to F = 1$ transition, then the linear polarization degree would be about 33%, well within the uncertainty of the measured polarization as in Fig. 10. We point out that if we include contribution from the spectrally unresolved $F = 1 \to F' = 1 \to F = 2$ Raman transition, the linear polarization would be reduced to 3/11 (~ 27 %). In Fig. 11 we present theoretical calculations of the time resolved fluorescence. The results are calculated for the same control field Rabi frequency as the experimental data; the relative control and probe beam sizes are also matched to the experimental results. We see in the theoretical results good correspondence with respect to the measurements when the probe field remains on. However, once the field is extinguished the experimental results show that the remaining transient fluorescence signals are unpolarized, within the experimental uncertainty, whereas there remains a clear polarized component in the theoretical results. We can understand these results by considering that the fluorescence signals must arise from components of the probe field that do not spectrally overlap the transparency window. With reference to Fig. 5, for the selected control field Rabi frequency of $\Omega_c = 0.69 \; \Gamma_p$, a significant fraction of the probe field satisfies this condition. This signal would be expected to have a linear polarization degree close to the estimates provided above for single atom scattering.

The measured time-resolved fluorescence signals with the *control field* vertically polarized, and for two orthogonal states of detected light linear polarization, are

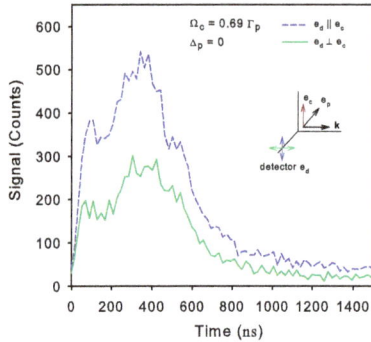

Fig. 12. Experimental results for the time-dependent polarized fluorescence under similar conditions. For these results the control field linear polarization is vertical while the probe field linear polarization is horizontal. As indicated, two results are shown; in one case the linear polarization of the detected light is in the same direction as the control field linear polarization, while in the other they are orthogonal.

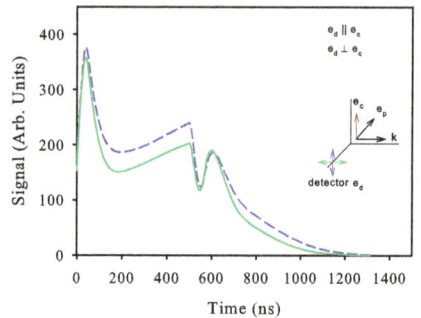

Fig. 13. Theoretical calculations of the time-dependent polarized fluorescence under similar conditions. For these results the control field linear polarization is vertical while the probe field linear polarization is horizontal. As indicated, two results are shown; in one case the linear polarization of the detected light is in the same direction as the control field linear polarization, while in the other they are orthogonal.

shown in Fig. 12. As for the data in Fig. 10, the time resolution is limited by the bin size of 20 ns. In the figure, the overall evolution of the experimental intensity is more structured than in the previous case. We also see that the fluorescence is linearly polarized over the entire time evolution of the signals, even after the probe and control fields are shut off; for the whole span of the data, the average linear polarization degree is about 35%. Theoretical results are calculated for the same control field Rabi frequency as the experimental data; the relative control and probe beam sizes are also matched to the experimental results. As in Fig. 11, we see in Figure 13 that the theoretical results have a qualitative correspondence with respect to the measurements. In contrast to the results of Fig. 10 and Fig. 11, we observe in Figs. 12–13 that the experimental and theoretical polarization of the fluorescence signals are quite closely matched, even when the probe and control fields are turned off.

2.1.6. *Conclusion*

In this section, we have discussed several aspects of near resonance lambda scheme optical excitation of a cold sample of ^{87}Rb atoms. Forward scattering measurements are used to characterize the optical response of the dressed atomic sample. Further experiments in a right angle fluorescence geometry explore the linear polarization degree of scattered light for two different spatial configurations of the pump and control fields with respect to the fluorescence detector. These measurements show a significant linear polarization degree that is essentially constant while the probe and control fields are both incident on the atomic sample. Once the fields are extinguished the initial polarization degree of the polarized probe fluorescence rapidly decays to zero, while that of the polarized control field fluorescence is maintained. These dynamical polarization effects are in reasonable agreement with theoretical modeling of the processes.

2.2. A High-Density Optical Atomic Trap for Study of Cooperative Interactions in Light Scattering

2.2.1. Brief Introduction

In this portion of the chapter, we give an overview of how we produce a quasi one dimensional configuration of high density and cold ^{87}Rb. Our approach builds on earlier work associated with a CO_2 laser based quasielectrostatic trap.[38,39] This modified trap may be used for a number of experiments including measurement of the forward coherent and incoherently scattered light, and experiments directed towards Anderson Localization of light in quasi one dimension. Here we focus on how we characterize the atomic sample, and illustrate with one method how a quasi one dimensional geometry can be created.

2.2.2. The Magneto Optical Trap

We begin with atoms collected and cooled in a magneto optical trap (MOT) well described by a density distribution $n(r)$ having a peak density n_0 and radius r_0 given by

$$n(r) = n_0 \exp\left(-\frac{r^2}{2r_0^2}\right). \tag{1}$$

In order to measure the number of atoms in the MOT, we determine the absorption of light from a near-resonance probe beam with intensity I_0 that is incident on the sample. From the Beer-Lambert Law, the transmitted light through the sample is

$$I_T = I_0 e^{-b}, \tag{2}$$

with b being the optical depth. For a not-too-dense atomic gas sample

$$b(r, \delta) = \sqrt{2\pi} r_0 n(\rho) \sigma(\delta) \tag{3}$$

where

$$\sigma(\delta) = \frac{\sigma_0}{1 + \left(\frac{2\delta}{\gamma}\right)^2} \tag{4}$$

is the scattering cross section with a laser detuning δ from resonance; γ is the lifetime of the excited state. Here, ρ is the distance transverse to the propagation direction of the probe beam. The expression for the cross section applies when the density is not too large, and when interatomic interactions may be ignored. Initially, a charge coupled device (CCD) camera takes an image of the transverse spatial probe intensity distribution. Then, an image is taken with the atomic sample present, yielding an absorption image. After background correction, the natural log of the ratio of these two images yields

$$\ln \frac{I_T(r,\delta)}{I_0} = -b(r,\delta). \tag{5}$$

This image is well fit to a Gaussian curve, with the amplitude equal to the peak optical depth and the $1/e^{\frac{1}{2}}$ radius giving r_0 of the sample. Integrating Equation (1), we find that the total number of atoms in the MOT is

$$N_{MOT} = (2\pi)^{3/2} n_0\, r_0^3 \tag{6}$$

Using an on-resonance probe ($\delta = 0$), substitution of Eqs. (3) and (4) into Eq. (6) yields

$$N = 2\pi \frac{r_0^2\, b_0}{\sigma_0} \tag{7}$$

$$\sigma_0 = \frac{2F'+1}{2F+1} \frac{\lambda^2}{2\pi} \tag{8}$$

Probing the $5^2 S_{1/2}, F = 2 \to 5^2 P_{3/2}, F' = 3$ transition with a laser of wavelength of 780 nm, we measured that the MOT typically has $\sim 5 \cdot 10^7$ atoms.

2.2.3. High Density Far Off Resonance Trap; Loading and Characterization

In order to reach the regime of higher atomic density, the atoms collected in the MOT are partially transferred to a far off resonance trap (FORT). This trap is formed at the focus of a fiber laser beam operating around 1.06 μm and having a power of several watts. To obtain optimal loading, good spatial overlap between the FORT beam and MOT atoms must be assured. This is done by controlling a wide parameter space that includes repumper beam power and turn-off time, MOT beam detuning, FORT power and alignment, and loading time. The most important of these is repumper power during loading. It is extremely sensitive to the optimal setting due to the need to reduce radiation pressure within the MOT and the ability to pump the atoms down into the F=1 ground state, which has a small inelastic collisional cross-section. Typically, the repumper power is reduced to $\sim$30 μW and has a turn-off time that precedes that of the MOT beam by 3 ms to help with the pumping process to the F=1 state. The MOT beam is detuned to also help with the compression of the MOT atoms around the FORT beam and a detuning of $\sim$10γ is generally optimum. The fiber laser that provides the FORT potential, which can be increased up to 35 W, is operated around 2 W. This power setting is sufficient as the beam is focused to $\sim$18 μm giving a trap depth of 670 μK. As referenced in Ref. 40, the FORT beam focus should be offset from the center of the MOT

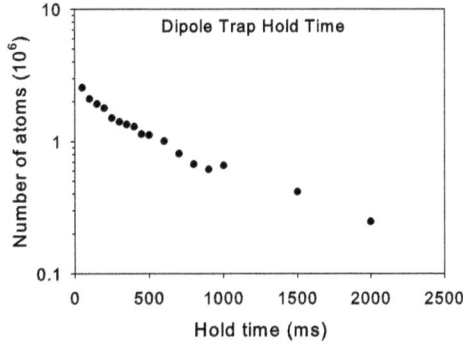

Fig. 14.   Number of atoms remaining in the dipole trap as a function of time. The atom number decreases due to collisional ejection of the atoms from the trap, mainly by hot Rb atoms in the chamber.

atoms as the best loading is achieved in the throat region of the trap. This entire experimental protocol lasts for 70 ms, typical of most FORTs,[41] at which point the MOT beams are shut off along with the magnetic field and a fraction of the MOT atoms thermalize by collisions in the dipole potential trap.

By comparing the fluorescence of the MOT and FORT under optically thin conditions, we can then find the transfer efficiency and number of atoms successfully shifted from the MOT to the FORT. With a typical transfer efficiency of $\sim 5\%$ the FORT has $\sim 2.5 \cdot 10^6$ atoms. The number of atoms in the FORT decreases with longer hold times, as seen in Fig. 14. This is due to collisions between atoms in the FORT and background gas atoms in the vacuum chamber.

After finding the number of atoms in the FORT, it is necessary to know the temperature of the atom sample. The FORT has a bi-Gaussian distribution of atoms

$$n(r, z) = n_0 \exp\left(-\frac{r^2}{2r_0^2} - \frac{z^2}{2z_0^2}\right) \tag{9}$$

and, once thermalized, a Maxwell-Boltzmann distribution of velocities[35]

$$f(v)d^3v = \left(\frac{2\pi k_B T}{m}\right)^{-3/2} \exp\left(-\frac{mv^2}{2k_B T}\right) d^3v. \tag{10}$$

Here, m is the mass of an individual atom, k_B is the Boltzmann constant, and T is the average temperature of the sample. When atoms in the FORT are released from the trap, their position will change with time according to

$$\vec{r} = \vec{r}\,' + \vec{v}\,t. \tag{11}$$

The time-dependent distribution of atoms, which is itself a bi-Gaussian distribution, is given to a good approximation by

$$n(\vec{r}, t) = \int\!\!\!\int\!\!\!\int_{-\infty}^{\infty} d^3v \, f(v) \, n(\vec{r} - \vec{v}\,t). \tag{12}$$

If we allow the cloud of atoms to ballistically expand, we can find the two characteristic radii of the sample as a function of time. The temperature is then proportional to the rate of change of the radii.

$$r^2 = r_0^2 + \frac{k_B T}{m} t^2$$
$$z^2 = z_0^2 + \frac{k_B T}{m} t^2 \tag{13}$$

The typical temperature of the FORT is $\sim 100\ \mu K$ and that of our MOT is $\sim 25\ \mu K$.

Finally, to find the initial radii of the FORT potential, and thus the effective volume, we utilize parametric resonance. Because of the low atom temperature in comparison with the trap depth, the distribution of atoms in the trap is concentrated near the bottom of the confining dipole potential. Then the radial and axial radii for the sample are significantly smaller than the beam waist and Rayleigh range, and the potential can be approximated as a harmonic oscillator

$$U(r, z) \approx -U_0 \left(1 - \left(\frac{z}{z_r^2} \right)^2 - 2 \left(\frac{r}{w_0^2} \right)^2 \right). \tag{14}$$

This can be rewritten to include a radial oscillation frequency ω_r and transverse frequency ω_z

$$U(r, z) = -U_0 + \frac{1}{2} m\, \omega_r^2\, r^2 + \frac{1}{2} m\, \omega_z^2\, z^2 \tag{15}$$

with harmonic oscillation frequencies of

$$\omega_r = \sqrt{\frac{4U_0}{m\omega_0^2}}$$
$$\omega_z = \sqrt{\frac{2U_0}{mz_r^2}}. \tag{16}$$

We begin with a trapping beam with an intensity described by a Gaussian distribution, and then apply a small sinusoidal modulation to this intensity for a fixed time while the atoms are held in the potential. The strength of the modulation is h, and the frequency is ω.

$$I(r, z, t) = I(r, z) + I_0\, h \cos(\omega\, t). \tag{17}$$

The one dimensional Mathieu equation describes the motion of atoms within the trap

$$\ddot{r} + \kappa\, \dot{r} + \omega_r^2\, r\, (1 + h\, \cos(\omega\, t)) = 0 \tag{18}$$

with a relaxation rate κ. A resonance occurs when

$$\omega = 2\omega_r/n, \qquad n = 1, 2, \ldots. \tag{19}$$

We can detect when we are on resonance by recording the intensity of the fluorescence as a function of frequency. The strongest response occurs at $\omega = 2\omega_r$. For our FORT, the radial frequency is on the order of $2\pi \cdot 4$ kHz and the transverse frequency is $2\pi \cdot 45$ Hz. With these frequencies and the temperature determined from ballistic expansion, we can calculate the radial and transverse radii of the atomic sample.

$$r_0 = \sqrt{\frac{k_b\, T}{m\omega_r^2}}; \qquad z_0 = \sqrt{2}\pi\, r_0\, \omega_0/\lambda \qquad (20)$$

While the trap has a beam waist of ~ 17.6 μm and Rayleigh range of ~ 0.92 mm, the FORT atom distribution has a radial Gaussian radius of ~ 3.4 μm and a transverse Gaussian radius of ~ 250 μm. Thus, the peak density of our sample of atoms is $6 \cdot 10^{13}$ atoms/cm^3.

2.2.4. *Creating and Probing a Dense and Cold Quasi-One-Dimensional Ensemble*

Cold atomic samples with high densities have numerous valuable and interesting properties. Our immediate interests are in three areas; (a) spectral and density dependence of the forward scattered light intensity, (b) demonstration and characterization of Anderson localization of light in a quasi one dimensional (Q1D) geometry, and (c) study of random lasing in Q1D. These experiments require creation of a Q1D channel through the atomic sample. One way to accomplish this is described below, and an image of such a realization is shown in Fig. 15.

We begin with two laser beams, one near the 795 nm D1 line (the light shift (LS) laser) and one near the 780 nm D2 line (the probe). The beams are nearly spatially mode matched and propagate along the same axis. They are focused so that their beam waist is approximately 15 μm with a Rayleigh range of approximately 900 μm. The LS laser is used to create a channel using the AC Stark shift. In

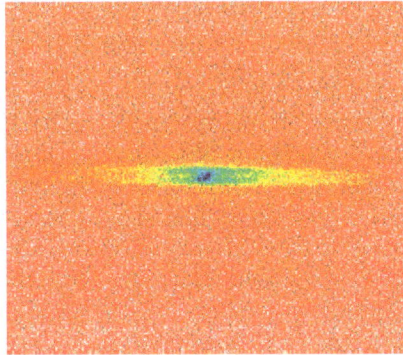

Fig. 15.   Head-on image of the channel created by a laser beam traversing the center of an optical dipole trapped atom sample.

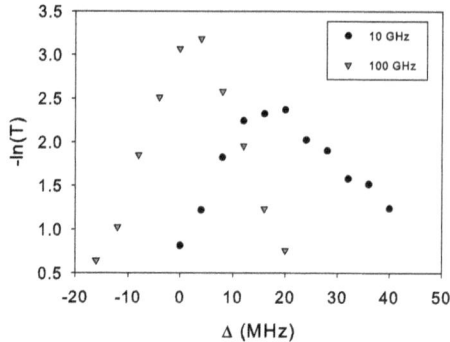

Fig. 16. Spectral shifts and Zeeman broadening of atomic resonance due to the ac Stark shift generated by the light shift laser. Data for two different detunings and a light shift laser power of about 5 mW is shown.

general, shifts in the ground and excited states must be considered so as to obtain the effective differential atomic resonance shift between the states. However, in performing calculations motivated by Safranova et al.[42] and Griffin,[43] it was found that to a good approximation only the ground state is shifted around 795 nm. The light-shift can then be estimated as

$$\Delta E = \frac{\alpha_0 P}{\pi \epsilon_0 c w_0{}^2} \tag{21}$$

where α_0 is the scalar polarizability, P is the power of the beam, ϵ_0 is the permitivitty of free space, c the speed of light, and w_0 the beam waist. This expression is written in SI units; division by Planck's constant h gives the shift in conventional frequency units.

The AC Stark shift can be measured by probing the spectral shift of hyperfine components of the D2 line. By changing the detuning or intensity of the LS laser, the ground state energy is changed, and this effect is seen by measuring the shift of resonance of the $F = 2 \rightarrow F' = 3$ transition of the D2 hyperfine multiplet. As seen in Fig. 16, detuning the LS laser 10 GHz below resonance with 5 mW of power leads to a shift of ~20 MHz. Increasing the detuning to 100 GHz reduces the shift by nearly a factor of 10 to 2.4 MHz.

Atoms within the LS beam path experience a resonance shift and the optical depth within this channel for light at the bare resonance for $5^2 S_{1/2} \rightarrow 5^2 P_{3/2}$ is reduced. The atoms outside of this channel create a dielectric wall which is optically deep and difficult for photons to penetrate. Thus, scattering is reduced in the transverse direction and mostly limited to forwards or backwards scattering, creating a quasi-1D system.

2.2.5. *Conclusion*

In this chapter section, we have given an overview of one method by which to create in the laboratory a quasi one dimensional configuration for light propagation through a high density and cold atomic gas. We have demonstrated and partially characterized such a configuration for an optical dipole trapped gas of ^{87}Rb atoms. The overall scientific aim is to generate a soft cavity type arrangement for study of random lasing and Anderson light localization in reduced spatial geometry.

Acknowledgments

We also appreciate financial support by the National Science Foundation (Grant Nos. NSF-PHY-0654226 and NSF-PHY-1068159), the Russian Foundation for Basic Research (Grant No. RFBR-CNRS 12-02-91056). D.V.K. would like to acknowledge support from the External Fellowship Program of the Russian Quantum Center (Ref. Number 86). We also acknowledge the generous support of the Ministry of Education and Science of the Russian Federation (State Assignment 3.1446.2014K).

References

1. See, for example, T. Bienaime, R. Bachelard, N. Piovella and R. Kaiser, *Cooperativity in light scattering by cold atoms*, Fortschr. Phys. **61**, 377 (2013).
2. R. Friedberg, S.R. Hartmann, and J.T. Manassah, *Frequency shifts in emission and absorption by resonant systems of two-level atoms*, Physics Reports **7**, 010 (1973).
3. J.T. Manassah, *Cooperative radiation from atoms in different geometries; decay rate and frequency shift*, Advances in Optics and Photonics **4**, 108 (2012).
4. J. Keaveney, A. Sargsyan, U. Krohn, I.G. Hughes, D. Sarkisyan, and C.S. Adams, *Cooperative Lamb Shift in an Atomic Vapor Layer of Nanometer Thickness*, Phys. Rev. Lett. **108**, 173601 (2012).
5. H. van Kampen, V.A. Sautenkov, C.J.C. Smeets, E.R. Eliel, and J.P. Woerdman, *Measurement of the excitation dependence of the Lorentz local-field*, Phys. Rev. A **59**, 271 (1999).
6. Juha Javanainen, Janne Ruostekoski, Yi Li, and Sung-Mi Yoo, *Shifts of a Resonance Line in a Dense Atomic Sample*, Phys. Rev. Lett. **112**, 113603 (2014).
7. Marlan O. Scully, *Collective Lamb Shift in Single Photon Dicke Superradiance*, Phys. Rev. Lett. **102**, 143601 (2009).
8. Ralf Rohlsberger, Kai Schlage, Balaram Sahoo, Sebastien Couet, and Rudolf Ruffer, *Collective Lamb Shift in Single-Photon Superradiance*, Science **328**, 1248 (2010).
9. R. H. Dicke, *Coherence in Spontaneous Radiation Processes*, Phys. Rev. **93**, 99 (1954).
10. P.W. Anderson, *Absence of Diffusion in Certain Random Lattices*, Phys. Rev. **109**, 1492 (1958).
11. Q. Baudouin, N. Mercadier, V. Guarrera, W. Guerin, R. Kaiser, *A cold-atom random laser*, Nature Physics **9**, 357 (2013).
12. Hui Cao, *Lasing in Disordered Media*, in Progress in Optics **45**, (2003).
13. V.S. Letokhov, *Generation of light by a scattering medium with negative resonance absorption*, Sov. Phys. JETP **16**, 835-840 (1968).

14. L. V. Gerasimov, V. M. Ezhova, D. V. Kupriyanov, Q. Baudouin, W. Guerin, and R. Kaiser, *Raman process under condition of radiation trapping in a disordered atomic medium*, arXiv:1401.6641. To appear, Phys. Rev. A (2014).

15. Ping Sheng, *Introduction to Wave Scattering, Localization, and Mesoscopic Phenomena* (Academic Press, San Diego, 1995).

16. Eric Akkermans and Gilles Montambaux, *Mesoscopic Physics of Electrons and Photons* (Cambridge University Press, Cambridge, UK, 2007).

17. G. Labeyrie, F. de Tomasi, J.-C. Bernard, C.A. Muller, C. Miniatura, and R. Kaiser, *Coherent Backscattering of Light by Cold Atoms*, Phys. Rev. Lett. **83**, 5266 (1999).

18. Y. Bidel, B. Klappauf, J.C. Bernard, D. Delande, G. Labeyrie, C. Miniatura, D. Wilkowski, and R. Kaiser, *Coherent backscattering of light by resonant atomic dipole transitions*, Phys. Rev. Lett. **88**, 203902-1 (2002).

19. D.V. Kupriyanov, I.M. Sokolov, N.V. Larionov, P. Kulatunga, C.I. Sukenik, and M.D. Havey, *Spectral dependence of coherent backscattering of light in a narrow-resonance atomic system*, Phys. Rev. A **69**, 033801 (2004).

20. M.D. Lukin, T*rapping and manipulating photon states in atomic ensembles*, Rev. Mod. Phys. **75**, 457 (2003).

21. M.D. Lukin and A. Imamoglu, *Controlling photons using electromagnetically induced transparency*, Nature **413**, 273 (2001).

22. T. Chaneliere, D.N. Matsukevich, S.D. Jenkins, S.-Y. Lan, T.A.B. Kennedy, and A. Kuzmich, *Storage and retrieval of single photons transmitted between remote quantum memories*, Nature **438**, 833 (2005).

23. D. N. Matsukevich and A. Kuzmich, *Quantum state transfer between matter and light*, Science 306, **663** (2004).

24. B. Darquie, M.P.A. Jones, J. Dingjan, J. Beugnon, S. Bergamini, Y. Sortais, G. Messin, A. Browaeys, P. Grangier, *Controlled Single-Photon Emission from a Single Trapped Two-Level Atom*, Science **309**, 454 (2005).

25. S.E. Harris, J.E. Field, and A. Kasapi, *Dispersive properties of electromagnetically induced transparency*, Phys. Rev. A **46**, R29 (1992).

26. Danielle A. Braje, Vlatko Balic, G.Y. Yin and S. E. Harris, *Low-light-level nonlinear optics with slow light*, Phys. Rev. A **68**, 041801(R) (2003).

27. Danielle A. Braje, Vlatko Balic, Sunil Goda, G.Y. Yin, and S. E. Harris, *Frequency Mixing Using Electromagnetically Induced Transparency in Cold Atoms*, Phys. Rev. Lett. **93**, 183601 (2004).

28. H. Kang, G. Hernandez, Y. Zhu, *Slow-light six-wave mixing at low light intensities*, Phys. Rev. Lett. **93**, 073601 (2004).

29. H. Kang, Y. Zhu, *Observation of large Kerr nonlinearity at low light intensities*, Phys, Rev. Lett. **91**, 93601 (2003).

30. D. Budker, W. Gawlik, D.F. Kimball, S.M. Rochester, V.V.Yashchuk, A. Weiss, *Resonant nonlinear magneto-optical effects in atoms*, Rev. Mod. Phys. **74**, 1153 (2002).

31. D. Budker, D.F. Kimball, S. M. Rochester, V.V. Yashchuk, *Nonlinear Magneto-optics and Reduced Group Velocity of Light in Atomic Vapor with Slow Ground State Relaxation*, Phys. Rev. Lett. **83**, 1767 (1999).

32. I.M. Sokolov, D.V. Kupriyanov, and M.D. Havey, *Coherent backscattering under conditions of electromagnetically induced transparency*, J. Modern Optics **58**, 1928 (2011).

33. V.M. Datsyuk, I.M. Sokolov, D.V. Kupriyanov, and M.D. Havey, *Diffuse light scattering dynamics under conditions of electromagnetically induced transparency*, Phys. Rev. A **74**, 043812 (2006).

34. V.M. Datsyuk, I.M. Sokolov, D.V. Kupriyanov, and M.D. Havey, *Electromagnetically induced optical anisotropy of an ultracold atomic medium*, Phys. Rev. A **77**, 033823 (2008).
35. H. J. Metcalf, P. van der Straten, *Laser Cooling and Trapping*, (Springer-Verlag, New York, 1999).
36. L. Giner, L. Veissier, B. Sparkes, A. S. Sheremet, A. Nicolas, O. S. Mishina, M. Scherman, S. Burks, I. Shomroni, D. V. Kupriyanov, P. K. Lam, E. Giacobino, J. Laurat, *Experimental investigation of the transition between Autler-Townes splitting and electromagnetically-induced-transparency models*, Phys. Rev. A **87**, 013425 (2013).
37. S. Balik, R.G. Olave, C.J. Sukenik, M.D. Havey, V. M. Datsyuk, I.M. Sokolov, and D.V. Kupriyanov, *Alignment dynamics of slow light diffusion in ultracold atomic Rb*, Phys. Rev. A **72**, 051402(R) (2005).
38. S. Balik, A.L. Win, M.D. Havey, I.M. Sokolov and D.V. Kupriyanov, *Near-resonance light scattering from a high-density ultracold atomic Rb gas* Phys. Rev. A **87**, 053817 (2013).
39. A.S. Sheremet, I.M. Sokolov, D.V. Kupriyanov, S. Balik, A.L. Win and M.D. Havey, *Light scattering on the F = 1 − F '= 0 transition in a cold and high density Rb vapor*, J. Mod. Optics **61**, 77 (2014).
40. S.J. Kuppens, K.L. Corwin, T.E. Chupp, and C.E. Wieman, *Loading an optical dipole trap*, Phys. Rev. A **62**, 013406 (2000).
41. M. Minarni, *Investigation of Ultracold Rb Atoms in a Pulsed Far Off Resonance Trap*, Old Dominion University (2005).
42. M.S. Safranova, C.J. Williams, and Charles W. Clark, *Relativistic many-body calculations of electric-dipole matrix elements, lifetimes, and polarizabilities in rubidium*, Phys. Rev. A **69**, 022509 (2004).
43. Paul F. Griffin, (2005) *Laser cooling and loading of Rb into a large period, quasi-electrostatic, optical lattice*, Durham theses, Durham University.

Chapter 3

Seeing Spin Dynamics in Atomic Gases

Dan M. Stamper-Kurn

Department of Physics, University of California, Berkeley, California 94720, USA
Materials Sciences Division, Lawrence Berkeley National Laboratory, Berkeley,
California 94720, USA

The dynamics of internal spin, electronic orbital, and nuclear motion states of atoms and molecules have preoccupied the atomic and molecular physics community for decades. Increasingly, such dynamics are being examined within many-body systems composed of atomic and molecular gases. Our findings sometimes bear close relation to phenomena observed in condensed-matter systems, while on other occasions they represent truly new areas of investigation. I discuss several examples of spin dynamics that occur within spinor Bose-Einstein gases, highlighting the advantages of spin-sensitive imaging for understanding and utilizing such dynamics.

1. Atomic Materials Science

It is conventional to begin a book chapter such as this with a coherent framing of the progress within a scientific field, perhaps portrayed as setting up the next clear question that must be answered or an opportunity that must be exploited. However, it has always struck me that such a presentation belies the true and exciting nature of scientific work, in which discoveries arise from unframed curiosity and in which the intellectual random walks of many individuals cause the bounds of our knowledge and our sense of what is worth knowing to diffuse outward. Read as a whole, the chapters of this book support this view of science as a diffusing endeavor. One may try to tie all the work described herein as a coherent body of work (about coherence). But maybe it is best to appreciate the diverse manners in which the field of AMO science is evolving, all the fun that people are having in their laboratories and offices, and all the places we may go in the coming years.

This Chapter is about spin dynamics in quantum degenerate gases. I shall emphasize the role of high-resolution direct imaging to visualize such dynamics. Direct imaging provides a great deal of detailed information about the gases being probed. While we have learned a few ways in which some portion of that information can be analyzed, there still remains a wealth of information to be mined. Above all, however, the direct images of spin dynamics have the value of being beautiful, of portraying the science at hand as an appealing subject to study.

Now, to keep with convention, let me reframe the study of spin dynamics in quantum gases in a coherent framework.

Simple, discrete quantum systems, along with methods that make use of resonance and coherence to manipulate and measure such systems, have been the bread and butter of atomic physics for a century. In experiments on atomic beams and vapor cells, and then certainly on trapped ions, the focus has often been on the internal dynamics within single atoms. Precise measurements of these dynamics have allowed us to test the principles by which atoms are constructed (e.g. quantifying high-order processes in quantum electrodynamics), or to characterize the conditions to which the atoms are exposed (e.g. sensing magnetic fields).

Increasingly, however, atomic physics is giving access to many-body systems in which the atoms evolve no longer in isolation, but rather as a collection of mutually interacting quantum objects. In other words, we are moving from the study of atoms to the study of materials. The focus of materials science is typically different than that of atomic physics: whereas the atomic physicist may appreciate a precise characterization of the full dynamics of a simple quantum system, the material scientist may focus instead on general characterizations of thermodynamic ground states and of small excitations atop such equilibria. Precise characterization of materials is often unwarranted since material samples are imperfect and differ from one another. Characterizing far-from-equilibrium states is unnecessary since most materials cannot typically be driven very far from equilibrium.

Now that atomic physicists are creating materials-like quantum systems under highly controlled conditions, both scientific approaches are beneficial and applicable. It is believed that atomic systems, serving as "quantum emulators" of paradigmatic many-body quantum systems, making use of the tremendous control, cleanliness, and powerful detection capabilities offered by atomic physics, may allow us to determine better the principles underlying the materials science of both existing and theorized matter. Conversely, one expects that the properties of complex materials realized by these atomic systems will make it possible for us to improve our performance in the typical tasks of an atomic physicist, vis a vis precision measurement and sensing.

This duality is exhibited by the study of magnetism in quantum gases, which is the topic of this Chapter. In particular, I focus on magnetic phenomena in spinor Bose-Einstein gases, ones composed of bosonic atoms that are free to occupy the various Zeeman sublevels of a manifold of spin states. Following the example of materials science, I describe the propagation of small-amplitude spin modulations in a ferromagnetic condensate (Sec. 3) in terms of the dispersion of magnon excitations, and the spin dynamics of non-equilibrium quantum fluids (Sec. 4) in terms of spontaneous symmetry breaking and phase transitions. Following the example of atomic physics, I describe the propagation of magnons (Sec. 3) as a promising form of atomic interferometry, the spin-mixing instability (Sec. 4) in terms of the generation of spin-nematic squeezed states, and the combination of Larmor precession and high-resolution, high-sensitivity imaging (Secs. 2, 5) as a potentially powerful magnetic field sensor.

This Chapter is not meant to be a comprehensive review of studies of magnetic phenomena in quantum gases, or even specifically of research on spinor Bose-Einstein gases. Such reviews are provided elsewhere (e.g. in Refs. 1–3). Rather, the intention is to highlight some of the new research directions in materials science and atomic physics that have been identified in recent years and that should, I believe, be taken up (by the reader!) in the coming years.

2. Imaging Methods

One of the standard ways to diagnose an ultracold gas is to let it out of its trapping container, allow it to expand a while, and then to take a picture that reveals the spatial distribution of the expanded cloud. This technique is favored because the expanded gas can be made far larger than the imaging resolution, so as to avoid imaging aberrations. Also, the expanded gas can be made to have a low optical density. One can illuminate the sample with light that is resonant with an atomic transition, and then quantify the resulting shadow image to determine faithfully the column density distribution. This distribution carries information on an amalgam of properties of the gas – such as the one-particle distribution and two-particle correlations in position and momentum space, the interaction energy, angular momentum, vorticity, etc. – depending on the time of flight. That is, the spatial distribution at zero time of flight is obviously the in-trap spatial distribution, but understanding which property is revealed by which feature in the time-of-flight distribution for longer expansion times depends on assumptions about the behavior of the gas upon expansion. For gases that are structured and correlated with greater complexity, the connection between time-of-flight spatial distributions and in-situ properties is less straightforward.

The time-of-flight imaging method can be made spin sensitive by applying a magnetic field gradient during the time of free expansion, emulating Stern and Gerlach's early experiment in which atoms in different Zeeman states were deflected to different spatial regions on a detector.[4] Separating the different Zeeman components from one another requires a non-zero time of flight, so that, aside from the overall population distribution among the Zeeman sublevels, all other information about the gas is ambiguous. In particular, the spatial spin distribution can be determined only with poor spatial resolution.

My group has developed imaging methods that allow for the spin distribution of a quantum gas to be measured directly, in-situ, with high spatial resolution and high spin sensitivity. Both methods make use of the atomic physics of single atoms to characterize the magnetic structure of a many-body system: The first method relies on optical birefringence, while the second relies on the spin-selectivity of microwave transitions between different hyperfine states within the electronic ground state.

2.1. Spin-sensitive Dispersive In-situ Imaging

Birefringent materials are ones for which the real optical susceptibility is a tensor, rather than a scalar; in other words, light passing through the material acquires a phase shift that depends on the polarization of the light, and, for this reason, may emerge from the material with a different polarization than it had before entering the material.

One distinguishes between linear and circular birefringence. In the case of linear birefringence, if we represent the polarization of light in a particular basis of linear polarizations, the optical susceptibility is a diagonal matrix, meaning that light entering the material with either of these linear polarizations acquires a polarization-dependent phase shift but does not change in polarization. For example, the birefringent materials used in optical waveplates are linearly birefringent, defined by orthogonal "fast" and "slow" axes. In the case of circular birefringence, the basis polarizations that diagonalize the susceptibility tensor are the left- and right-handed circular polarizations. Circular birefringence occurs in situations that break time-reversal symmetry, e.g. in optical isolators that use strong magnetic fields and the Faraday effect.

A spin-polarized atomic gas can exhibit both linear and circular birefringence. For example, consider a gas fully polarized in the $|m_J = 1/2\rangle$ sublevel (with respect to some spatial quantization axis) of a $J = 1/2$ ground state. The oscillator strength for an optical transition to either a $J' = 1/2$ or $J' = 3/2$ excited state differs for σ^+ and σ^- polarized light; this can be seen by examining the values of Clebsch-Gordan coefficients. Thus, the dispersive phase shift imposed on light that passes through such an atomic gas, and that is detuned from the optical transition, is different for the two circular optical polarizations. This difference quantifies the spin polarization of the atomic gas. As another example, consider a gas fully polarized in the $|m_J = 0\rangle$ sublevel of a $J = 1$ ground state, and probed near a transition to a $J' = 1$ excited state. This gas does not couple to π-polarized light (linear along the quantization axis), but does couple to the orthogonal linear polarization. Thus, such a gas exhibits linear birefringence which, in this case, identifies the spin quantization axis.

It can be shown that circular birefringence reveals the spin-vector moments while linear birefringence reveals the spin-quadrupole moments of the atoms being probed.[5] These moments yield the magnetization vector and nematicity tensor of the material comprised of such atoms. The spin-vector moments are associated with coherences between Zeeman sublevels that differ by $\Delta m = 1$ in their magnetic quantum number, while the spin-quadrupole moments are associated with $\Delta m = 2$ coherences. That is why I could use the example of a $J = 1/2$ ground state to

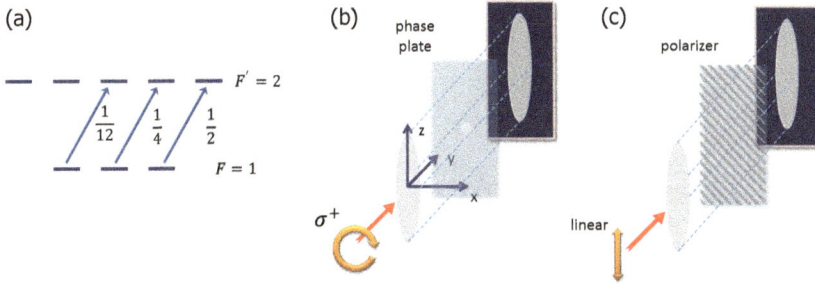

Fig. 1.   We employ circular birefringence to image the magnetization of an $F = 1$ spinor gas. (a) The relative oscillator strengths for circular polarized light driving an optical transition from the $F = 1$ ground state to an $F' = 2$ excited state are shown. Atoms polarized in the $|m_F = +1\rangle$ sublevel, quantized along the direction of the optical helicity, interact more strongly with the light. A dispersive image taken with light near this transition therefore contains information on the magnetization of the gas along that direction. The dispersive signal is turned into an image in one of two standard ways. (b) One way is to convert the phase shift on a circularly polarized light field into an intensity image using a phase plate placed in the Fourier plane of the image, this being a form of phase contrast imaging.[8] (c) A second way is to illuminate the sample with linear polarized light. The circular birefringence causes a rotation of the linear polarization, and this rotation is analyzed by passing the light through a linear polarizer before the camera.

discuss circular birefringence, but needed a $J = 1$ ground state, which can have $\Delta m = 2$ coherences, to exemplify linear birefringence.

My group made use of the circular birefringence of ^{87}Rb atoms in the $F = 1$ hyperfine ground state in order to image their magnetization.[6] Polarized light detuned by several hundred MHz from the D1 optical transitions was sent along the thin axis of a scalene ellipsoidal optical trap containing a gas of several million atoms, which were typically deep within the quantum degenerate regime (Fig. 1). In one approach, we imaged with a brief pulse of circular polarized light, and used phase-contrast imaging to convert the spatially varying phase shift on the probe light into an intensity image. A single image yielded information on both the density and the column-integrated magnetization of the gas along the probe axis. To isolate the magnetization signal, and also characterize the projections of the magnetization along all axes, we used a combination of rf pulses and Larmor precession to rotate the magnetization (Fig. 2). In a second approach, we used linearly polarized light and quantified its polarization rotation using a linear polarizer before the camera. The image now measured only the magnetization. Again, a sequence of rf-pulses was used to bring each of the magnetization components into view.[7]

The quality of data obtained by this method is exemplified by Fig. 3, which reveals the response of a quantum degenerate gas to being quenched across a paramagnet-to-ferromagnet phase transition.[9] The physics revealed by these images is discussed in Sec. 4. Here, let me just emphasize how eye-opening these images were. Prior studies of spin-mixing dynamics in rubidium spinor gases,[10,11] relying on Stern-Gerlach time-of-flight imaging, had revealed damped oscillations in

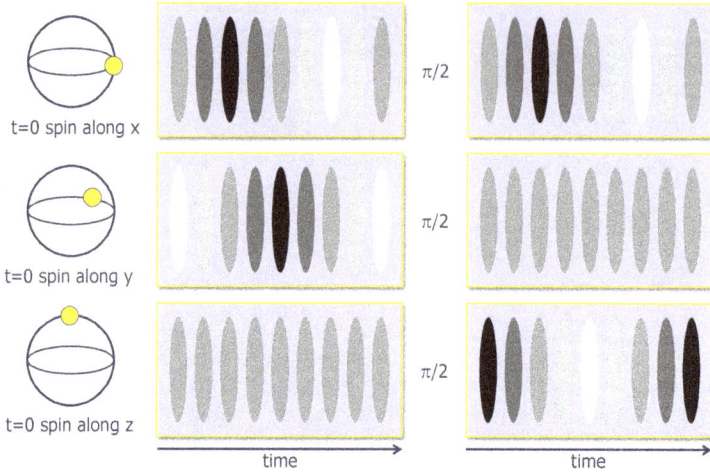

Fig. 2. All three components of the vector magnetization can be quantified by taking a sequence of imaging frames while the gas magnetization precesses in an applied magnetic field, and then is re-oriented using rf pulses. Shown is the expected image frame sequence for a gas initially magnetized along each of the cardinal directions. The first set of panels shows the expected signal during nine equally spaced image frames taken during Larmor precession. The image signal from transversely magnetized sample shows a temporal oscillation, the phase of which determines the direction of the magnetization within the transverse plane. Following a $\pi/2$ rf pulse, which causes the longitudinal magnetization to be rotated into the transverse plane, a second set of several image frames can be analyzed to quantify that transverse magnetization strength. A two-dimensional image of the vector magnetization is reconstructed from such an imaging sequence. The entire sequence occurs within just a few ms, faster than the timescales for spin transport and mixing dynamics.

the Zeeman-state populations, with very little spatial resolution. Theorists guessed that the damping was due to some sort of multi-mode spatial dynamics,[12] but it was impossible to tell from the data whether that was the proper explanation. In any case, such multi-mode effects did not seem like something interesting to examine further, but rather were seen as something to try to eliminate by working with smaller and more tightly confined samples.[a] In contrast, our high-resolution imaging of spin distributions revealed a rich dynamics: the development of spin domains with magnetization lying in all directions transverse to the magnetic field direction, separated by snaking domain walls and pierced by spin vortices.

Indeed, the progress of ultracold atomic physics over the past couple of decades has shown, time and time again, that improved imaging methods allow us to examine interesting physical phenomena whose study was previously inaccessible, and, indeed, inconceivable. Examples of such leaps include the first realization of

---

[a]The strategy of restricting the sample geometry to achieve strictly single-mode dynamics has been pursued and has turned out to be very beneficial for isolating and controlling the many-body-quantum nature of spin-mixing dynamics; see Refs. 13–15 for some examples.

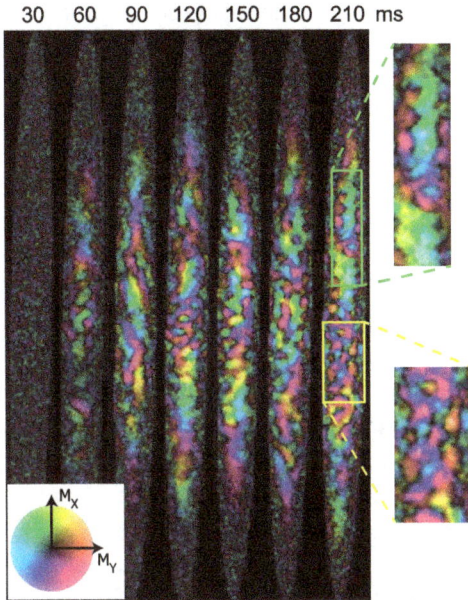

Fig. 3. The transverse magnetization of a spinor Bose-Einstein condensate was measured at a variable time (indicated at top of image) after it was quenched across a paramagnet-to-ferromagnet phase transition. The quantum degenerate gas was initially allowed to equilibrate in the $|m_F = 0\rangle$ state at a high quadratic Zeeman shift q where that state is the equilibrium spin state. The quench was performed by rapidly lowering the quadratic Zeeman energy to near zero, encouraging the formation of a transversely magnetized ferromagnetic condensate. The data are encoded in a color scheme (see color wheel at lower left) where the transverse magnetization orientation is indicated by the hue, and its magnitude by the color saturation. The image of each separate condensate has a width of about 30 microns. The magnetization appears over 10's of ms after the quench, resulting in the formation of some regions with small neighboring spin domains of opposite orientation (highlighted by the yellow box and magnified for view for the latest time image) and also large domains of common orientation (green box). The figure is adapted from Ref. 9.

non-destructive imaging[16] that allowed for studies of bosonic stimulation in the formation of a Bose-Einstein condensate,[17] high-resolution objectives that allowed for the realization of Josephson junctions formed by optical potentials[18] and observations of scale invariance[19] and quantum criticality[20] in two-dimensional gases, and the recent studies of dynamics and magnetism in the single-site-resolving quantum gas microscope.[21,22] I am certain this trend will continue.

2.2. *Absorptive Spin-sensitive In-situ Imaging*

Dispersive imaging has been billed as a form of "non-destructive imaging," implying that imaging the gas does not deposit so much energy so as to utterly destroy it. Perhaps a better moniker is "less-destructive imaging," for the act of imaging necessarily disturbs the gas. There is a tradeoff between how much information one wants to extract from the gas and how much one wants to preserve the gas for continued evolution and imaging. The off-resonant probe light used in dispersive imaging perturbs the gas in several ways. Characterizing the vector magnetization, and even more so the tensor nematicity, of a gas requires a sequence of several images, and, thus, that each image be non-destructive enough to leave enough signal for the subsequent image. Therefore, there is an upper bound on the signal to noise with which these characteristics can be quantified. These bounds were explored in my group's work on spatially resolved magnetic field sensing, which is discussed further in Sec. 5.

The limitations we identified led us to consider an alternative imaging scheme: absorptive spin-sensitive in-situ imaging (ASSISI). The idea here is to analyze the population of atoms in one set of atomic levels $|A\rangle$ by promoting a small fraction of them to unoccupied atomic levels $|B\rangle$, and then to image those promoted atoms selectively with probe light that is resonant with levels $|B\rangle$ but far detuned from levels $|A\rangle$. The promoted atoms are allowed to scatter very many photons, allowing one to obtain a low-noise image of their distribution. After the image is taken, the imaged atoms are expelled from the trap, leaving the remaining sample of atoms relatively unperturbed. The application of this scheme to imaging the $|F = 1\rangle$ hyperfine states of ^{87}Rb is described in Fig. 4. The signal-to-noise ratio of this imaging method is found to be limited by shot-noise fluctuations in the number of promoted atoms, and vastly improved over the quality of dispersive imaging.

Another important benefit of ASSISI is that one can fully characterize the atomic spin state. In particular, the dispersive imaging method described previously permits measurements of the spin-quadrupole moment only through the linear birefringence of the atomic gas. For alkali atoms, this linear birefringence is weak except for light very close to atomic resonances, where the absorption of the probe light causes severe problems. In contrast, ASSISI permits a complete in-situ "Stern-Gerlach" analysis of the Zeeman sublevels along any axis. Performing such analyses along several axes permits one to determine the atomic spin state completely, including the $\Delta m = 2$ coherences associated with the spin-quadrupole moments, and even higher-order coherences in the case of high-spin gases.

3. Coherent Propagation of Magnons

Now I turn to the first of three examples where high-resolution in-situ imaging has been useful for elucidating magnetic phenomena in quantum gases, and where such phenomena have been examined both from the viewpoint of materials science and of atomic physics. We start with understanding how magnetic excitations propagate in such gases.

Ultracold spinor Bose-Einstein gases will condense into a state that is both magnetically ordered and also superfluid. After all, it is the magical nature of Bose-Einstein condensation to cause particles to occupy the lowest-energy single particle

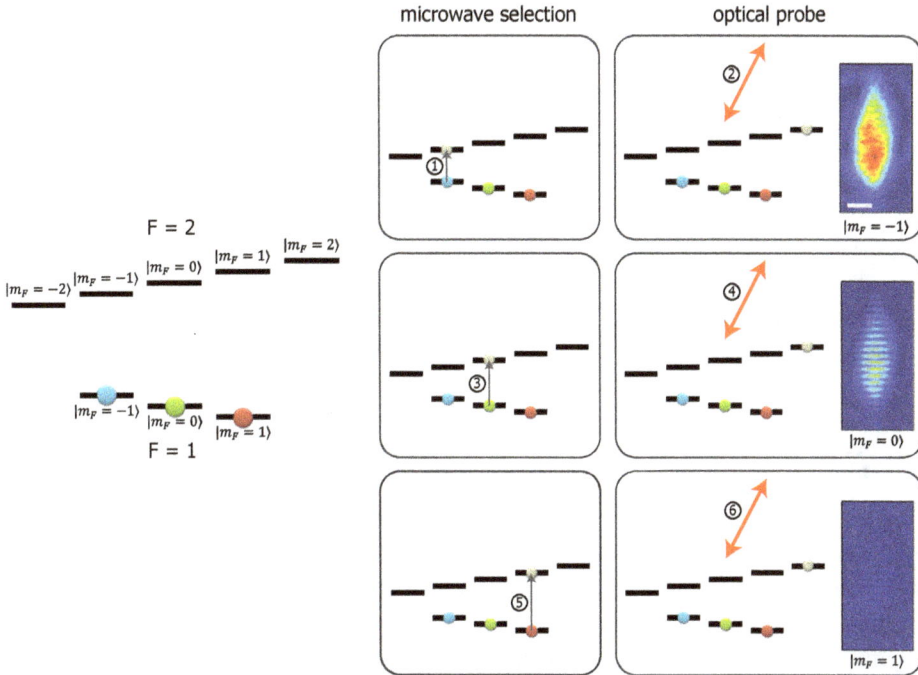

Fig. 4.   Absorptive spin-selective in-situ imaging of an $F = 1$ rubidium gas. Left: The ground electronic state has two hyperfine states, denoted by quantum numbers $F = 1$ and $F = 2$. Here the Zeeman sublevels within each manifold are separated by the linear Zeeman shift, which is shown out of scale to the very large energy separation between hyperfine spin manifolds. Our task is to image a gas that occupies the $F = 1$ levels. This is accomplished in six steps. First, a short microwave pulse is used to drive a small fraction of atoms from one of the $F = 1$ magnetic sublevels to the unoccupied $F = 2$ state. Second, an optical probe measures the in-situ column density of the selected atoms. In the process the atoms are optically pumped, so that the absorption image is quantitative, and also the atoms are heated sufficiently so that they are expelled from the trap after imaging. This selection/probe sequence is repeated three times to gather three image frames. The images show a standing wave of magnons in a ferromagnetic initial state. The scale bar indicates a 50 μm distance.

states with a macroscopic occupancy, even at temperatures T where the energy difference between the lowest and next-lowest energy is much smaller than the thermal energy available $k_B T$. Given this tendency, one expects spinor Bose-Einstein condensates to adopt the single-particle spin state with the lowest energy. Absent any external spin-dependent influences such as applied magnetic fields, the dominant spin-dependent energy that remains is the contact interaction, which describes the effect of the short-range interaction potential between the neutral atoms colliding at very low incident energy. This interaction can be expressed by the operator

$C(\mathbf{F}_i, \mathbf{F}_j)\delta^3(\mathbf{r}_i - \mathbf{r}_j)$ where i and j are the indexes of the two atoms colliding, $\mathbf{F}_{i,j}$ are their spin operators and $\mathbf{r}_{i,j}$ their positions. We know that $C(\mathbf{F}_i, \mathbf{F}_j)$ must be rotationally symmetric. Here I have neglected the magnetic dipole interactions, which are treated as a long-range interaction and are rotationally symmetric only under rotations both in position and spin space.

To keep the discussion focused, I will concentrate on the specific example of the $F = 1$ spinor gas of ^{87}Rb, for which one finds $C(\mathbf{F}_i, \mathbf{F}_j) = c_0^{(1)} + c_1^{(1)}\mathbf{F}_1 \cdot \mathbf{F}_2$, with $c_1^{(1)} < 0$.[b] Such an interaction favors ferromagnetic ordering, in which all atoms' spins are commonly oriented so as to minimize the contact interaction energy. For this reason, we may denote the $F = 1$ ^{87}Rb quantum gas as a ferromagnetic superfluid.

3.1. *Nambu-Goldstone Bosons of Magnetically Ordered Superfluids*

Systems that spontaneously break a continuous symmetry commonly possess gapless bosonic collective excitations, known as Nambu-Goldstone bosons. To visualize such excitations, we consider an equilibrium state in which a continuous symmetry S is broken; e.g. a ferromagnet with uniform magnetization (Fig. 5a). If one applies an infinitesimal operation within S throughout the system, the energy is unchanged. Now consider that one transforms the system by applying an infinitesimal symmetry operation that varies in space with a well-defined wavevector $\mathbf{k}$. Such a transformation defines a bosonic collective excitation: the Nambu-Goldstone boson. This excitation may be expected to raise the energy of the system, since now the order parameter is nonuniform. However, in the limit $\mathbf{k} \to 0$, the transformed system is nearly uniform, so the rise in energy should tend to zero, implying the boson is gapless. The Nambu-Goldstone theorems do not hold for systems with long-range interactions, e.g. for charged superfluids and for the broken electroweak symmetry.

In recent years, the Nambu-Goldstone theorems have been re-examined in an attempt to reconcile the phenomena of symmetry breaking in relativistic and non-relativistic scenarios: In particle physics, excepting the Higgs mechanism, each broken symmetry leads to a gapless linearly dispersing bosonic mode. In contrast, in material systems, not every broken symmetry produces a new boson, and many of these bosons disperse quadratically, rather than linearly. The mathematical conditions leading to these differences have only recently been identified.[23,24]

Now we return to considering the ferromagnetic superfluid realized by the $F = 1$ ^{87}Rb gas. It breaks three symmetries: a combination of gauge symmetry and rotational symmetry, and the symmetries of rotations about the transverse $\mathbf{x}$ and $\mathbf{y}$ axes. While one might naively expect the ferromagnetic state to carry three

---

[b]The quantity $c_1^{(1)}$ is determined by differences in s-wave scattering lengths for collisions in different angular momentum channels, which are in turn determined by the particular form of the atom-atom interaction potential, which is in turn determined by comparison to a variety of measurements.

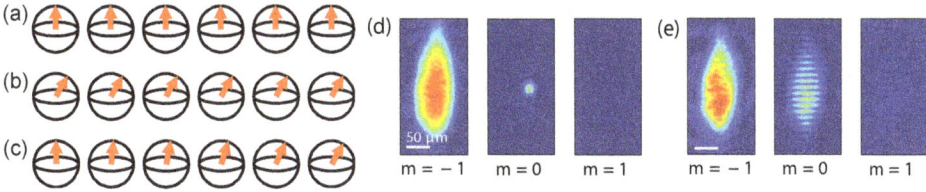

Fig. 5. Magnon schematic (left): (a) A material equilibrates in a uniform broken symmetry state (e.g. a ferromagnet with magnetization pointing up). (b) Applying a uniform symmetry transformation (e.g. rotating the magnetization) does not change the free energy. (c) This suggests that an inhomogeneous transformation of the system, which varies only over long length scales, costs little extra energy. This excitation is the Nambu-Goldstone boson, which, in the absence of long-range interactions, is gapless at the limit of zero wavevector. Magnon images (right): Zeeman-state populations are characterized in-situ by a sequence of microwave excitation and resonant optical absorption imaging. (d) A localized coherent wavepacket is created using the vector ac Stark shift of light focused tightly within the center of the condensate. (e) A spatial grating of magnons with well defined wavevectors is created using two beams of light that form a high-contrast interference pattern at the location of the trapped gas. For weak magnon excitations atop a longitudinally polarized gas, the $m_F = 0$ population is representative of the magnon population.

Nambu-Goldstone bosons, in fact there are just two. The breaking of gauge/rotation symmetry (rotation about $\mathbf{z}$ gives a phase shift) is the origin of the gapless, linearly dispersing phonon. The consequence of the breaking of transverse rotational symmetries is different. Because the generators of the two broken symmetries have a commutator with a non-zero expectation value in the ground state,[23] the two symmetries generate only a single Nambu-Goldstone boson, and the dispersion relation of this boson is quadratic with wavevector, i.e. $E(\mathbf{k}) = \hbar^2 k^2/(2m^*)$. This boson is the ferromagnetic magnon. It corresponds to rotating the uniform ground state by an infinitesimal angle θ about the transverse axis $\mathbf{n}_\phi = \mathbf{x}\cos\phi + \mathbf{y}\sin\phi$ with $\phi = \mathbf{k}\cdot\mathbf{r}$. Under mean-field theory, the effective mass m^* is equal to the atomic mass.

3.2. Coherent Atom Optics with Magnon Excitations

Magnons can be written onto a spinor Bose gas by realizing a spatially dependent rotation of the gas magnetization. Such rotation can be effected using the vector ac Stark shift of an off-resonant light field,[25] which imposes an effective magnetic field oriented along the helicity of the light field and is proportional in strength to a product of the light ellipticity and intensity.

To see how this light-induced energy shift creates magnon excitations, consider the effect of light with σ^+ circular polarization along the $\mathbf{x}$ axis, with an intensity pattern of $I(\mathbf{r})$. The vector ac Stark shift creates an effective inhomogeneity in the magnetic field $\Delta\mathbf{B}(\mathbf{r}) \propto I(\mathbf{r})\mathbf{x}$. Suppose a fully polarized ferromagnetic condensate is prepared with its magnetization uniform, and oriented transversely to a magnetic field along $\mathbf{x}$. The vector ac Stark shift now causes the condensate magnetization to precess by an additional angle $\Delta\phi \propto I(\mathbf{r})$ in the transverse plane. This inhomogeneous spin texture represents a coherent wavepacket of magnons (realized in Ref. 26). Similarly, consider that the ferromagnetic condensate is prepared with uniform longitudinal polarization along a magnetic field oriented perpendicular to $\mathbf{x}$. If the intensity or polarization of the incident light is now modulated at the Larmor precession frequency, the atoms will undergo Rabi oscillations due to the applied light field, again creating a coherent magnon excitation.

The magnons produced in this way were characterized by our ASSISI method, as shown in Figs. 5. To interpret these images, we recall that a rotation by the angle θ about the $\mathbf{n}_\phi$ axis transverse to the magnetization of a ferromagnetic spinor transforms the condensate wavefunction, to lowest order in θ, as

$$\psi = \sqrt{n} \begin{pmatrix} 1 \\ 0 \\ 0 \end{pmatrix} \rightarrow \sqrt{n} \begin{pmatrix} 1 \\ e^{i\phi}\theta/\sqrt{2} \\ 0 \end{pmatrix}. \tag{1}$$

To lowest order, a magnon excitation atop a condensate in the $|m_F = +1\rangle$ internal state is just a single atom in the $|m_F = 0\rangle$ state. Therefore, an image of the $m_F = 0$ atomic population is effectively an image of the (small) magnon density.

Consider that one prepares a spatial grating of magnons using the vector ac Stark shift of light beams intersecting with wavevector difference $\mathbf{q}$. One thereby prepares a coherent magnon wave with the initial wavefunction

$$\psi_M(t=0) \propto 1 + \eta \cos(\mathbf{q} \cdot \mathbf{r}) = 1 + \frac{\eta}{2} \left(e^{i\mathbf{q}\cdot\mathbf{r}} + e^{-i\mathbf{q}\cdot\mathbf{r}} \right) \tag{2}$$

where we have assumed the ferromagnetic condensate to be stationary and uniform. The factor η accounts for the visibility of the light interference pattern. This density-modulated magnon wave is a superposition of magnons at three wavevectors: $\mathbf{k} \in \{0, \mathbf{q}, -\mathbf{q}\}$. After evolving for a time t, the energy difference between stationary and moving magnons causes the magnon density distribution to become

$$|\psi_M(t)|^2 \propto 1 + \eta^2 \cos^2(\mathbf{q} \cdot \mathbf{r}) + 2\eta \cos\left(\frac{E(q) - E(0)}{\hbar}t\right)\cos(\mathbf{q} \cdot \mathbf{r}) \tag{3}$$

We observe the amplitude of the density modulation at wavevector $\mathbf{q}$ is temporally modulated at the Bohr frequency $(E(q) - E(0))/\hbar$, providing an interferometric measurement of the magnon dispersion relation. Such modulations are shown in Fig. 6. By Galilean invariance, the contrast modulation frequency is Doppler free. Contrast interferometry was demonstrated in a scalar Bose-Einstein condensate, where the signal was strongly sensitive to density-dependent shifts of the Bragg resonance frequency.[27]

From a materials science perspective, a precise measurement of the magnon dispersion relation is a precise test of quantitative many-body theories that describe the interacting ferromagnetic Bose gas. A recent theoretical calculation[29] that goes beyond mean-field theory to consider the effects of phonon-magnon interactions predicts an increase in the magnon mass above the bare atomic mass by a fractional

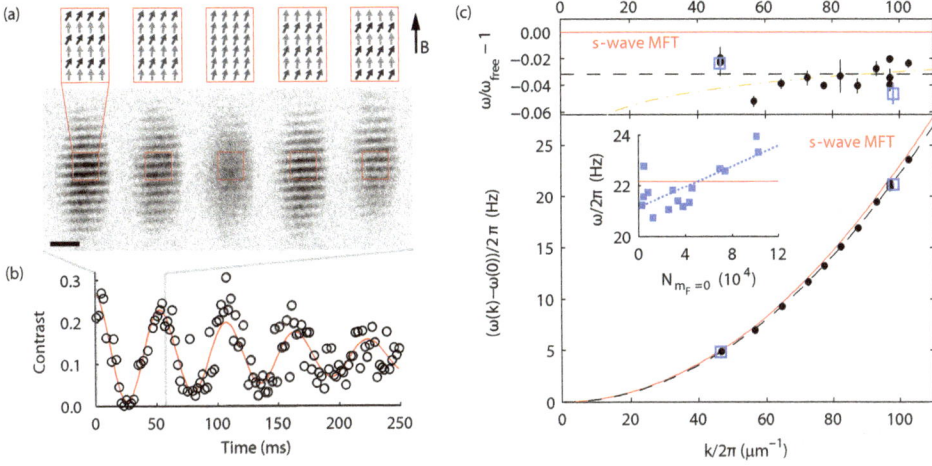

Fig. 6.   In our magnon interferometer, a standing wave of magnons is imprinted onto the condensate, being composed of magnon waves at three distinct wavevectors. Free propagation of the magnon waves causes the contrast of the magnon density modulation to oscillate at the frequency $2(E(q) - E(0))/\hbar$. (a) Images of the $m_F = 0$ population during this free propagation show high initial contrast, diminished contrast after half a temporal cycle, and the high contrast obtained after a full temporal cycle (with fringes π out of phase with the initial pattern). (b) The contrast shows clear temporal oscillations. (c) From these oscillations, we obtain the magnon recoil energy, shown in closed black circles. The recoil frequency is corrected for a frequency shift with the number of magnons produced in the interferometer, with the measured recoil energy being the value obtained by interpolation to zero density. Data are compared to the theoretical prediction of mean field theory (red line) that the magnon recoil energy should equal the free-atom recoil energy. Our data show the magnon mass to be slightly larger than the bare atomic mass. Figure reproduced from Ref. 28.

amount of about 3×10^{-3}. In our experiment, we find the magnon mass to be heavier than the atomic mass by a fractional amount of 3×10^{-2}. The origin of this discrepancy is unknown, though one should note that the theoretical calculations were performed for a uniform zero-temperature gas with no magnetic dipole interactions, whereas the samples used in our experiment were inhomogeneous, finite-sized, at non-zero temperature, and possessed discernable magnetic dipolar interactions (as were also seen in the experiment). Given the precision of the magnon contrast interferometer and the flexibilty of the experimental setup, it is clear there will be scientific payoffs in continuing this investigation: probing gases of variable size and geometry, isolating the effects of dipolar interactions, varying the temperature in a controlled manner, probing different initial states, and so on.

From an atomic physics perspective, the remarkable observation is how close is the magnon mass to the atomic mass: In a gas with a density on the order of 10^{14} cm^{-3}, interactions appear to shift the magnon recoil frequency only at the precent level, with the absolute value of the shift being less than a Hertz. In contrast, the density-induced shifts in the Bragg resonance frequency would be orders of magnitude larger.

Indeed, the properties of ferromagnetic magnons have enticing implications for atom interferometry. Their gapless nature implies that magnons are created without a density-dependent interaction energy shift. With the magnon mass being just about equal to the atomic mass, magnon excitations propagate just about like free particles. Thus, an interferometer based on the free propagation of magnons within a ferromagnetic spinor Bose-Einstein condensate at equilibrium (equal chemical potential across the condensate) is equivalent to an atom interferometer using free particles at zero gravity and in vacuum.

4. Spin-mixing Instability in a Spatially Multi-mode System

The Section above illustrates how the study of atomic materials can complement that of solid-state materials by allowing and justifying precise quantitative tests of theory. Another advantage offered to the study of materials science is the study of materials far from equilibrium, which is possible in atomic gases owing to the vast separation of timescales between the equilibration time (from just below a millisecond to much longer) and the shortest time (e.g. microseconds) in which the system Hamiltonian can be varied or the atomic internal state can be modified.

One scenario considered for such non-equilibrium dynamics is the response of an initially equilibrated many-body system to a rapid change of the system Hamiltonian. Similar to quenching a hot iron bar by submersing it in water, here we consider a quench of a quantum system to a condition that favors a different equilibrium state. The many-body system is thus uniformly in a non-equilibrium state. The deviations from equilibrium can propagate spatially across the system, allowing correlations and entanglement to span a distance that grows with time. The deviations can also serve to nucleate instabilities.[30]

Several works by my group have examined the quench of a degenerate spinor Bose-Einstein gas across a symmetry breaking phase transition. In particular, we consider the ^{87}Rb $F = 1$ spinor gas for which the spin-dependent contact interactions favor ferromagnetism, as discussed above. Atop these interactions, we add a single-particle energy term, a quadratic Zeeman energy qF_z^2 along an experimentally selected axis $\mathbf{z}$. For large positive q, this term favors a condensate fully polarized in the $|m_F = 0\rangle$ state quantized along the $\mathbf{z}$ axis. The Hamiltonian of the system is now symmetric under rotations about $\mathbf{z}$ and also under uniform gauge transformations (multiplication of the system wavefunction by a uniform complex phase). For large positive q, the non-magnetized equilibrium state breaks just the gauge symmetry, in establishing a complex, scalar superfluid order parameter. For small positive q, the ferromagnetic state is magnetized transverse to the $\mathbf{z}$ axis, and thus breaks the axisymmetry of the Hamiltonian as well as gauge symmetry. These states of different broken symmetry are separated by a phase transition, which can be traversed by changing q. An additional change of symmetry occurs for $q < 0$, where the favored longitudinally magnetized state breaks the discrete $\mathbb{Z}_2$ time-reversal symmetry but retains axisymmetry.

Let us focus on a quench from the non-magnetized, axisymmetric phase at large positive q to the symmetry-broken ferromagnetic phase at small positive q. How does a system quenched through this transition dynamically break the previously unbroken axisymmetry?

4.1. *The Kibble-Zurek Picture of Inhomogeneous Symmetry Breaking*

In the study of materials one encounters many cases of systems that undergo symmetry-breaking transitions, typically after being rapidly cooled (conventionally quenched). Considering systems as conventional as ferromagnets or supercooled water, we observe that symmetry breaking occurs locally. Small regions of the material undergo the phase transition independently, and these regions then serve to nucleate larger regions of commonly broken symmetry. Where these regions abut, we observe defects – domain walls in magnets or crystal defects in ice – that signify a conflict between the choices of broken symmetry in different regions, and that heal only very slowly as these conflicts are resolved.

The same phenomena are understood to occur in quantum materials undergoing quantum phase transitions. The picture often invoked to describe the symmetry breaking dynamics is called the "quantum Kibble-Zurek mechanism,"[31-33] following earlier considerations for classical thermal phase transitions.[34,35] This picture divides the quench dynamics into three periods in time. In the first, a quantum system quenched across a phase transition will form regions of commonly broken symmetry with a size that shrinks as the quench rate is increased, with the exact relation between size and rate determined by the character of the dispersion relation for excitations near the phase transition point. Second, the jumbled symmetry breaking will result in a tangle of defects of various types that depend on the geometry of the order parameter manifold. Third, these defects will slowly coarsen and heal over long times as the system becomes ordered at longer length scales.

So far, the predictions for the second and third periods of evolution have been examined in spinor Bose-Einstein condensates that are sufficiently large in either one or two dimensions so that it makes sense to invoke the Kibble-Zurek mechanism to describe their evolution. The spatially inhomogeneous symmetry breaking following a very rapid (essentially instantaneous) quench across the aforementioned paramagnet-to-ferromagnet transition is shown in Fig. 3.[9] Regions of transverse magnetization appear some tens of milliseconds after the quench. Regions of differently oriented magnetization are joined either by extended regions in which the magnetization gradually varies from one orientation to another, or by sharp lines of apparently unmagnetized gas (although it is possible the gas within these lines remains fully magnetized, with the magnetization varying over a length scale below our imaging resolution). The dynamics of symmetry breaking also give rise to topological defects: polar-core spin vortices that, in fact, turn out to be the only topologically distinct vortex for the $F = 1$ ferromagnetic superfluid. The quantum quench in a polar $F = 1$ spinor condensate has also been studied.[36]

As for the third period of the Kibble-Zurek chronology, several elements of coarsening dynamics have been examined. Early studies[37,38] shed light on the growth of spin domains in one-dimensional spinor Bose-Einstein condensates through two distinct mechanisms: Just as the normal and superfluid components of superfluid helium show distinct transport characteristics, here the non-condensed component of the gas allows for coarsening through thermal activation and re-equilibration, while the condensed component of the gas allows for domain growth through coherent quantum tunneling across domain walls.

My group examined the process of coarsening in two dimensional systems. We examined the evolution of an initially hot and unmagnetized gas after it was cooled across the Bose-Einstein condensation phase transition and allowed to equilibrate for several seconds. Regions of magnetized gas were observed after crossing the transition.[c] In contrast with the ~10 ms timescale for the initial formation of spin domains, here we observe a seconds-long timescale for domain coarsening. One disappointing conclusion of this study is that the equilibration timescales for the ^{87}Rb $F = 1$ spinor Bose-Einstein gas are so long as to be experimentally inaccessible for spatially extended gases.

4.2. *Symmetry Breaking through Parametric Amplification of Spin Fluctuations*

Let us return to the question of how an isolated system spontaneously breaks symmetry. If the initial state of the system is symmetric, and the Hamiltonian of the system conserves that symmetry at all times in the evolution, then the final state of the system is perforce symmetric. In that case, it stands to reason that the states we observe at late times after the quench are indeed symmetric superpositions of states of macroscopically broken symmetry.

Quantum optics offers a candidate for such a superposition state: squeezed vacuum. Applying a phase-sensitive parametric amplifier to the vacuum state causes one quadrature of the electromagnetic field to be amplified. Both the vacuum state and the squeezed vacuum state are symmetric, in that quadrature measurements are equally likely to give positive or negative values. However, measurements on

---

[c] As far as we could tell, the condensation of the gas and the formation of magnetized spin domains occurred simultaneously. However, according to our experimental protocol, for each gas sample we were able to measure either the presence of a condensate (in time-of-flight images) or the presence of transverse magnetization (in in-situ images). It would be interesting to perform both probes in a single experimental realization so that one can resolve better whether Bose-Einstein condensation occurs temporally before the formation of magnetized domains. I suspect that it does.

the squeezed state are likely to yield either a large positive or large negative value, deviating from the vacuum levels by potentially macroscopic amounts.

Perhaps, then, the dynamics of a many-body quantum system following a quantum quench may be described as the parametric amplification of initially symmetric and microscopic vacuum fluctuations. Questions arise: What is being amplified? What is being squeezed? Is the amplifier quantum-limited? What is the gain of this amplifier? If this amplifier is applied to a spatially extended condensate, the amplifier might be expected to act simultaneously but separately on fluctuations in different portions of the condensate. If so, how are different spatial patterns of fluctuations affected by the amplifier? In other words, what is the spatial gain spectrum of this amplifier?

Answers have been provided to several of these questions. It is understood that the spin fluctuations in this case – amplification through the spin-mixing instability of a gas initially in the $|m_F = 0\rangle$ quantum state – exist in two independent planes, representing two different polarizations of magnon excitations of the initial state.[39,40] Each plane is spanned by two observables: one component of the tranverse spin and one component of the spin-quadrupole tensor, serving as the independent and non-commuting quadratures equivalent to the electric field quadratures of light or the position and momentum quadratures of a mass on a spring. The transverse spin quadrature can be measured directly using circular birefringence dispersive imaging.[6,9,39] The spin-quadrupole moment can be measured by first rotating it onto the transverse-spin quadrature by briefly applying a strong quadratic Zeeman shift, as has been demonstrated in single-mode experiments probing this instability.[13]

For spatially extended systems, one can characterize the gain of this amplifier in momentum space, just as one characterizes an amplifier that acts on a continuous signal by its gain spectrum in frequency space. Through a linear stability analysis of the Gross-Pitaevskii equation that describes the coherent evolution of an initially uniform spinor Bose-Einstein condensate, one finds the magnon dispersion relation for the $|m_F = 0\rangle$ initial state, given as $E_s^2 = (\epsilon_k + q)(\epsilon_k + q - |c_1^{(1)}|n)$ where $\epsilon_k = \hbar^2 k^2 / 2m$ is the free particle kinetic energy at wavenumber k and n is the condensate density. For $q < q_0 = 2|c_1^{(1)}|n$, there is a range of wavenumber where $E_s^2 < 0$ indicating a regime of parametric instability with a temporal gain given by $\sqrt{|E_s^2|}/\hbar$. The colored spectrum of this amplifier can be interpreted as defining particular localized spatial modes that are amplified with highest gain.[40,41] Rough features of this gain spectrum, and how it tunes with the quadratic Zeeman shift, were observed in Ref. 39 (Fig. 7).

The noise characteristics of this parametric amplifier have been examined in single-mode spinor systems, where it has been shown to yield highly spin squeezed states.[13] My group attempted to characterize the noise limits of this parametric amplification in a spatially extended spinor gas, using our high-resolution imaging to quantify the amplifier output after a variable time of amplification.[39] Unfortunately,

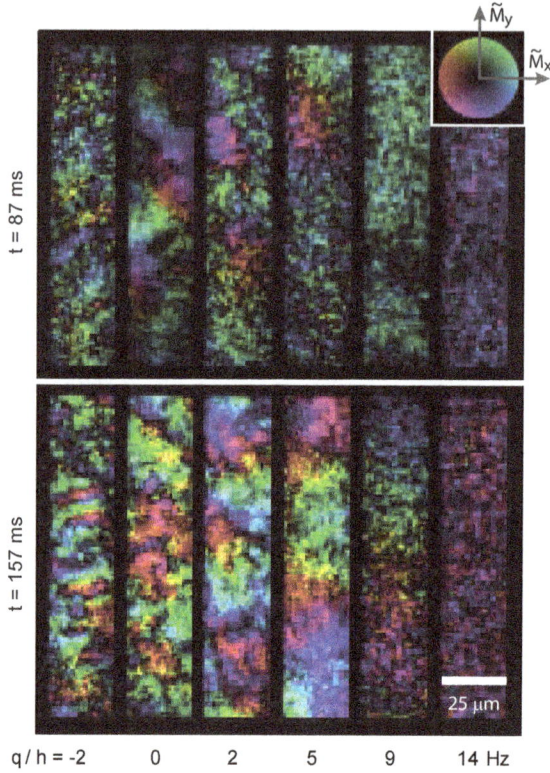

Fig. 7.   An electronic amplifier has a spectrum, describing the fact that gain at particular temporal frequencies is stronger than at other frequencies. Similarly, in a spatially extended spinor Bose-Einstein condensate, quenching to conditions that favor symmetry breaking into a ferromagnetic states causes the condensate to act as a spin-fluctuation amplifier that has a gain that depends on the spatial mode structure of the fluctuation. In particular, quenching the system to lower values of the quadratic Zeeman shift (images further to the left) causes spin fluctuations with smaller wavelength to be amplified more strongly. This spatial gain spectrum is visible in images taken at variable times after the quench. The initial spin noise in the spinor condensate, which we presume to be present at equal strength at all wavevectors (spatially white) is amplified by a wavelength-dependent spin amplifier. The macroscopic magnetization spectrum produced by this quench shows the spatial lengths most favored by the amplifier. Each image here is a map of the tranverse magnetization in a single condensate, at one of two different times after the quench (indicated on the left) at several different values of the quadratic Zeeman shift (indicated on the bottom). Figure taken from Ref. 39.

our uncertainty in the temporal gain of the amplifier, stemming from uncertainty in the scattering lengths of ^{87}Rb, did not permit us to place tight constraints on the amplifier noise.

5. Spatially Resolved Magnetometry with a Quantum Fluid Sensor

Finally, let me consider the use of cold atomic gases for spatially resolved magnetometry to study materials. At present, several technologies are used for spatially resolved magnetometry at micron length scales, including superconducting quantum interference devices (SQUIDs), scanning Hall probe microscopes, magnetic force microscopes and magneto-optical imaging techniques.[42] Among these, the technologies offering the highest field sensitivity for single-pixel measurements are the SQUID microscope and the scanning Hall probe microscope. SQUID sensors excel at measuring high-frequency signals, e.g. in the GHz range, where quantum-limited performance has been achieved,[43] equivalent to roughly $S = 1\,\mathrm{pT}^2/\mathrm{Hz}$ sensitivity in a $100\,\mu\mathrm{m}^2$ measurement area. For low frequency signals, the performance of SQUID microscopes is compromised by $1/f$ flux noise, achieving $S \simeq (40\,\mathrm{pT})^2/\mathrm{Hz}$ in a $100\,\mu\mathrm{m}^2$ measurement area.[44,45] The scanning Hall probe microscope achieves a worse sensitivity of $S \simeq (100\,\mathrm{nT})^2/\mathrm{Hz}$, but its high spatial resolution (optimal around 200 nm) makes it competitive with SQUID sensors at length scales at or just below one micron.

Both these technologies require the sensor to be scanned sequentially across the field of view of the microscope, reducing the amount of measurement time available at each resolved pixel. Thus, magnetic field images with many resolved pixels are acquired with correspondingly worse measurement sensitivity. This disadvantage is circumvented by the use of magneto-optical imaging, which relies on the linear Faraday effect (birefringence) of solid-state materials. Magneto-optical imaging allows for the simultaneous acquisition of a magnetic-field image across a wide field of view, with spatial resolution around 1 μm matching the wavelength of light used for imaging. The sensitivity of this technique in present implementations is lower, at single pixels, than the SQUID and scanning-Hall probes described above. However, this worsening of sensitivity is made up by the parallelism of measuring across a wide field of view.

These magnetic microscopy methods are used in an extensive range of applications. Scientific applications include imaging vortex structures in superconductors[42,46] and resolving their temporal dynamics[47] and probing for spontaneous magnetization in p-wave superconductors.[48] A major technological application is non-destructive evaluation of metallic objects and microfabricated devices by means of detecting eddy currents or bound currents in stressed ferromagnetic materials, susceptibility imaging, and detailed mapping of current flow.[49]

A cold-atom magnetometer is compatible with many of the applications discussed above, and yet achieves a single-pixel sensitivity that is orders of magnitude better than the SQUID and scanning Hall probe sensors while maintaining the advantage of detecting over a wide field of view in a single measurement. The possibility of using spinor Bose-Einstein gases for high precision magnetometry was established by prior work of my group.[26] In that work, a dense, spin-polarized

Bose-Einstein condensed gas of ^{87}Rb was confined in a near-planar geometry in a far-off-resonant optical dipole trap. A measurement was begun by tipping the spin polarization to lie transverse to an applied magnetic field, with a spatially uniform initial orientation. The gas was then allowed to undergo Larmor precession, at an angular frequency determined by the local magnetic field magnitude $B(\rho)$ as $\omega_L(\rho) = \gamma B(\rho)$ where $\gamma = g_F \mu_B / \hbar$ is the atomic gyromagnetic ratio[d] and ρ indicates position in the imaged 2D plane. After a measurement time τ, the Larmor precession phase $\phi(\rho)$ was measured by high-resolution magnetization-sensitive imaging.

Let me highlight several features of that work. First, the cold atom gas represents an extended sensing field, providing a simultaneous spatial (2D) map of the field magnitude. Compared with a scanning single sensor, such as a scanning SQUID microscope, this simultaneous mapping rejects long-spatial-wavelength temporal field fluctuations, which appear as a spatially uniform shift of the Larmor phase $\bar{\phi}$, to reveal static short-wavelength spatial inhomogeneities, which are measured as $B(\rho) - \bar{B} = \frac{1}{\gamma\tau}(\phi(\rho) - \bar{\phi})$. Such common-mode rejection bestows the practical advantage of achieving superior magnetic field sensitivity even in the absence of any magnetic shielding.

Second, the sensitivity of this proof-of-principle experiment already approached the standard quantum limit for magnetic sensing. This limit derives from atomic-shot-noise limit to phase measurements, $(\Delta\phi)^2 = N^{-1}$, where the number of atoms in a resolved pixel of area A is given as $N = n_{2D} A$ with n_{2D} being the areal atom number density. The measurement sensitivity is then given as

$$S(A) = \frac{1}{\gamma^2} \frac{1}{A} \frac{1}{n_{2D}\tau D} \tag{4}$$

The sensitivity, with units $[\mathrm{T}^2/\mathrm{Hz}]$, scales inversely with the areal resolution, and improves with higher atom column density (n_{2D}), single-measurement time τ, or duty cycle $D \leq 1$ which is the ratio of the single-measurement time to the cycle time between measurements.

Now, consider implementing this measurement with a layer of ultracold atoms trapped at micron distances from a mirror surface, where they serve as a sensor for magnetic materials below, on, or above the surface (Fig. 8). For example, the atoms might be trapped within an optical lattice trap, formed by light reflecting at shallow angles off the surface of the material being probed or off a thin "cover slip" placed above the material. The quality of the optical trap is not critical: distortions of the potential by scattered light will distort the atomic density distribution, but will not affect the Larmor precession of the trapped atoms.

Once placed within the surface-supported optical lattice trap, the spin-polarized atoms can serve as a sensor of steady (dc) or alternating (ac) magnetic fields. For

---

[d]Conventionally, γ would be defined with a minus sign. We use γ here to relate the Larmor frequency directly to the magnetic field.

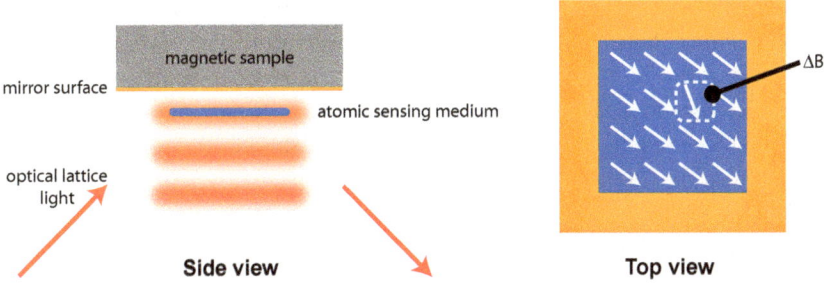

Fig. 8. Magnetometery scheme. Side view: A mirror surface on a magnetic sample is used to produce a vertical optical lattice from laser light reflected off the mirror surface. The incident angle of the laser light controls the lattice spacing. Top view: Following Larmor precession for the single-shot measurement time, the orientation of the transverse magnetization or nematicity is imaged at high spatial resolution across the atomic sample. Local inhomogeneities of the magnetic field are visible in the inhomogeneous orientation of the atomic spin.

this, the atomic spin is prepared in a spatially uniform, transversely polarized, initial state, allowed to evolve for a measurement time τ in the presence of the field produced by the sample, an applied bias field, and rf pulses. Last, the atoms are probed using state-sensitive imaging to reveal the phase of Larmor precession with high spatial resolution, and the imaging data processed to derive the target magnetic field.

The quantum-limited sensitivity to be expected from this device is derived from Eq. 4. Consider that atoms are trapped in the optical-lattice site closest to the reflective material surface, at a distance $d = (\lambda/4)\csc\theta$ from the mirror surface, where λ is the wavelength of the light forming the lattice and θ is its angle of incidence. With λ on the order of 1 μm, $d \geq 250$ nm. Let us assume a three-dimensional density of $n_{3D} = 3 \times 10^{14}$ cm^{-3}, a thickness of $d/4$ for the lattice-confined atomic layer, and single-shot measurement time of $\tau = 3$ s. One then obtains

$$S(A) \simeq \left(1.5 \frac{\text{pT}^2}{\text{Hz}}\right) \times \left(\frac{\mu\text{m}^2}{A}\frac{\mu\text{m}}{d}\frac{1}{D}\right) \tag{5}$$

This measurement sensitivity is far superior to existing dc SQUID magnetometers (limited at $\sim (10\,\text{nT})^2/\text{Hz}$) which are regarded presently as the most powerful magnetometers at the micron length scale. Additional benefits of such a cold-atom magnetometer include the following:

Absolute calibration: The magnetic sensitivity for this cold-atom device is based directly on atomic properties. In contrast, microfabricated SQUID magnetometers are subject to calibration errors and drifts. Magnetometers based on solid-state spin impurities are strongly influenced by their inhomogeneous environments, so that each impurity must be individually diagnosed.

Simultaneous measurement across a wide spatial region: The cold-atom magnetometer envisioned here excels at differential measurements of the magnetic field obtained in parallel in many resolved pixels of the sensor surface. This differential sensitivity provides three important advantages. First, long-wavelength magnetic disturbances result in common-mode shifts of the Larmor phase and are thus rejected. Thus, the influence of any source of stray magnetic fields further than 100's of microns from the sensor is essentially eliminated, obviating the need for strict controls or shielding of the magnetic environment. Second, this feature obviates the need to register precisely the positions of a magnetic object and the sensor; rather, the magnetic object is simultaneously located and measured by the cold-atom sensor. Third, making simultaneous measurements over a total sensor area A_{tot} at many resolved pixels, each of area A, provides an additional improvement in sensitivity, reducing S in Eq. 4 by a factor A/A_{tot}, compared to a single scanning sensor.

Operation at high temperature: Atoms can be maintained at ultralow temperature even within micron distances of high-temperature objects.[50] Thus, a cold-atom magnetometer can measure magnetic objects over a wide temperature range, e.g. for the study of magnetic phase transitions.

Operation at high magnetic field: While high fields disturb the operation of superconductor-based magnetic sensors, the immunity of our cold-atom magnetometer to uniform bias fields allows high sensitivity to be achieved even in a strong uniform applied magnetic field. This allows for sensitive detection of magnetic susceptibility.

Let us quantify the ideal sensitivity of such a cold-atom sensor for several applications.

Detecting currents: First, consider the task of detecting currents on or within the sample, an essential component in many applications of magnetometry for non-destructive evaluation of materials and devices, including detection of conducting materials by the generation of eddy currents, the identification of disturbances in current flow through conductors, the assessment of material defects by the hysteretic magnetization of stress-modulated materials, and the characterization of failures in integrated circuits.[49,51] A line current I running parallel to the mirror surface at a depth of r produces a magnetic field of magnitude $B = \mu_0 I/(2\pi r)$. We compare this field to the single-shot atom-noise-limited measurement sensitivity $S(A)_{single} = \frac{1}{\gamma^2}\frac{1}{An_{2D}\tau^2}$. We consider the task of resolving features of the field produced by this wire (or many wires) with areal spatial resolution of $A = d^2$ set by the distance of the atoms from the mirror surface. As above, we take $n_{2D} = n_{3D} \times d/4$ and $n_{3D} = 3 \times 10^{14}\,\mathrm{cm}^{-3}$, and single-shot measurement time $\tau = 3\mathrm{s}$. This gives a signal-to-noise ratio of unity for a current of

$$I = 4\,\mathrm{pA} \times \left(\frac{\mu\mathrm{m}}{d}\right)^{3/2} \times \left(\frac{r}{\mu\mathrm{m}}\right) \tag{6}$$

Detecting permeable materials: Second, consider the task of detecting a spherical structure of radius R, composed of material with magnetic susceptibility χ_m, and buried within the substrate at a distance $r > R + d$ from the atomic sensor layer. Let us assume the surrounding material has susceptibility $\chi_m = 0$. In an applied magnetic field $\mathbf{B_0}$, the spherical structure acquires a uniform magnetization of $\mathbf{M} \simeq (\chi_m/\mu_0)\mathbf{B_0}$. Outside the sphere, the induced magnetization produces a field equivalent to a magnetic dipole of moment $\mathbf{m} = (4/3)\pi R^3 \mathbf{M}$, i.e. the field produced at the atomic gas above the mirror surface has magnitude $B \simeq (\mu_0/4\pi)|\mathbf{m}|/r^3$. This induced field is nearly uniform over an area $A \sim r^2$ across the atomic gas; taking this as the resolving area of the magnetometer, we conclude a single-shot measurement will detect a susceptibility of order

$$\chi_m = 3 \times 10^{-8} \times \left(\frac{\mu m}{d}\right)^{1/2} \times \left(\frac{r}{\mu m}\right)^2 \times \left(\frac{\mu m}{R}\right)^3 \times \left(\frac{G}{B_0}\right) \tag{7}$$

Thus, at just a few Gauss applied field, micron-scale structures with susceptibility below 10^{-8} should be detectible; this is well below the typical susceptibility of diamagnetic solids (10^{-5}) and equal to those of diamagnetic gases (10^{-8}).

Detecting surface spins: Finally, we consider the task of locating and quantifying magnetic particles on the mirror surface, at distance d from the atomic sensor. Let these particles have a magnetic moment equal to $N_e \mu_B$ (i.e. N_e commonly aligned electron spins). Using the areal resolution $A = d^2$, the single-shot sensitivity is equivalent to

$$N_e = 0.9 \times \left(\frac{d}{\mu m}\right)^{3/2} \tag{8}$$

Thus, even nanoscale ferromagnetic particles should be detectible.

6. Conclusion

There remain many things to be learned about magnetic phenomena in quantum gases, and methods of direct imaging may allow us to investigate them. One area that is clearly ripe for investigation is the detailed dynamics of spin textures, domains, domain walls, and vortices. Our methods for imaging are now sufficiently refined that we should be able to investigate such dynamics by taking a set of images of a single gaseous sample as it evolves in time. It will be interesting to track the motion of magnetic features so as to understand how they change and influence one another. Such investigations will reveal the basic elements of superfluid hydrodynamics as it occurs in magnetic quantum gases, where the flow of mass and spin currents are interrelated. It will also be valuable to measure precisely the temporal spin correlation function, which can be quantified by examining a pair of consecutive images taken of an evolving gas. For systems in thermal equilibrium, this correlation function is related to the nature of spin excitations through the fluctuation-dissipation theorem.

Another open area of investigation is the effect of non-zero temperature on magnetic order and dynamics. Much of the literature on spinor Bose-Einstein condensates considers their zero-temperature properties. At non-zero temperature, we know that additional phenomena will arise, such as "spin-locking"[52] and spin waves driven by the exchange interaction.[53] Both these phenomena have been observed with spatial resolution in pseudo-spin-1/2 gases, but detailed investigations in higher-spin spinor gases are lacking. There are several types of experiments where very high quality data are being produced on lowest-temperature gases, such as the studies of quantum spin mixing dynamics and our recent precision measurement of the magnon dispersion relation. Extending these studies carefully to higher temperature gases might help reveal essential differences stemming from the dynamics of non-condensed atoms.

Finally, we still await applications of spinor-gas magnetometry to the studies of magnetic materials. Our measurement of the magnetic dipolar interactions in the rubidium spinor Bose gas, identified as a gap in the magnon excitation spectrum in Ref. 28, was an application of spinor-gas magnetometry to measure magnetic gases through the magnetic field they generate. However, as sketched in this Chapter, there are excellent prospects for studying solid-state materials by placing a spinor-gas sensor nearby. These applications will be boosted by quantum effects such as spin-nematic squeezing and by improvements in spin-sensitive imaging.

Acknowledgments

I thank the members of my research group who worked with me on spinor Bose-Einstein gases over the years. In particular, I acknowledge the "E4 team" of G. Edward Marti, Ryan Olf, Andrew MacRae, Fang Fang, and Sean Lourette for inspiring a new set of ideas about how to improve spin-dependent imaging and how to characterize magnon excitations. This work is supported by the NSF, NASA and by the AFOSR's MURI program on advanced quantum materials.

References

1. I. Bloch, J. Dalibard, and W. Zwerger, Many-body physics with ultracold gases, *Rev. Mod. Phys.* **80**, 885, (2008).
2. I. Bloch, J. Dalibard, and S. Nascimbène, Quantum simulations with ultracold quantum gases, *Nature Physics.* **8**, 267, (2012).
3. D. M. Stamper-Kurn and M. Ueda, Spinor Bose gases: Symmetries, magnetism, and quantum dynamics, *Rev. Mod. Phys.* **85**, 1191, (2013).
4. W. Gerlach and O. Stern, The directional quantisation in the magnetic field, *Ann. Phys.-Berlin.* **74**, 673, (1924).
5. D. Suter, *The Physics of Laser-Atom Interactions.* Cambridge Studies in Modern Optics, (Cambridge University Press, Cambridge, 2005).

6. J. Higbie, L. Sadler, S. Inouye, A. P. Chikkatur, S. R. Leslie, K. L. Moore, V. Savalli, and D. M. Stamper-Kurn, Direct, non-destructive imaging of magnetization in a spin-1 Bose gas, *Phys. Rev. Lett.* **95**, 050401, (2005).

7. J. Guzman, G. B. Jo, A. N. Wenz, K. W. Murch, C. K. Thomas, and D. M. Stamper-Kurn, Long-time-scale dynamics of spin textures in a degenerate F=1 ^{87}Rb spinor Bose gas, *Phys. Rev. A.* **84**, 063625, (2011).

8. E. Hecht, *Optics.* (Addison-Wesley, Reading, 1989), 2nd edition.

9. L. Sadler, J. Higbie, S. Leslie, M. Vengalattore, and D. Stamper-Kurn, Spontaneous symmetry breaking in a quenched ferromagnetic spinor Bose condensate, *Nature.* **443**, 312, (2006).

10. M.-S. Chang, C. D. Hamley, M. D. Barrett, J. A. Sauer, K. M. Fortier, W. Zhang, L. You, and M. S. Chapman, Observation of spinor dynamics in optically trapped Rb Bose-Einstein condensates, *Phys. Rev. Lett.* **92**, 140403, (2004).

11. H. Schmaljohann, M. Erhard, J. Kronjäger, M. Kottke, S. van Staa, L. Cacciapuoti, J. J. Arlt, K. Bongs, and K. Sengstock, Dynamics of F = 2 spinor Bose-Einstein condensates, *Phys. Rev. Lett.* **92**, 040402, (2004).

12. J. Mur-Petit, M. Guilleumas, A. Polls, A. Sanpera, M. Lewenstein, K. Bongs, and K. Sengstock, Dynamics of F=1 ^{87}Rb condensates at finite temperatures, *Phys. Rev. A.* **73**, 013629, (2006).

13. C. D. Hamley, C. S. Gerving, T. M. Hoang, E. M. Bookjans, and M. S. Chapman, Spin-nematic squeezed vacuum in a quantum gas, *Nat Phys.* **8**, 305, (2012).

14. B. Lücke, J. Peise, G. Vitagliano, J. Arlt, L. Santos, G. Tóth, and C. Klempt, Detecting multiparticle entanglement of Dicke states, *Phys. Rev. Lett.* **112**, 155304, (2014).

15. H. Strobel, W. Muessel, D. Linnemann, T. Zibold, D. B. Hume, L. Pezzè, A. Smerzi, and M. K. Oberthaler, Fisher information and entanglement of non-Gaussian spin states, *Science.* **345**, 424, (2014).

16. M. Andrews, M.-O. Mewes, N. van Druten, D. Durfee, D. Kurn, and W. Ketterle, Direct, non-destructive observation of a Bose condensate, *Science.* **273**, 84, (1996).

17. H.-J. Miesner, D. Stamper-Kurn, M. Andrews, D. Durfee, S. Inouye, and W. Ketterle, Bosonic stimulation in the formation of a Bose-Einstein condensate, *Science.* **279**, 1005, (1998).

18. S. Levy, E. Lahoud, I. Shomroni, and J. Steinhauer, The a.c. and d.c. Josephson effects in a Bose-Einstein condensate, *Nature.* **449**, 579, (2007).

19. C.-L. Hung, X. Zhang, N. Gemelke, and C. Chin, Observation of scale invariance and universality in two-dimensional Bose gases, *Nature.* **470**, 236, (2011).

20. X. Zhang, C.-L. Hung, S.-K. Tung, and C. Chin, Observation of quantum criticality with ultracold atoms in optical lattices, *Science.* **335**, 1070, (2012).

21. W. S. Bakr, J. I. Gillen, A. Peng, S. Folling, and M. Greiner, A quantum gas microscope for detecting single atoms in a Hubbard-regime optical lattice, *Nature.* **462**, 74, (2009).

22. J. F. Sherson, C. Weitenberg, M. Endres, M. Cheneau, I. Bloch, and S. Kuhr, Single-atom-resolved fluorescence imaging of an atomic Mott insulator, *Nature.* **467**, 68, (2010).

23. H. Watanabe and H. Murayama, Unified description of Nambu-Goldstone bosons without Lorentz invariance, *Phys. Rev. Lett.* **108**, 251602, (2012).

24. H. Watanabe and H. Murayama, Redundancies in Nambu-Goldstone Bosons, *Phys. Rev. Lett.* **110**, 181601, (2013).

25. C. Cohen-Tannoudji and J. Dupont-Roc, Experimental study of Zeeman light shifts in weak magnetic fields, *Phys. Rev. A.* **5**, 968, (1972).

26. M. Vengalattore, J. M. Higbie, S. R. Leslie, J. Guzman, L. E. Sadler, and D. M. Stamper-Kurn, High-resolution magnetometry with a spinor Bose-Einstein condensate, *Phys. Rev. Lett.* **98**, 200801, (2007).
27. S. Gupta, K. Dieckmann, Z. Hadzibabic, and D. Pritchard, Contrast interferometry using Bose-Einstein condensates to measure h/m and alpha, *Phys. Rev. Lett.* **89**, 140401, (2002).
28. G. E. Marti, A. MacRae, R. Olf, S. Lourette, F. Fang, and D. M. Stamper-Kurn, Coherent magnon optics in a ferromagnetic spinor Bose-Einstein condensate, *Phys. Rev. Lett.* **113**, 155302, (2014).
29. N. T. Phuc, Y. Kawaguchi, and M. Ueda, Beliaev theory of spinor Bose-Einstein condensates, *Annals of Physics.* **328**, 158, (2013).
30. A. Lamacraft and J. Moore, *Potential insights into non-equilibrium behavior from atomic physics*, In eds. K. Levin, A. Fetter, and D. Stamper-Kurn, *Ultracold Bosonic and Fermionic Gases.* Contemporary Concepts of Condensed Matter Science. Elsevier, (2012).
31. J. Dziarmaga, Dynamics of a quantum phase transition: Exact solution of the quantum Ising model, *Phys. Rev. Lett.* **95**, 245701, (2005).
32. A. Polkovnikov, Universal adiabatic dynamics in the vicinity of a quantum critical point, *Phys. Rev. B.* **72**, 161201, (2005).
33. W. H. Zurek, U. Dorner, and P. Zoller, Dynamics of a quantum phase transition, *Phys. Rev. Lett.* **95**, 105701, (2005).
34. T. W. B. Kibble, Topology of cosmic domains and strings, *J. Phys. A.* **9**, 1387, (1976).
35. W. H. Zurek, Cosmological experiments in superfluid helium?, *Nature.* **317**, 505, (1985).
36. E. M. Bookjans, A. Vinit, and C. Raman, Quantum phase transition in an antiferromagnetic spinor Bose-Einstein condensate, *Phys. Rev. Lett.* **107**, 195306, (2011).
37. H.-J. Miesner, D. Stamper-Kurn, J. Stenger, S. Inouye, A. Chikktur, and W. Ketterle, Observation of metastable states in spinor Bose-Einstein condensates, *Phys. Rev. Lett.* **82**, 2228, (1999).
38. D. Stamper-Kurn, H.-J. Miesner, A. Chikkatur, S. Inouye, J. Stenger, and W. Ketterle, Quantum tunneling across spin domains in a Bose-Einstein condensate, *Phys. Rev. Lett.* **83**, 661, (1999).
39. S. Leslie, J. Guzman, M. Vengalattore, J. D. Sau, M. L. Cohen, and D. Stamper-Kurn, Amplification of fluctuations in a spinor Bose Einstein condensate, *Phys. Rev. A.* **79**, 043631, (2009).
40. J. D. Sau, S. R. Leslie, M. D. Cohen, and D. M. Stamper-Kurn, Spin squeezing of high-spin, spatially extended quantum fields, *New Journal of Physics.* **12**, 085011, (2010).
41. A. Lamacraft, Quantum quenches in a spinor condensate, *Phys. Rev. Lett.* **98**, 160404, (2007).
42. S. J. Bending, Local magnetic probes of superconductors, *Adv Phys.* **48**, 449, (1999).
43. M. Mück, J. B. Kycia, and J. Clarke, Superconducting quantum interference device as a near-quantum-limited amplifier at 0.5 GHz, *App. Phys. Lett.* **78**, 967, (2001).
44. J. R. Kirtley, M. B. Ketchen, K. G. Stawiasz, J. Z. Sun, W. J. Gallagher, S. H. Blanton, and S. J. Wind, High-resolution scanning SQUID microscope, *App. Phys. Lett.* **66**, 1138, (1995).
45. T. S. Lee, E. Dantsker, and J. Clarke, High-transition temperature superconducting quantum interference device microscope, *Rev. Sci. Instrum.* **67**, 4208, (1996).

46. J. W. Guikema, H. Bluhm, D. A. Bonn, R. X. Liang, W. N. Hardy, and K. A. Moler, Two-dimensional vortex behavior in highly underdoped YBa2Cu3O6+x observed by scanning Hall probe microscopy, *Phys. Rev. B.* **77**, 104515, (2008).

47. B. L. T. Plourde, D. J. Van Harlingen, R. Besseling, M. B. S. Hesselberth, and P. H. Kes, Vortex dynamics in thin superconducting strips observed by Scanning SQUID microscopy, *Physica C: Superconductivity.* **341-348**, 1023, (2000).

48. J. R. Kirtley, C. Kallin, C. W. Hicks, E. A. Kim, Y. Liu, K. A. Moler, Y. Maeno, and K. D. Nelson, Upper limit on spontaneous supercurrents in Sr2RuO4, *Phys. Rev. B.* **76**, 014526, (2007).

49. J. Kirtley and J. Wikswo, Scanning SQUID microscopy, *Annual Review of Material Science.* **29**, 117, (1999).

50. J. M. Obrecht, R. J. Wild, M. Antezza, L. P. Pitaevskii, S. Stringari, and E. A. Cornell, Measurement of the temperature dependence of the Casimir-Polder force, *Phys. Rev. Lett.* **98**, 063201, (2007).

51. W. G. Jenks, S. S. H. Sadeghi, and J. P. Wikswo, SQUIDs for nondestructive evaluation, *J Phys D Appl Phys.* **30**, 293, (1997).

52. J. M. McGuirk, D. M. Harber, H. J. Lewandowski, and E. A. Cornell, Normal-superfluid interaction dynamics in a spinor Bose gas, *Phys. Rev. Lett.* **91**, 150402, (2003).

53. J. McGuirk, H. Lewandowski, D. Harber, T. Nikuni, J. Williams, and E. Cornell, Spatial resolution of spin waves in an ultracold gas, *Phys. Rev. Lett.* **89**, 090402, (2002).

II

Atom-Like Coherent Solid State Systems

Chapter 4

Precision Magnetic Sensing and Imaging Using NV-Diamond

R. L. Walsworth

Harvard University, Department of Physics
Harvard-Smithsonian Center for Astrophysics
Cambridge, MA 02138, USA

I provide an overview of precision magnetic sensing and imaging using Nitrogen-Vacancy (NV) color centers in diamond, with particular attention to imaging of magnetic fields from biological cells under ambient conditions using a wide-field-of-view diamond magnetic imager. NV-diamond provides an unprecedented combination of magnetic field sensitivity and spatial resolution in a room-temperature solid due to the remarkable properties of NV centers, including long electronic spin coherence times, optical spin polarization and read-out, a large Zeeman shift of the spin transitions, and the robust and biocompatible physical properties of diamond in a wide variety of forms (bulk crystals, films, nanocrystals, etc.).

1. Overview of NV-Diamond Magnetometry

Nitrogen vacancy (NV) color centers in diamond are quantum defects that enable nanoscale magnetic sensing and imaging under ambient conditions.[1-3] As recently shown using a variety of methods,[4-7] NV centers within room-temperature diamond can be brought into several nm proximity of magnetic field sources of interest while maintaining long NV electronic spin coherence times ($\sim$millisec), a large ($\sim$Bohr magneton) Zeeman shift of the NV spin states, and optical preparation and readout of the NV spin. Recent demonstrations of NV-diamond magnetometry include precision sensing and nanoscale imaging of externally applied and controlled magnetic fields,[4-7] as well as the detection of electron[8,9] and nuclear[10] spins within the diamond crystal and on the diamond surface.[11-14] This section provides an overview of the physics of NV centers and how they enable precision magnetic sensing and imaging. The second section of this chapter discusses one very successful modality for NV-diamond magnetic imaging, known as the "diamond magnetic imager," which provides unique capabilities for wide field-of-view magnetic imaging of biological cells under ambient conditions (see Fig. 1).[15] The final section of this chapter reviews several other notable achievements with NV-diamond quantum sensors.

The negatively-charged NV color center in diamond consists of a substitutional nitrogen atom adjacent to a vacancy in the diamond lattice, with an associated

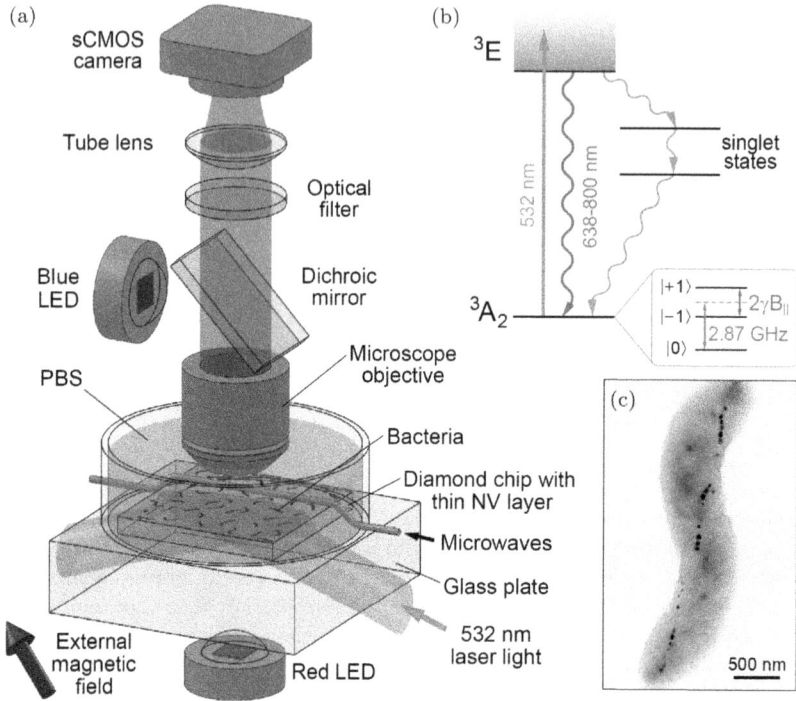

Fig. 1.   Diamond magnetic imager. (a) Bio samples such as live magnetotactic bacteria (MTB) in phosphate-buffered saline (PBS) are placed on the surface of a diamond chip implanted with NV centers. Vector magnetic field images are derived from optical interrogation of NV centers excited by a 532 nm totally-internally-reflected laser beam, with red NV fluorescence imaged onto an sCMOS camera; and spatially correlated with bright field images from red LED illumination. (b) NV center energy-level diagram, with ground spin states ($S = 1$) in the inset. (c) Transmission electron microscope (TEM) image of an AMB-1 MTB. Magnetite nanoparticles appear as dark spots of high electron density. Adapted from Ref. 16.

electronic spin (S = 1) that can be optically polarized and measured. The ground state spin triplet has a 2.87 GHz zero-field splitting between the $m_S = |0>$ and $|\pm 1>$ spin states (Fig. 1b). The NV can be optically excited into a phonon sideband of the electronic excited state using a 532 nm wavelength laser, followed by rapid phonon decay and spontaneous emission of a fluorescence photon into the 640–800 nm phonon sideband of the electronic ground state. This excitation and emission process primarily conserves the $|0>$ spin state, whereas the electronically excited $|\pm 1>$ states decay non-radiatively via metastable singlet states into the $|0>$ ground spin state with ~50% probability. This spin-state-selective decay allows rapid optical pumping of the NV spin into the $|0>$ ground state after several optical excitation cycles, with the polarization preserved for $T_1 > 5$ ms in the room temperature diamond lattice. The spin-state-dependent decay also provides a means of optically distinguishing the $|0>$ and $|\pm 1>$ spin states, since fewer fluorescence photons are emitted when the NV center is in the $|\pm 1>$ spin states than when it is in the $|0>$ state.

The magnetic field at the NV location is determined by measuring the corresponding Zeeman shift in the $|\pm 1>$ energy levels. The zero-field splitting of the $|0>$ and $|\pm 1>$ spin states quantizes the spin along the NV center's symmetry axis. In the presence of an external magnetic field, the $|0> \Leftrightarrow |\pm 1>$ spin-flip transition frequencies shift from the zero-field splitting by $\Delta f = \pm \gamma B_O$, where $\gamma = 2.8\,MHz/G$ is the gyromagnetic ratio of the NV electronic spin, and B_O is the magnetic field projection along the NV symmetry axis. One method of measuring the shift in the NV spin-transition frequency is using optically-detected electron-spin resonance (ESR), by monitoring the NV fluorescence rate during continuous optical excitation as the frequency of a continuously-applied microwave field is swept through resonance. When the microwave frequency passes through either of the $|0> \Leftrightarrow |\pm 1>$ transition resonances, some of the NV state population is transferred from the $|0>$ state to a mixed state, and the fluorescence rate decreases. The magnitude of the magnetic field projection along the NV symmetry axis is then determined from the frequencies corresponding to these dips in the fluorescence signal. Such ESR measurements provide a simple and robust means of measuring static or slowly varying magnetic fields with sub-micron spatial resolution and at room temperature, where the resolution is given by the diffraction limit of the optics used for fluorescence detection.

Pulsed measurement schemes commonly used in nuclear magnetic resonance (NMR) research can also be applied to NV magnetometry.[1,2,4] For example, static (DC) magnetic fields may be measured using a Ramsey pulse sequence, with sensitivity $\approx 1\mu\,T/\sqrt{Hz}$ for a single NV, limited by the inhomogeneously broadened transverse spin dephasing time $T_2^* \approx 1\mu s$. Oscillating (AC) magnetic fields can be measured with higher sensitivity using Hahn-echo or CPGM pulse sequences, with sensitivity $\approx 30\,nT/\sqrt{Hz}$ for a single NV,[2] limited by the intrinsic transverse coherence $T_2 \approx 1\,ms$ with static inhomogeneity effects eliminated. In general, the magnetic field sensitivity of a single NV can be improved by $\sqrt{N}$ by averaging the signal from multiple NV centers, where N is the number of NVs within the averaging volume, which provides a $\sqrt{N}$ improvement in the shot-noise limited optical magnetometer sensitivity.[1]

Compared to other state-of-the-art techniques, NV-diamond magneometry provides a unique combination of wide field-of-view ($>1\,mm^{16}$), nanoscale spatial resolution ($\sim 1\,nm$ with strong gradients[9]), excellent magnetic field sensitivity (<1 nT/Hz$^{1/2}$ using optimized fluorescence collection techniques[17]), and operation under a wide range of conditions. In addition, NV-diamond has other enabling properties for both physical and life science applications, including: fluorescence that typically does not bleach or blink; ability to be fabricated into a wide range of forms such as nanocrystals, atomic force microscope (AFM) tips, and bulk samples with NVs a few nanometers from the surface or uniformly distributed at high density;

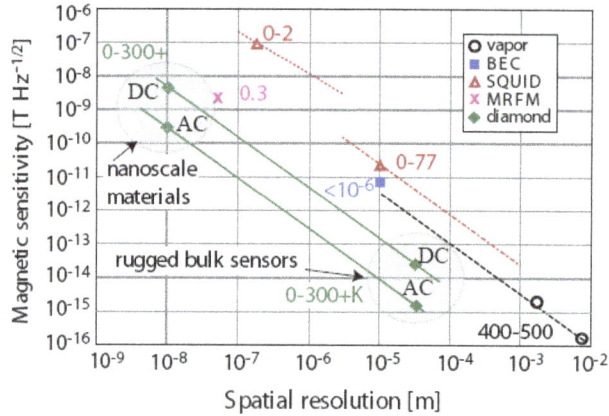

Fig. 2.   Reproduction of figure from the book *Optical Magnetometry*,[18] which provides a comparison of magnetic field sensitivity and spatial resolution of several ultra-sensitive techniques. Note that NV-diamond can provide a unique combination of sensitivity and resolution, along with bio-compatibility.

compatibility with most materials (metals, semiconductors, liquids, polymers et al.) including biological samples under ambient conditions (temperature, pressure, wet, etc.[15]); benign chemical properties; and good endocytosis and no cytotoxicity for diamond nanocrystals and other structures used in sensing and imaging of living biological cells and tissues. Existing magneometry techniques (e.g., MRI, scanning Hall probe microscopy, atomic magnetometers, SQUID microscopy, electron holography, magnetic force microscopy, and magnetic resonance force microscopy) have worse spatial resolution and/or magnetic sensitivity, or entail operating conditions (cryogenics, vacuum, etc.) that preclude application to many samples of interest such as living biological samples. A summary comparison of the magnetic field sensitivity and spatial resolution of leading ultra-sensitive technologies, including NV-diamond, is found in a recently published book entitled *Optical Magnetometry*.[18] Figure 2 reproduces a figure from this book that compares the sensitivity and resolution of several leading magnetic sensing techniques.

2.   Studies of Biological Samples with the Wide-Field Diamond Magnetic Imager

As shown in the above schematic of the diamond magnetic imager (Fig. 1a), the biological sample being studied is placed on a planar diamond chip that contains a dense layer of NV centers in close proximity to the diamond surface. This NV surface layer can be produced by CVD growth or by implanting nitrogen ions in a pure diamond chip followed by annealing.[4] The NV centers are optically excited using a 532 nm laser beam switched on and off using an acousto-optic modulator. The NV red fluorescence is collected by a microscope objective and imaged onto an sCMOS camera, which allows rapid parallel readout of the NV fluorescence within a wide field-of-view, with each camera pixel corresponding to a different magnetic field imaging area on the diamond surface. The NV spin states are coherently manipulated using a spatially uniform microwave field, which is generated by applying the amplified output of a microwave synthesizer to a loop-antenna mounted on the diamond chip. The laser and microwave fields are controlled using a pulse generator, allowing a variety of NV magnetometry protocols for sensing either static or time-varying magnetic fields. Bright-field optical images of the sample on the diamond chip are obtained using separate illumination from a red LED ($\lambda = 660$ nm) imaged onto the same camera, providing spatially-correlated optical and magnetic images. Importantly, the 532 nm laser beam is coupled into the diamond chip from below using a glass prism. This enables total internal reflection of the laser beam from the diamond-air (or diamond-water) interface where the sample is located, significantly reducing the net optical power transfer, e.g., to living biological cells, and consequent potential for undesirable photochemistry. Similarly, the applied microwave fields are too weak to damage biological samples, as confirmed by live-dead assays.[15]

In work published in *Nature* in 2013, we applied the diamond magnetic imager to measure the vector magnetic field patterns from both live and fixed *magnetotactic bacteria* MTB on the surface of the diamond sensor chip with ≈ 400 nm spatial resolution and ~ 100 micron field-of-view.[15] MTB form membrane-bound organelles containing magnetic nanoparticles (magnetosomes) arranged in chains with a net dipole moment (Fig. 1c), allowing them to orient and travel along geomagnetic field lines (magnetotaxis).[19] MTB are of considerable interest as a model system for the study of molecular and biophysical mechanisms involved in magnetic biomineralization.[19] Also, the nanoparticles synthesized by MTB have exceptional chemical purity and well controlled species-specific crystal morphologies and size distributions, making them of great interest for biomedical applications.[20]

We acquired correlated optical bright-field and magnetic field images (Fig. 3) of populations of wild-type *Magnetospirillum magneticum* AMB-1. This MTB strain forms magnetic nanoparticles with cubo-octahedral morphology and an average diameter of 50 nm; see Fig. 1c for a characteristic TEM image showing the size and distribution of nanoparticles in a single AMB-1. Measurements were performed on both living and dried MTB on the diamond surface, with consistent results. Such wide-field magnetic images enable rapid, simultaneous measurement of magnetic particle distributions in many individual cells. Previous assays of magnetic nanoparticles in MTB relied on electron microscopy to resolve the size and location of individual nanoparticles, but did not provide correlated, direct measurements of magnetic fields from large numbers of individual cells. The magnetic field sensitivity and spatial resolution of the diamond magnetic imager are sufficient both to localize nanoparticle chains within the MTB and to quantify their magnetic moment from the acquired field magnetic images. To verify that the diamond magnetic field images are consistent with those predicted from precise knowledge of the

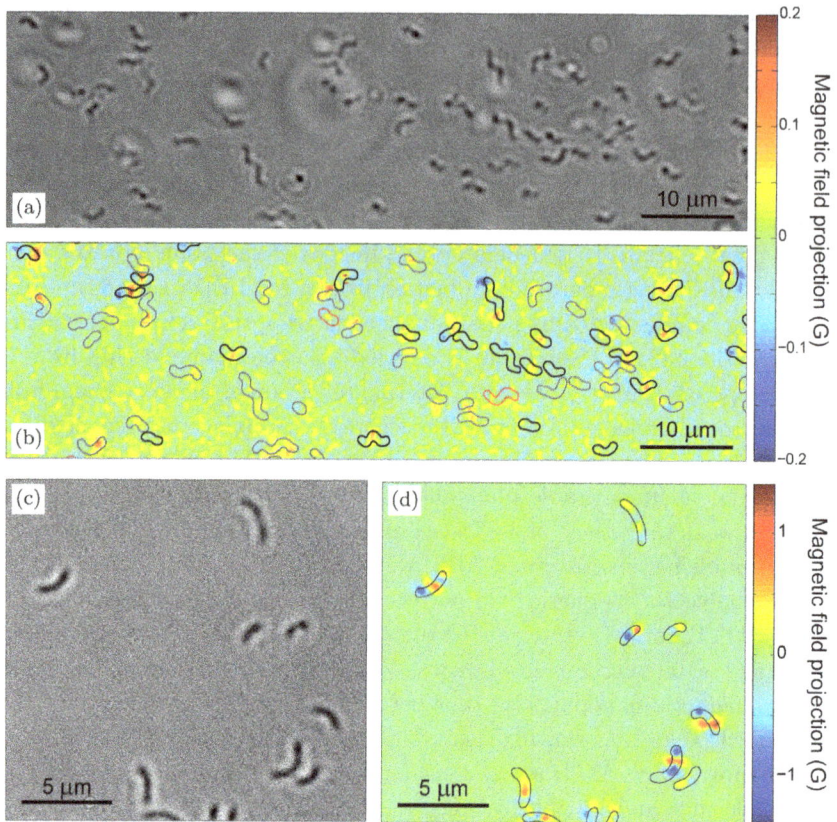

Fig. 3. Correlated bright-field and magnetic images of magnetotactic bacteria. (a) Bright-field image of live MTB on diamond surface immersed in PBS. (b) NV-diamond image of one vector component of magnetic field for same region as (a), showing characteristic dipole magnetic field pattern from MTB magnetosome chains. Superimposed outlines indicate MTB locations determined from (a). Outline colors indicate results of a live-dead assay performed after magnetic imaging (black for living, red for dead, and grey for indeterminate). (c) Bright-field image of dried MTB on diamond chip. (d) NV-diamond image of one vector component of the magnetic field for the same region, with outlines indicating MTB locations determined from (c). Adapted from Ref. 15.

nanoparticle configurations determined by electron microscopy, we carried out scanning electron microscope (SEM) imaging of a small number of MTB on the surface of the diamond chip. Positions and relative sizes of the magnetic nanoparticles within each MTB were determined from the backscattered electron SEM images, and used to calculate the expected vector magnetic field pattern. The directly measured field patterns (from the diamond magnetic imager) were in excellent agreement with the calculated patterns (using SEM data) for a variety of nanoparticle distributions within the MTB, one example of which is shown in Fig. 4.

In follow-up work, we applied the diamond magnetic imager to studies of wild-type and mutant RS-1 MTBs;[21] increased the field-of-view to ~1 mm;[16] and used

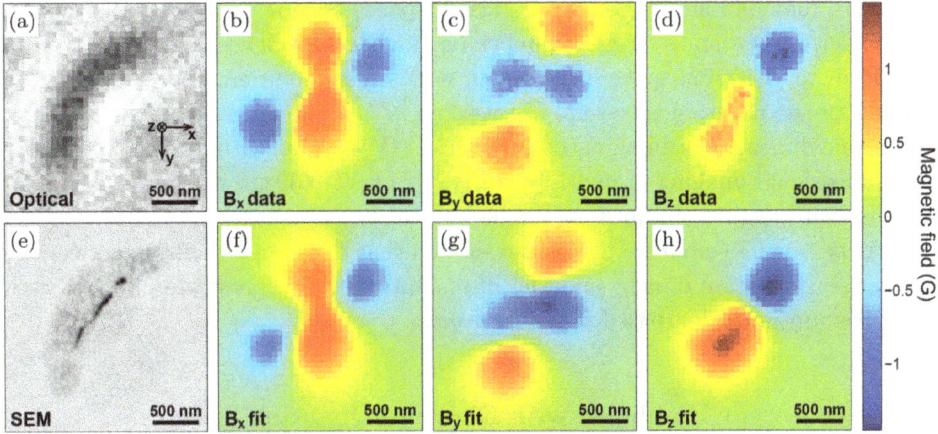

Fig. 4.   Good agreement is found for MTB magnetic fields that are optically imaged (by NV-diamond) and simulated (from SEM images). (a) Bright-field optical image of an AMB-1 bacterium. (b)-(d) NV-diamond magnetic field images for vector components along x,y, and z axes within the same field-of-view as the bright field image. (e) Scanning electron microscope (SEM) image of the same bacterium, acquired in backscattered electron mode. (f)-(h) Simulated magnetic field projections along x, y, and z axes within the same field-of-view as the SEM image. The total magnetic moment of the nanoparticles was determined to be $(1.2 \pm 0.1) \times 10^{-16}$ A m^2 by fitting the calculated field distribution to the measurement, with the moment and the standoff distance left as free parameters. Adapted from Ref. 15.

Fourier imaging techniques to extend the resolution to < 30 nm.[22]   This next-generation system will enable rapid measurement of the magnetic properties of individuals in a given MTB population under ambient laboratory conditions, a measurement capability that was previously unavailable; and will allow time-resolved magnetic measurements on single MTB with the ability to resolve the magnetic fields from individual magnetic nanoparticles and magnetosomes. Future planned work includes detailed studies of magnetic nanoparticle growth in various MTB species and biomineralization mutants.

In other recent work, we extended the performance and biological applicability of the diamond magnetic imager to realize noninvasive magnetic imaging of cells immunologically targeted with magnetic nanoparticles, with an expanded 1 mm field-of-view. This demonstration has single cell resolution, excellent cell detection fidelity, quantitative measurement of biomarker expression on multiple cell lines, and applicability to opaque samples.[16] Due to low magnetic backgrounds in most bio-samples, targeting cells and sub-cellular structures with magnetic labels enables more accurate and robust identification than optical methods, which suffer from scattering, absorption, and intrinsic autofluorescence.

To improve the spatial resolution of the diamond magnetic imager, we developed a Fourier imaging technique analogous to conventional magnetic resonance imaging (MRI).[22] Pulsed magnetic field gradients, generated by microfabricated

coils on a diamond substrate, are applied to phase-encode spatial information on
the NV electronic spins in wavenumber or "k-space;" followed by a fast Fourier
transform to yield real-space images with nanoscale resolution, wide field-of-view,
and compressed sensing speed-up. We used an NV-diamond test sample to demonstrate one-dimensional imaging of individual NV centers with <5 nm resolution;
two-dimensional imaging of multiple NV centers with <30 nm resolution; and two-
dimensional imaging of nanoscale magnetic field patterns, with magnetic gradient
sensitivity $\sim 14\,\mathrm{nT/nm/Hz}^{1/2}$. We also showed that compressed sensing accelerates
the image acquisition time by an order-of-magnitude using sparse sampling followed
by convex-optimization-based signal recovery.

3. Other Recent Highlights with Diamond Quantum Sensors

Applications of NV-diamond quantum sensors are rapidly advancing and diversify-
ing. In addition to magnetic sensing and imaging, emphasized here, NV-diamond
can provide nanoscale electric field sensing via a linear Stark shift in the NV ground-
state spin levels induced by interactions with the crystalline lattice;[23] as well as
nanoscale temperature sensing via a change in the zero-magnetic-field splitting
between the NV spin levels.[24] Furthermore, NV optical measurements are ide-
ally suited for applications of superresolution imaging techniques, which provide
sub-wavelength resolution by exploiting the nonlinear response of optical emitters.
In particular, the group of Stefan Hell used NV centers to achieve imaging resolution
below 10 nm — a world record based on far field optics;[25] while my group and Harvard collaborators worked with Hell and colleagues to realize two-dimensional NV
optical magnetic imaging with resolution of about 30 nanometers.[26] Some other
recent highlights using NV-diamond quantum sensors, all performed at ambient
laboratory conditions, are summarized below.

The magnetic field from a vortex produced by a thin ferromagnetic film was
imaged using a scanning NV-diamond AFM by the group of Vincent Jacques and
collaborators.[27] In related work, the group of Amir Yacoby and collaborators used
a shallow NV centers to map the magnetic-field dependence of spin-wave excitations in a ferromagnetic microdisc, by detecting the associated local reduction in
the disc's longitudinal magnetization.[28] In addition, they characterized the spin-
noise spectrum by NV-spin relaxometry, finding excellent agreement with a general analytical description of the stray fields produced by spin-spin correlations
in a 2D magnetic system. In future work such ultra-high-resolution NV-diamond
magnetic imaging may be applied to probe advanced materials where quantum
spins and their associated magnetic fields play a fundamental role in determining the material properties. Examples include frustrated magnetic systems with
skyrmionic ordering, spin liquids where quantum spin fluctuations prevent the system from ordering, and topological insulators with quantized spin-carrying surface
states.

Sensing of the local temperature within living human cells using NV centers in nanodiamonds was performed by the groups of Mikhail Lukin and Hongkun Park.[29] This technique allows precision monitoring of thermally-induced cell death from targeted laser power absorption by gold nanoparticles within the cells, which may provide a useful tool for guiding thermoablative therapy to kill tumor cells selectively. Other potential applications of functionalized *in vivo* nanodiamond quantum sensors include: guiding temperature-induced control of gene expression and development; identifying local tumor activity by mapping atypical thermogenesis at the single-cell level; and nanoscale functional MRI within cells, e.g., from changes in free radical concentrations or currents induced by ion channel dynamics.

References

1. J. M. Taylor, P. Cappellaro, L. Childress, L. Jiang, D. Budker, P. R. Hemmer, A. Yacoby, R. Walsworth and M. D. Lukin, "High-sensitivity diamond magnetometer with nanoscale resolution," *Nat. Phys.* **4**, 810–816 (2008).
2. J. R. Maze, P. L. Stanwix, J. S. Hodges, S. Hong, J. M. Taylor, P. Cappellaro, L. Jiang, M. V. G. Dutt, E. Togan, A. S. Zibrov, A. Yacoby, R. L. Walsworth, and M. D. Lukin, "Nanoscale magnetic sensing with an individual electronic spin in diamond," *Nature* **455**, 644–647 (2008).
3. G. Balasubramanian, I. Y. Chan, R. Kolesov, M. Al-Hmoud, J. Tisler, C. Shin, C. Kim, A. Wojcik, P. R. Hemmer, A. Krueger, T. Hanke, A. Leitenstorfer, R. Bratschitsch, F. Jelezko, and J. Wrachtrup, "Nanoscale imaging magnetometry with diamond spins under ambient conditions," *Nature* **455**, 648–651 (2008).
4. L. M. Pham, D. Le Sage, P. L. Stanwix, T. K. Yeung, D. Glenn, A. Trifonov, P. Cappellaro, P. R. Hemmer, M. D. Lukin, H. Park, A. Yacoby, and R. L. Walsworth, R. L. Walsworth, "Magnetic field imaging with nitrogen-vacancy ensembles," *New J. Phys.* **13**, 045021 (2011).
5. P. Maletinsky, S. Hong, M. S. Grinolds, B. Hausmann, M. D. Lukin, R. L. Walsworth, M. Loncar, and A. Yacoby, "A robust scanning diamond sensor for nanoscale imaging with single nitrogen-vacancy centres," *Nature Nanotech.* **7**, 320–323 (2012).
6. S. Steinert, F. Dolde, P. Neumann, A. Aird, B. Naydenov, G. Balasubramanian, F. Jelezko, and J. Wrachtrup, "High sensitivity magnetic imaging using an array of spins in diamond," *Rev. Sci. Instrum.* **81**, 043705 (2010).
7. B. J. Maertz, A. P. Wijnheijmer, G. D. Fuchs, M. E. Nowakowski, and D. D. Awschalom, "Vector magnetic field microscopy using nitrogen vacancy centers in diamond," *Appl. Phys. Lett.* **96**, 092504 (2010).
8. M. S. Grinolds, S. Hong, P. Maletinsky, L. Luan, M. D. Lukin, R. L. Walsworth, and A. Yacoby, "Nanoscale magnetic imaging of a single electron spin under ambient conditions," *Nat. Phys.* **9**, 215–218 (2013).
9. M. S. Grinolds, M. Warner, K. De Greve, Y. Dovzhenko, L. Thiel, R. L. Walsworth, S. Hong, P. Maletinsky, and A. Yacoby. "Sub-nanometer resolution in three-dimensional magnetic-resonance imaging of individual dark spins," *Nat. Nanotech.* **9**, 279–282 (2014).
10. L. Childress, M. V. G. Dutt, J. M. Taylor, A. S. Zibrov, F. Jelezko, J. Wrachtrup, P. R. Hemmer, and M. D. Lukin, "Coherent Dynamics of Coupled Electron and Nuclear Spin Qubits in Diamond," *Science* **314**, 281–284 (2006).

11. H. J. Mamin, M. Kim, M. H. Sherwood, C. T. Rettner, K. Ohno, D. D. Awschalom, and D. Rugar, "Nanoscale nuclear magnetic resonance with a nitrogen-vacancy spin sensor," *Science* **339**, 557–560 (2013).

12. T. Staudacher, F. Shi, S. Pezzagna, J. Meijer, J. Du, C. A. Meriles, F. Reinhard, and J. Wrachtrup, "Nuclear magnetic resonance spectroscopy on a (5-nanometer)^3 sample volume," *Science* **339**, 561–563 (2013).

13. S. J. DeVience, L. M. Pham, I. Lovchinsky, A. O. Sushkov, N. Bar-Gill, C. Belthangady, F. Casola, M. Corbett, H. Zhang, M. D. Lukin, H. Park, A. Yacoby, and R. L. Yacoby, and R. L. Walsworth, "Nanoscale NMR Spectroscopy and imaging of multiple nuclear species," *Nat. Nanotech.* **10**, 129–134 (2015).

14. A. O. Sushkov, I. Lovchinsky, N. Chisholm, R. L. Walsworth, H. Park, and M. D. Lukin, "Magnetic resonance detection of individual proton spins using quantum reporters," *Phys. Rev. Lett.* **113**, 197601 (2014).

15. D. Le Sage, K. Arai, D. R. Glenn, S. J. DeVience, L. M. Pham, L. Rahn-Lee, M. D. Rahn-Lee, M. D. Lukin, A. Yacoby, A, Komeili, and R. Walsworth, "Optical cells," *Nature* **496**, 486–489 (2013).

16. D. R. Glenn, K. Lee, H. Park, R. Weissleder, A. Yacoby, M. D. Lukin, H. Lee, R. L. Walsworth, C. B. Connolly, "Single cell magnetic imaging using a quantum diamond microscope," *Nat. Methods* (in press).

17. D. Le Sage, L. M. Pham, N. Bar-Gill, C. Belthangady, M. D. Lukin, A. Yacoby, and R. L. Walsworth, "Efficient photon detection from color centers in a diamond optical waveguide," *Phys. Rev. B* **85**, 121202(R) (2012).

18. D. Budker and D. F. Jackson Kimball (Eds.), *Optical Magnetometry*, Cambridge: Cambridge University Press (2013).

19. A. Komeili, "Molecular mechanisms of compartmentalization and biomineralization in magneto-tactic bacteria," *FEMS Microbiol. Rev.* **36**, 232–255 (2012).

20. D. Faivre and D. Schüler, "Magnetotactic Bacteria and Magnetosomes," *Chem. Rev.* **108**, 4875–4898 (2008).

21. L. Rahn-Lee, M. E. Byrne, D. Le Sage, D. R. Glenn, T. Milbourne, R. L. Walsworth, H. Vali, and A. Komeili, "A genetic strategy for probing the functional diversity of magnetosome formation," *PLOS-Genetics* **11**, e1004895 (2015).

22. K. Arai, C. Belthangady, H. Zhang, N. Bar-Gill, S. J. DeVience, P. Cappellaro, A. Yacoby, and R. L. Walsworth, "Fourier Magnetic Imaging with Nanoscale Resolution and Compressed Sensing Speed-up using Electronic Spins in Diamond," submitted to *Nat. Nanotech.*, arxiv:1409.2749.

23. F. Dolde, H. Fedder, M. W. Doherty, T. Nöbauer, F. Rempp, G. Balasubramanian, T. Wolf, F. Reinhard, L. C. L. Hollenberg, F. Jelezko, and J. Wrachtrup, "Electric field sensing using single diamond spins," *Nat. Phys.* **7**, 459–463 (2011).

24. V. Acosta, E. Bauch, M. P. Ledbetter, A. Waxman, L.-S. Bouchard, and D. Budker, "Temperature dependence of the nitrogen-vacancy magnetic resonance in diamond," *Phys. Rev. Lett.* **104**, 070801 (2010).

25. E. Rittweger, K. Y. Han, S. E. Irvine, C. Eggeling, and S. W. Hell, "STED microscopy reveals crystal color centres with nanometric resolution," *Nat. Photon.* **3**, 144–147 (2009).

26. P. C. Maurer, J. R. Maze, P. L. Stanwix, L. Jiang, A. V. Gorshkov, A. A. Zibrov, B. Harke, J. S. Hodges, A. S. Zibrov, A. Yacoby, D. Twitchen, S. W. Hell, R. L. Walsworth, and M. D. Lukin, "Far-field optical imaging and manipulation of individual spins with nanoscale resolution," *Nat. Phys.* **6**, 912–918 (2010).

27. L. Rondin, J. P. Tetienne, S. Rohart, A. Thiaville, T. Hingant, P. Spinicelli, J. F. Roch, V. Jacques, "Stray-field imaging of magnetic vortices with a single diamond spin," *Nat. Comm.* **4**, 2279 (2013).

28. T. van der Sar, F. Casola, R. L. Walsworth, and A. Yacoby, " Nanometre-scale probing of spin waves using single electron spins," submitted to *Nat. Comm.*, arxiv:1410.6423.

29. G. Kucsko, P. C. Maurer, N. Y. Yao, M. Kubo, H. J. Noh, P. K. Lo, H. Park, and M. D. Lukin, "Nanometre-scale thermometry in a living cell," *Nature* **500**, 54–58 (2013).

Chapter 5

Entanglement and Quantum Optics with Quantum Dots

A. P. Burgers, J. R. Schaibley, and D. G. Steel

Harrison M. Randall Laboratory, The University of Michigan
Ann Arbor, MI 48118, USA

Quantum dots (QDs) exhibit many characteristics of simpler two-level (or few level) systems, under optical excitation. This makes atomic coherent optical spectroscopy theory and techniques well suited for understanding the behavior of quantum dots. Furthermore, the combination of the solid state nature of quantum dots and their close approximation to atomic systems makes them an attractive platform for quantum information based technologies. In this chapter, we will discuss recent studies using direct detection of light emitted from a quantum dot to investigate coherence properties and confirm entanglement between the emitted photon and an electron spin qubit confined to the QD.

1. Introduction

Quantum dots (QDs) have emerged as a viable option in the quest for scalable quantum technologies. In particular, InAs/GaAs self-assembled QDs are shown to satisfy many of the criteria necessary for a scalable quantum computer.[1] The ability to confine a single electron spin to the QD provides us with a quantum bit (qubit) for quantum information processing. Initialization to a pure spin state, spin rotation and read-out have all been demonstrated in this system showing its viability as a qubit for quantum computation.[2-5] Recent work, by our group and others, demonstrates that the spontaneously emitted photon from the QD is entangled with the electron spin state.[6-8] Many quantum information architectures utilize entangled photons to coherently transport information between qubits, so this result makes many of these schemes possible with QDs.[9-11]

The dots investigated in this chapter are self-assembled InAs/GaAs QDs, grown by molecular beam epitaxy. The 7% lattice mismatch between InAs and GaAs causes InAs islands to form which gives the 3 dimensional confinement when a GaAs capping layer is deposited on the islands. Unlike electrically defined dots, these dots confine both the electron and the hole in the same location, enabling a strong optical coupling to transitions, often in the visible or near IR. The dots are embedded in a Schottky diode heterostructure which allows selective charging of the QD with single electrons by varying sample voltage.[12,13] When a charged QD is optically excited a trion or negative exciton, X$^-$, is created. The trion charge states are determined using photoluminescence (PL) spectroscopy, and further identified with voltage modulation spectroscopy.[14] Here the electron is swept in and out of the dot by modulating the voltage across the Schottky diode, and a narrowband (<1MHz) continuous wave (CW) laser scans across the resonance to determine the exact frequency and linewidth.

The energy level structure for the trion at zero magnetic field is shown in Fig. 1a, using the growth axis as the axis of quantization. The two allowed transitions are excited by circularly polarized light as show in Fig. 1a. The cross transitions e.g. $|z-\rangle \leftrightarrow |T_z+\rangle$, are forbidden making optical manipulation of the spin state impossible in the current quantization. By breaking the symmetry with a transverse magnetic field (perpendicular to the growth axis or the Voigt geometry) the states become mixed and yields the level diagram in 1b. Now the cross transitions are allowed and the selection rules become linear with the ground state split according to the electron g-factor, g_e, and the excited state by the hole g-factor, g_h. The

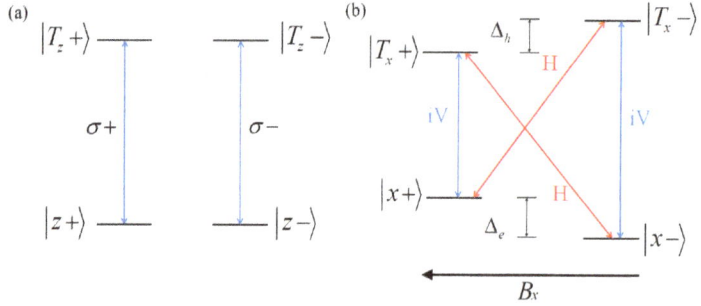

Fig. 1.   The trion level diagram a) at zero magnetic field and b) with a magnetic field applied in the transverse direction, which lifts the degeneracy and gives rise to a new basis (x) with linear selection rules.

2 Λ systems give us the ability to manipulate the electron spin via the accessible trion states. We also note that spontaneous emission from one of the excited states will lead to an entangled state between the photon and the electron spin.[15] The resulting wave function from a decay from $|T_x-\rangle$ would be of the following form,

$$|\Psi\rangle = \frac{|\omega_r\rangle|H\rangle|x+\rangle - i|\omega_b\rangle|V\rangle|x-\rangle}{\sqrt{2}} \qquad (1)$$

Where $|H\rangle$ (and $|V\rangle$) and $|\omega_r\rangle$ (and $|\omega_b\rangle$) denote the polarization state and frequency, respectively. Verification of this entanglement will be discussed in detail in section 4. The following sections will discuss the experimental techniques needed to directly detect the QD radiation and time tagging procedures that are vital to demonstrating entanglement and further studies with QDs. The first step is to directly detect the photons scattered off the trion transitions when a CW laser is tuned on resonance.

2. Resonant Rayleigh Scattering from a QD

The trion system described above provides us with a degenerate two level system with circularly polarized selection rules as shown in Fig. 1. Experiments involving photon counting with QDs, such as entanglement protocols, require the suppression of the excitation laser to allow the detection of photons that are emitted or scattered by the QD exciton (or trion). Previous measurements utilize lock-in detection and avalanche photodiodes (APDs) to observe the absorption (a measure of the change in transmission when the laser is on resonance vs off, or DT/T) of laser light by the QD.[14] In those experiments, the APDs measures the transmitted laser intensity instead of the direct single photon scatter from the QD. Here, we use polarization rejection to filter out the excitation laser so that we are only sensitive to single photons scattered from the QD (i.e. the resonant Rayleigh scattering).

For the degenerate two-level system described above linear polarization couples both of the transitions, so we set our input laser to vertical (V). Two aspheric lenses are placed on either side of the QD sample, the front for excitation and the back for collection. The QD will scatter in 4π, but we are only sensitive to the 2% collected by the back asphere. The laser light however is almost completely collected and therefore must be removed to observe the scatter photons. A polarizer set to H in the transmission geometry suppresses the excitation laser by a factor of 10^5. Unfortunately, this level of suppression is insufficient so the use of an optical fiber for spatial filtering gives an extra factor of 100. A single photon avalanche detector (SPAD), which is an APD operated in geiger mode (operated above the breakdown voltage), is sensitive to single photon events that cause the device to register a single event and reset. With the SPAD and a 10^7 level of suppression, scanning the laser

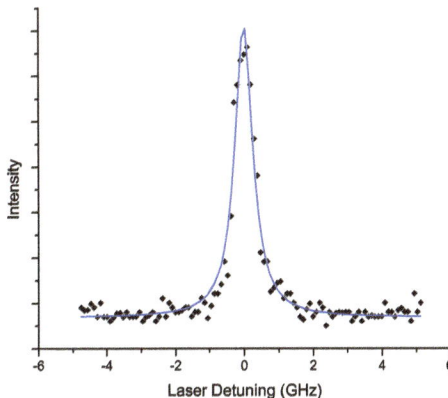

Fig. 2. An example CW resonance fluorescence scan with a Lorentzian fit yielding a line width of 623 ± 25 MHz.

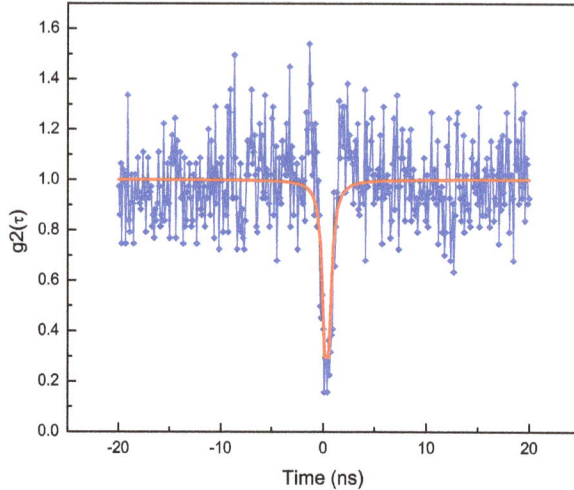

Fig. 3. The intensity correlation spectrum, $g^{(2)}(\tau)$, for a trion system at zero magnetic field. The dip in the spectrum below 0.5 indicates the non-classical nature of the emitter.

across a QD resonance reveals a Lorentzian shape signal shown in Fig. 2. The state linewidth for this QD is 623±25 MHz.

A final important step before moving to the time-resolved data involving QDs is to determine if the trion system of interest is indeed a single photon emitter. The most common method to determine if a system is a single emitter is to preform a temporal intensity correlation measurement, $g^{(2)}(\tau)$. The arrival times of the photons are recorded in a Hanbury Brown and Twiss (HBT) type set-up where a dip at zero delay indicates a non-classical signature.[16,17] Data indicating the single emitter nature of the QDs studied here is found in Fig. 3.

The excitation laser rejection and the ability to recover the resonant Rayleigh scattered spectrum are essential to photon counting with QDs. The observation of many of the coherent properties of the QD require the ability to do single photon counting as seen in the $g^{(2)}(\tau)$ measurements. In the next section more sophisticated time tagging techniques, required to performed advanced correlation measurements, are introduced.

3. Time Domain Fluorescence

Here we present data probing the nature of coherent population oscillations, Rabi oscillations , and discuss excitation techniques that are useful for other coherent control experiments using QDs. The data has been modified from a previous publication.[18]

In the previous section we reported on observations in the frequency domain using CW lasers for excitation. A further tool we can use in characterizing the coherence properties of QDs is to observe the fluorescence in the time domain. By time gating the excitation of a QD, we can observe the dynamics during and after excitation that are not available in CW measurements. By using an electro-optic modulator (EOM), we time gate a resonant CW laser, synced to a picosecond event timer. The QD light is time tagged using a single photon detector with 40 ps timing jitter and a RF pulse generator sets the master clock and provides the pulse train for the EOM, which intensity modulates the laser. Fig. 4 shows a typical diagram for this technique.

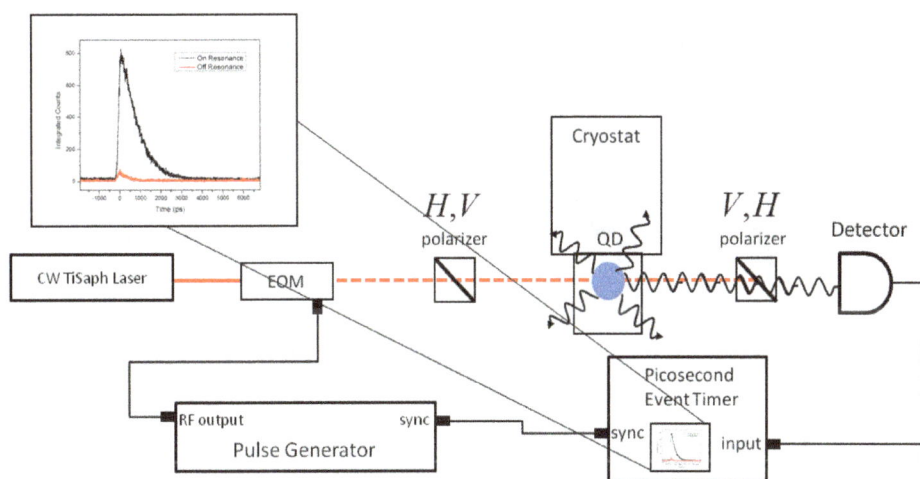

Fig. 4. Diagram of a typical time tagged fluorescence set-up. The excitation laser is rejected using polarization allowing detection of the QD fluorescence. The pulse generator sends RF signals to the EOM gating the CW laser at specified intervals. The inset is an example histogram using 250 ps pulses on (black) and off (red) resonance.

A direct measurement of the QD's transient fluorescence follows the dynamics of the excited state for this 2 level system. Using pulse widths less than the radiative lifetime, we observe the characteristic exponential decay of the excited state population (ρ_{22}), shown in Fig. 5. From these data, we extract a radiative lifetime of 640 ± 25 ps which is the inverse of the population decay rate. As the incident power increases, the exponential decay will increase until we reach a π pulse meaning all the population has been transferred to the excited state. Increasing the power past the π power drives the system back to the ground state before emission and thus should decrease the counts present in the QD emission spectrum. Fig. 6a shows that as the power increases the peak emission along the decay curve (the part of the histogram after 250 ps) decreases. A more pronounced visual of this effect is seen when we integrate over the emission and plot the integrated counts

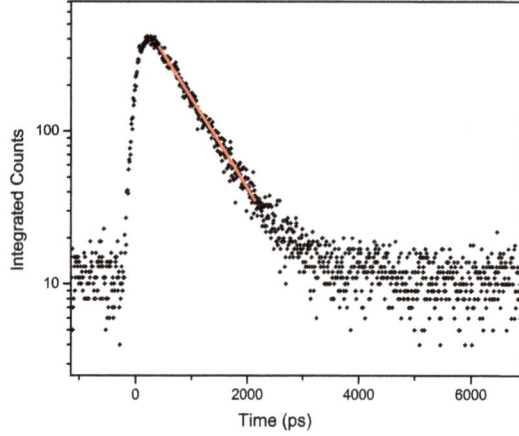

Fig. 5. An integrated histogram of the time tagged emission from the QD on a log scale with the exponential fit. Excitation is preformed with a 250 ps resonant pulse. The single exponential fit gives a lifetime of 640 ± 25 ps

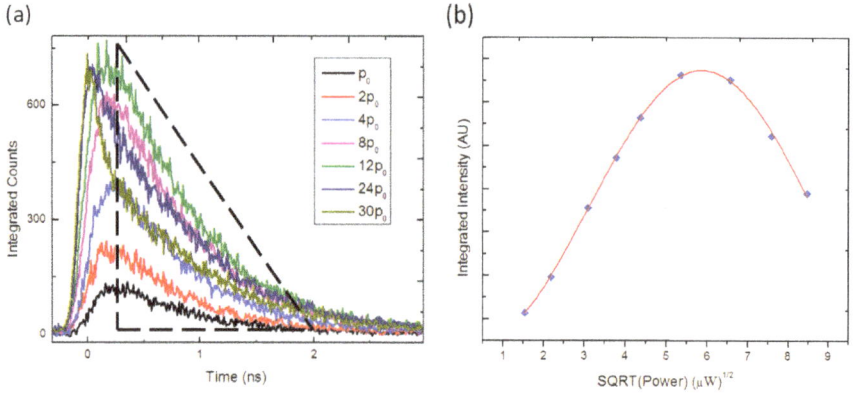

Fig. 6. (a) Histogram data from a resonant 250 ps pulse with increasing powers showing the increase and decrease in emission as the power surpasses the π power ($12P_0$). (b) Integrated counts taken from the black dotted triangle in (a), are plotted as a function of $\sqrt{(Power)}$ to show the Rabi turnover.

as a function of square root of power, Fig. 6b. These are Rabi oscillations and are evident in the formula for the excited state population under the influence of a driving field E_0,

$$\rho_{22}(t) = sin^2 \left(\frac{\mu E_0 t}{2\hbar} \right). \qquad (2)$$

Where μ is the optical dipole moment. This formula, however, neglects effects from spontaneous emission and decoherence, which are vital to understanding the complete dynamics of the system. Including these effects and solving for the excited state population we find,

$$\rho_{22}(t) = \frac{\Omega_0^2}{2(\Omega_0^2 + \gamma\gamma_2)} \left[1 - \left(cos(\lambda t) + \frac{\gamma + \gamma_2}{2\lambda} sin(\lambda t) \right) e^{-\frac{\gamma + \gamma_2}{2} t} \right] \tag{3}$$

with

$$\lambda = \sqrt{\Omega_0^2 - \frac{(\gamma - \gamma_2)^2}{4}}$$

$$\Omega_0 = \frac{\mu E_0}{\hbar} \tag{4}$$

Here, we assume resonant excitation, and the generalized Rabi frequency reduces to the usual Ω_0. The exponential in Eq. (3) shows that the oscillation term decays with the average of the longitudinal, γ_2, and transverse, γ, decay rates. An interesting physical interpretation to take from this is that as the optical Bloch vector rotates about the one axis, x, it spends an equal amount of time in both y and z. The magnitude along each axis will decay according to the separate decay rates where y decays at γ, and z at γ_2.

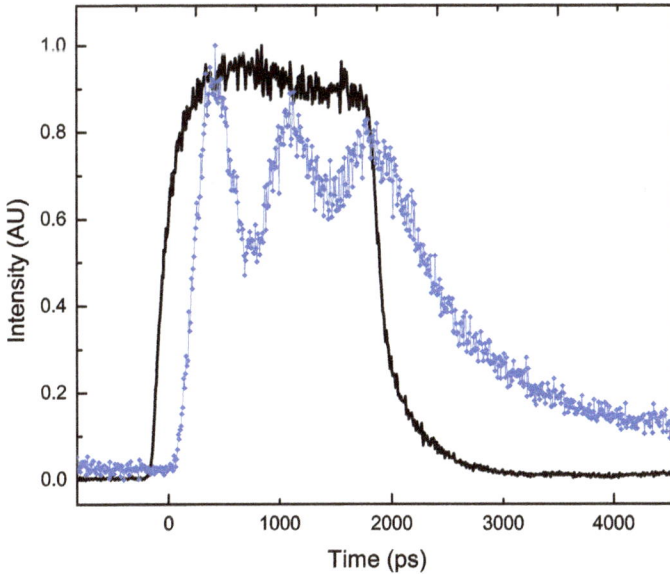

Fig. 7. A scaled histogram of the 2 ns excitation pulse (black) and the resultant QD fluorescence (blue) when the excitation pulse is suppressed using polarization rejection.

Equation (3) follows the excited state population for a constant field, rendering the analysis on the 250 ps pulse data incomplete when considering the time evolution of ρ_{22} under constant excitation. To further probe the behavior of the excited state, we should look at excitations with a duration larger than the radiative lifetime. Increasing the pulse width and continuing to monitor the fluorescence in the same way allows for the direct probe of Eq. (3).

As the pulse width and power are increased, we begin to see oscillations in the time tagged emission. A scaled example of the pulse and the quantum dot scattering are shown in Fig. 7. The oscillation frequency increases linearly with the square root of power as expected. Due to the previous determination of the population decay lifetime, $1/\gamma_2$, only the Rabi frequency and decoherence lifetime, γ, are left as unknowns. Using data for multiple input powers in Fig. 8, we can fit Eq. (3) during the excitation pulse leaving Ω_0 and γ as fitting parameters. From the fits we extract a decoherence lifetime, $1/\gamma$, of 1.22 ± 0.06 ns. In the absence of pure dephasing we would expect that $2\gamma = \gamma_2$, which is what we observe and is consistent with other studies done on InAs QDs.[19] Also, of interest is the comparison of this time domain data to the CW data shown in Fig. 2 of the previous section. The line width extracted from the CW laser scan is 623 ± 25 MHz, which would imply a decoherence lifetime of 511 ± 21 ps. Since the time domain measurements indicate a much longer coherence time we must conclude that the frequency domain measurements of the line width are broadened by some spectral wandering process, which has been previously proposed.[20,21]

Previous studies of Rabi oscillations in QDs utilize short pulses, 2 ps, to drive the oscillations and are observed as a function of the electric field by way of a CW readout laser.[22] Additional experiments show the onset of Rabi oscillation as they manifest in the intensity correlation function, $g2(t)$, of the QD emission.[17] However, our experiment is unique in that we are able to follow the time evolution of the system as predicted by the solution to the initial value problem of an exciting field, Eq. (3). By using both long (2 ns) and short (250 ps) pulse studies we have determined that the frequency domain response of the QD appears to be broadened beyond the natural linewidth by a spectral wandering process and that the optical decoherence rate is limited by the spontaneous emission rate. Verification of this has profound implications for QDs as nodes in a quantum memory since many architectures utilize photons as carriers of quantum information between nodes.[9-11] A short optical coherence time would significantly diminish the effectiveness of such systems.

4. Entanglement Studies with QDs

This section follows data and analysis on entanglement between a photon emitted from a QD and its electron spin qubit previously published by our group,[6] and others.[7,8] As was previously discussed in section 1, a single electron confined to a QD provides an exceptional candidate for quantum computing. To make this

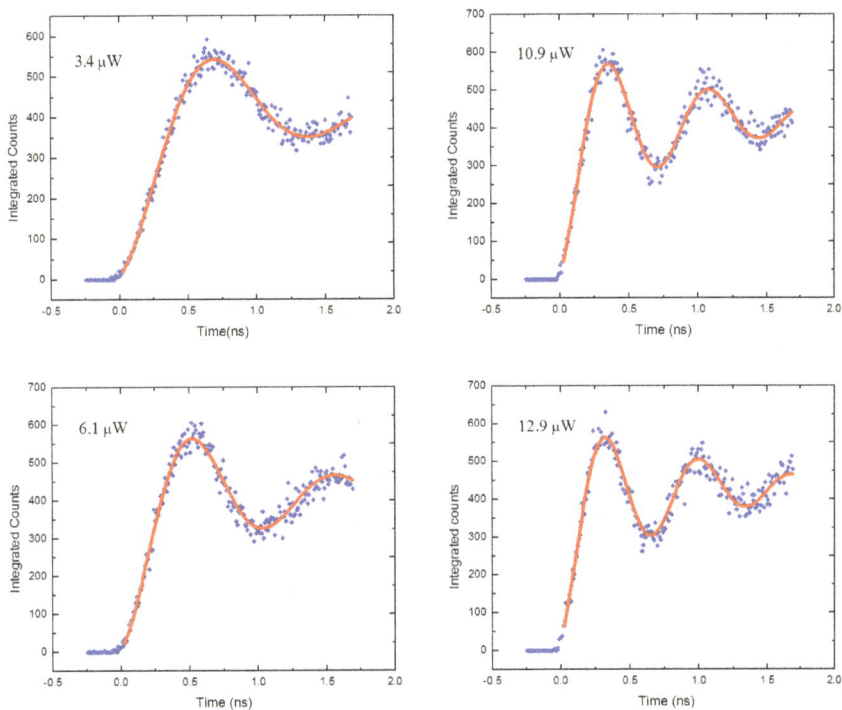

Fig. 8. Histograms of time resolved QD fluorescence resulting from a 2 ns excitation pulse. The time-dependent Rabi oscillations are seen to increase in frequency as the excitation power increases. The input powers are given on each graph.

system viable there must be some way to coherently transport information from one quantum node to another. A straightforward way to accomplish this is by using the spontaneously emitted photon from the QD to transmit the spin qubit information between nodes. Such a technique requires entanglement between the emitted photon and the spin qubit.[9–11] Experimental verification of this entanglement is vital to the feasibility of QDs for quantum computing.

  The protocol we use to verify the entanglement demonstrates correlation between the spontaneously emitted photon's polarization state and the final spin state. The correlation measurements are preformed in two orthogonal bases x and z for the spin state, which correspond to linear and circular polarization for the photon. This technique has been utilized in previous demonstrations of entanglement between a matter qubit and an emitted photon.[23–26] We will be working with the energy level structure shown in Fig. 1b and will be focusing on a photon emitted from the $|T_x-\rangle$ state resulting in the entangled state described in Eq. (1). As described above the entanglement of interest is between the spin state ($|x+\rangle$ and $|x-\rangle$) and the polarization state of the photon ($|H\rangle$ and $|V\rangle$). The frequency is also entangled with the spin and will destroy the entangled state if the frequencies are resolvable by elements in the set-up. To eliminate this problem we employ a fast detection scheme where the timing resolution of the single photon detector used to observe the entangled photon is (48 ps) faster than the bandwidth of the ground state splitting (set at $\Delta_e/2\pi = 7.35$ GHz) to destroy the "which-path" information.[27] The resulting entangled state of the system becomes,

$$|\Psi\rangle = \frac{|H\rangle|x+\rangle - i|V\rangle|x-\rangle}{\sqrt{2}} \qquad (5)$$

The time resolved fluorescence techniques developed in section 3 are essential when initializing the system, generating the entanglement, and finally reading out the spin state. To begin we must initialize to a pure state, $|x-\rangle$, using an EOM to gate a narrow bandwidth CW laser similar to the 2 ns pulse employed earlier to generate Rabi oscillations. This pulse is tuned to the $|x+\rangle \leftrightarrow |T_x+\rangle$ transition for 4 ns and results in optical pumping to the $|x-\rangle$ state. Following this initialization a pulse (250 ps), shorter than the radiative lifetime, excites to the $|T_x-\rangle$ state. In order to transfer the entire population to the excited state we employ the technique shown in Fig. 6 of section 3 to determine the π power. Once the pulse is turned off, the excited state will decay to one of two paths creating the entangled state of Eq. (5).[15] The experiment is performed at a repetition rate of 76 MHz, so the optical pumping pulse from the next shot of the experiment will act as a readout of the $|x+\rangle$ state since it is already tuned to that transition.

The linear basis measurement correlation is rather straightforward. We need to determine the conditional probabilities of a H or V photon emitted and being in $|x-\rangle$ or $|x+\rangle$. The use of polarization rejection makes this measurement easy to perform. However, in the z (rotated) basis, polarization selection happens using $\sigma+$ or $\sigma-$ polarized light, since we are correlating these polarizations with $|z+\rangle$ and $|z-\rangle$, which are linear combinations of the nondegenerate x basis states and hence evolve in time. Our detection scheme requires a collection polarization orthogonal to the input polarization (see section 2). This demands that the short pulse excitation laser be narrow enough in frequency to only excite to $|T_x-\rangle$, since the input polarization would not prohibit it from coupling to $|T_x+\rangle$. The EOM generated pulse provides us with the narrow band excitation required.

In the x basis, the conditional probabilities we want to find are, $P(x+|H), P(x-|H), P(x+|V)$, and $P(x-|V)$. For example, $P(x+|H)$ reports the probability that the system is in state $|x+\rangle$ if a H photon is emitted after excitation to $|T_x-\rangle$, this should be unity for perfect correlation. For the positive correlations, $P(x+|H)$ and $P(x-|V)$, the timing diagram is shown in Fig. 9. The following description is how we determine $P(x+|H)$, and the extension to $P(x-|V)$ is obvious. Using V polarized input light, so we can collect the H polarized emission, once the state is initialized to $|x-\rangle$ we excite with a π pulse to $|T_x-\rangle$ and measure the time of decay. If a photon is detected along the decay then the QD should be in $|x+\rangle$ and

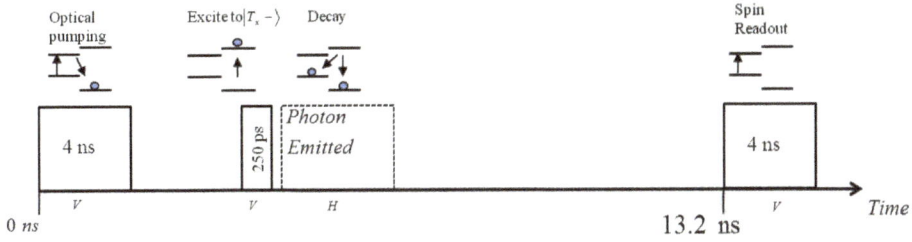

Fig. 9. Pulse sequence for determining positive correlations for $P(x + |H)$. A photon emitted in the dashed region will only be detected if it is a H photon and therefore scattering should occur with a probability of 1 during the spin readout. To normalize this measurement, correlations between detection of a photon in the dashed region and scattering from a spin readout of a distant shot are set to 1/2 since the entire system has been reset from shot to shot.

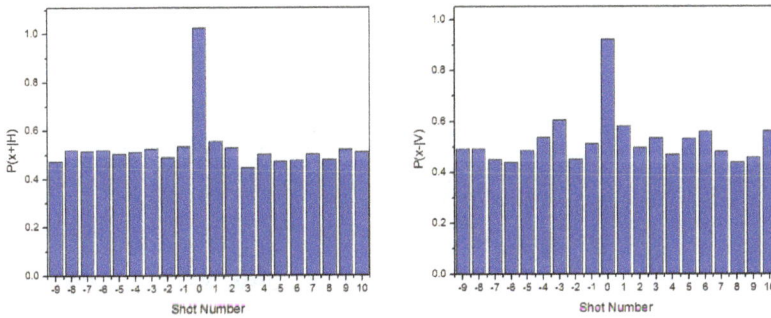

Fig. 10. Conditional probability data for $P(x+|H)$ and $P(x-|V)$ where the distant shot number is label and the average correlations between a photon in the emission region of shot 0 and a readout photon from all distant shots, have been set to 1/2.

the readout (initialization pulse from the next shot of the experiment) will scatter a photon. By correlating this with readout pulses in distant shots of the experiment, which should have no correlation, we are able to observe a strong positive correlation (Fig. 10).

For negative correlations, $P(x - |H)$ and $P(x + |V)$, we insert an extra 250 ps pulse (see Fig. 11). Again, we will describe the case for the detection of H polarized light, $P(x - |H)$. The initialization and first excitation are the same as before, however, the additional 250 ps pulse acts as a readout for $|x-\rangle$. If a detection occurs after the first 250 ps pulse, then we should expect nothing in the second since the QD is now in state $|x+\rangle$. The anti-correlation is seen by correlating detection in the first 250 ps pulse with the second and is normalized by correlating with second pulses from distant shots of the experiment. The anti-correlation signals are seen

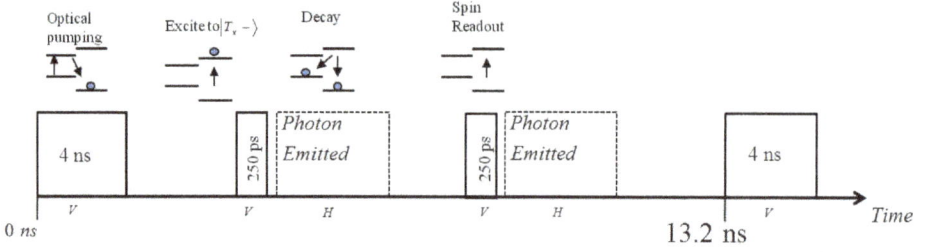

Fig. 11.   Pulse sequence for anti-correlation measurements of $P(x - |H)$ and $P(x + |V)$. The addition of an extra 250 ps pulse serves as a readout of the $|x-\rangle$ state. If a photon is found in the emission region of the first pulse then an H photon was emitted and there can be no population in $|x-\rangle$ so no emission should occur in both emission regions of the same shot. As before the normalization is determined by correlating with distant shots.

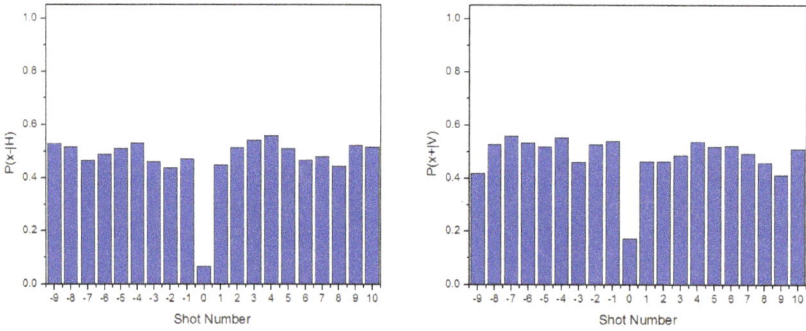

Fig. 12.   Conditional probability data for $P(x - |H)$ and $P(x + |V)$ where the distant shot number is label and the average correlations between a photon in the emission region of shot 0 and a spin readout photon from all distant shots, have been set to $1/2$.

in Fig. 12. A summary of all four x basis conditional probabilities is shown in Fig. 15a.

As stated before the z basis measurement is more complicated since it is time dependent, allowing the measurement of two conditional probabilities at the same time. Since we wish to look for $P(z \pm |\sigma+)$ and $P(z \pm |\sigma-)$ polarization projection along a circular axis (e.g. $\sigma+$) is required, and the orthogonal circular polarization (e.g. $\sigma-$) must be the input, so we can reject the excitation lasers. The experimental sequence begins in the same way as the x basis; a 4 ns optical pumping pulse initializes to $|x-\rangle$. Next, a 250 ps π pulse transfers the QD population to $|T_x-\rangle$ where the system decays generating the entangled photon. The photon is then polarization selected leaving the system momentarily in $|z+\rangle = (|x+\rangle - |x-\rangle)/\sqrt{2}$ or $|z-\rangle = (|x+\rangle + |x-\rangle)/\sqrt{2}$. However, the spin begins to precess under unitary time

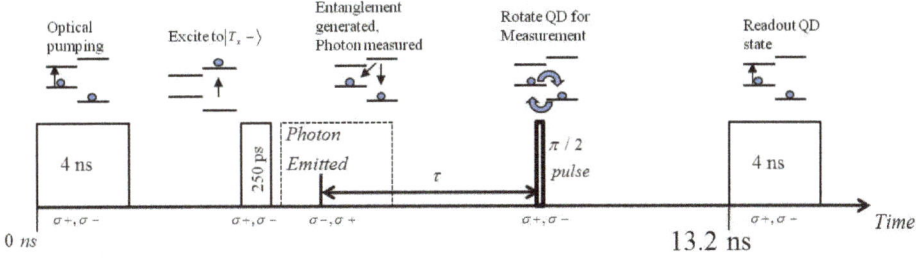

Fig. 13.   Timing diagram for the z (rotated) basis measurement. The entangled photon is emitted randomly along an exponential decay and some time τ later is rotated from a coherence into a population by a detuned $\pi/2$ Raman pulse. The population is $|x+\rangle$ is then readout by the following initialization pulse acting as a readout. Time binning of the emission with respect to the rotation creates an exponentially weighted fringe pattern, which signifies coherent electron precession.

evolution about the x axis until it is rotated into a population (into $|x+\rangle$ or $|x-\rangle$) by a $\pi/2$ Raman pulse. The spin rotation is achieved by using a detuned broadband pulse driving a two-photon transition. Since the rotation pulse is circularly polarized it non-destructively couples all the transitions allowing geometrical phase rotation of the QD spin, dependent on the pulse area. The rotation occurs about the optical axis, in our case z.[3,4] Finally, the optical pumping pulse from the next shot of the experiment reads out the population in $|x+\rangle$. The mathematical description of this sequence is shown below and a timing diagram appears in Fig. 13.

$$\langle \sigma \pm |\Psi\rangle = \frac{|H\rangle|x+\rangle \mp |x-\rangle}{2} \tag{6}$$

$$U(\tau)\langle \sigma \pm |\Psi\rangle = e^{-i\frac{\Delta_e}{\hbar}\tau}\frac{|H\rangle|x+\rangle \mp |x-\rangle}{2} \tag{7}$$

$$|\langle x + |R_{\sigma\mp}(\pi/2)U(\tau)\langle \sigma \pm |\Psi\rangle|^2 = \tfrac{1}{4}(1 + \sin \Delta_e \tau) \tag{8}$$

Here the projection $\langle x + |$ represents the readout of the $|x+\rangle$ state and the rotation is a $\pi/2$ rotation about the z axis,
$R_{\sigma\mp}(\pi/2) = \left[|x+\rangle\langle x + | + |x-\rangle\langle x - | \pm i(|x+\rangle\langle x - | + |x-\rangle\langle x + |) \right]/\sqrt{2}$.

From this analysis, the time, τ, between emission and detection is crucial, and to observe the frequency of precession, Δ_e (the spin difference frequency), we must utilize a low magnetic field to keep Δ_e smaller than the detector bandwidth. The added complication in choosing the correct magnetic field strength is in the excited state splitting. Since we are using polarization that will couple to all transitions, we must use frequency selectivity to only excite to $|T_x-\rangle$. Taking into account these complications, we use a 1.1T magnetic field giving an excited state and ground state splitting of 4.62GHz and 7.35GHz, respectively. The emission from the excited state will be randomly distributed along the exponential decay, so by time binning the emission with respect to the stationary rotation pulse a fringe pattern will emerge weighted by the exponential decay. This fringe pattern will peak when the time between emission and rotation is sufficient for the Bloch vector to be rotated into $|x+\rangle$ by a $\pi/2$ rotation, and dip when rotated into $|x-\rangle$. This oscillation occurs at the spin difference frequency, Δ_e. After correlating a photon scatter event in the 4 ns pulse with an observed decay after the 250 ps pulse and subtracting out the exponential we can see a fringe signal emerge, where the conditional probabilities can be read directly from the fringe contrast (see Fig. 14).

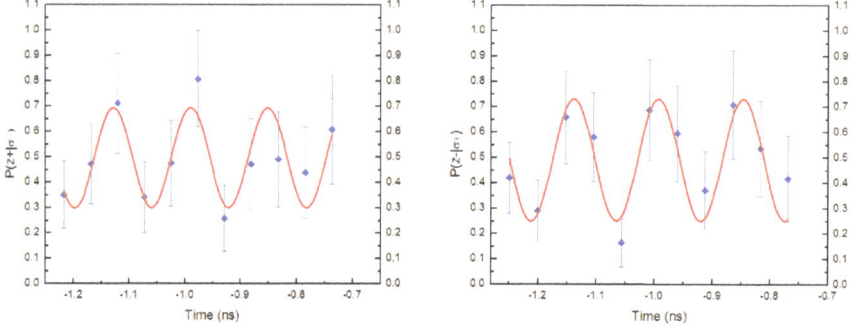

Fig. 14. The fringe pattern recovered by time binning the photon emission, following the 250 ps pulse when a readout photon is detected as well. This coincidence should only occur at certain times during the electron precession as evident from this fringe pattern. The exponential decay has been divided out to make the fringes easier to observe. A sinusoidal fit recovers the spin precession frequency and reveals the conditional probabilities from the contrast.

By requiring that the conditional probabilities sum to one we can extract the following for the x basis measurement: $P(x + |H) = 0.94 \pm 0.05, P(x - |H) = 0.06 \pm 0.01, P(x + |V) = 0.16 \pm 0.01, P(x - |V) = 0.84 \pm 0.04$. The z basis conditional probabilities are: $P(z + |\sigma-) = 0.69 \pm 0.04, P(z - |\sigma-) = 0.31 \pm 0.04, P(z - |\sigma+) = 0.70 \pm 0.05, P(z + |\sigma+) = 0.30 \pm 0.05$. For both the x and z bases the conditional probabilities combine (see Fig. 15) to give a lower bound on the entanglement fidelity, calculated using established techniques.[23]

$$F \geq 1/2 \quad (\rho_{Hx+,Hx+} + \rho_{Vx-,Vx-} - 2\sqrt{\rho_{Hx-,Hx-}\rho_{Vx+,Vx+}} +$$
$$\rho_{\sigma-z+,\sigma-z+} - \rho_{\sigma-z-,\sigma-z-} + \rho_{\sigma+z-,\sigma+z-} - \rho_{\sigma+z+,\sigma+z+}) \quad (9)$$

We find a lower bound on the fidelity of $F \geq 0.59 \pm 0.04$. It is important to note that this lower bound is mostly limited by the fringe contrast in the z basis measurement, which is diminished by the finite timing resolution of the detector. In many protocols for entangling multiple quits via interference of the entangled photons, the detector timing resolution is not vital.[28,29]

Here we have shown that the spontaneously emitted photon from a QD is entangled with its spin state, having profound implications to the world of quantum information. QDs have the advantage that they can be easily integrated into an industrial infrastructure designed around semi-conductor fabrication using on-chip

(a)

(b)

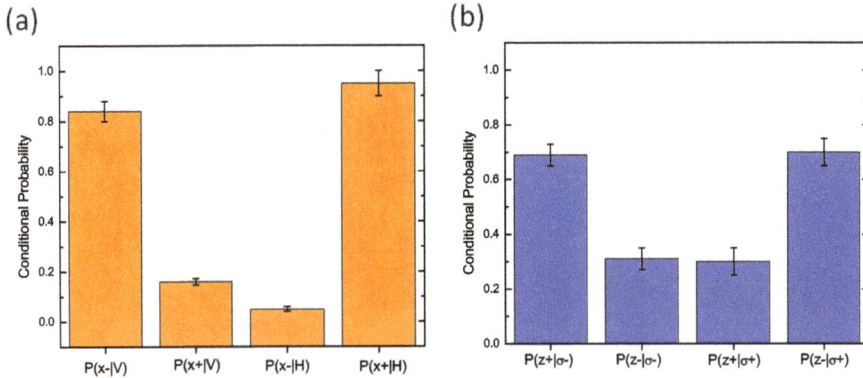

Fig. 15. The combined conditional probabilities for the (a) x basis, where the probabilities are extracted from previous graphs and constrained to add up to 1n and (b) z basis, where the probabilities are read directly from the fringe contrast.

photonics waveguides these entangled photons can be directed to many different quantum nodes making a scalable quantum architecture possible.

5. Summary and Future Directions

In this Chapter we have laid out the recent techniques developed to manipulate QDs in a more versatile way. The direct detection of the light scattered from a QD is a recent technological feat, and the use of EOMs to gate CW laser has already yielded results interesting to the field. The Rabi oscillations observed during a long excitation pulse give us insight to the optical decoherence nature of the QD and allow us to extract a decoherence time consistent with no pure de-phasing. The ability to follow the excited state population evolution is itself unique and has never been achieved in QDs before. One might take for granted that QDs give data similar to a well behaved two level system, but it is not obvious that such a system would be capable of suppressing the many-body physics to reveal a more elegant behavior.

The demonstration that a photon emitted from a QD is entangled with the ground state spin qubit has numerous implications for quantum information. We can now send these photons to any distant node and retain the spin information coded on them. The information transfer is typically mediated by the interference of entangled photons on a Hong-Ou-Mandel (HOM) interferometer, which projects the entangled matter qubits to a specific state upon coincidence causing them to coherently evolve together.[28-30] In this way, two separate QDs could be entangled. Also we can use the entangled nature of the emitted photon to teleport information onto the QD spin state.[31] The possible experiments greatly increase when you consider the capabilities for manipulation and coherent control of QDs . Experiments such as these further the case that QDs can indeed be considered viable and useful when considering the future of quantum technologies.

The authors would like to acknowledge useful discussions with P. Berman, L.-M. Duan, C. Chow, and L. Webster. This research is supported by NSF grants No. PHY0804114 and No. PHY1104446; ARO-MURI grant No. W911NF-1-0406; ARO grant No. W911NF-08-1-0487; DARPA grants No. FA8750-12-2-0333 and No. FA9550-10-1-0534; AFOSR grant No. FA9550-09-1-0457.

References

1. D. Loss and D. P. DiVincenzo, Quantum computation with quantum dots, *Phys. Rev. A.* **57**, 120–126 (Jan, 1998).
2. M. Atatüre, J. Dreiser, A. Badolato, A. Högele, K. Karrai, and A. Imamoglu, Quantum-dot spin-state preparation with near-unity fidelity, *Science.* **312**(5773), 551–553 (04, 2006).
3. D. Press, T. D. Ladd, B. Zhang, and Y. Yamamoto, Complete quantum control of a single quantum dot spin using ultrafast optical pulses, *Nature.* **456**(7219), 218–221 (11, 2008).
4. E. D. Kim, K. Truex, X. Xu, B. Sun, D. G. Steel, A. S. Bracker, D. Gammon, and L. J. Sham, Fast spin rotations by optically controlled geometric phases in a charge-tunable inas quantum dot, *Phys. Rev. Lett.* **104**, 167401 (Apr, 2010).
5. X. Xu, Y. Wu, B. Sun, Q. Huang, J. Cheng, D. G. Steel, A. S. Bracker, D. Gammon, C. Emary, and L. J. Sham, Fast spin state initialization in a singly charged inas-gaas quantum dot by optical cooling, *Phys. Rev. Lett.* **99**, 097401 (Aug, 2007).
6. J. R. Schaibley, A. P. Burgers, G. A. McCracken, L.-M. Duan, P. R. Berman, D. G. Steel, A. S. Bracker, D. Gammon, and L. J. Sham, Demonstration of quantum entanglement between a single electron spin confined to an inas quantum dot and a photon, *Phys. Rev. Lett.* **110**, 167401 (Apr, 2013).
7. W. B. Gao, P. Fallahi, E. Togan, J. Miguel-Sanchez, and A. Imamoglu, Observation of entanglement between a quantum dot spin and a single photon, *Nature.* **491**(7424), 426–430 (11, 2012).
8. K. De Greve, L. Yu, P. L. McMahon, J. S. Pelc, C. M. Natarajan, N. Y. Kim, E. Abe, S. Maier, C. Schneider, M. Kamp, S. Hofling, R. H. Hadfield, A. Forchel, M. M. Fejer, and Y. Yamamoto, Quantum-dot spin-photon entanglement via frequency downconversion to telecom wavelength, *Nature.* **491**(7424), 421–425 (11, 2012).
9. J. I. Cirac, P. Zoller, H. J. Kimble, and H. Mabuchi, Quantum state transfer and entanglement distribution among distant nodes in a quantum network, *Phys. Rev. Lett.* **78**, 3221–3224 (Apr, 1997).
10. L. M. Duan, M. D. Lukin, J. I. Cirac, and P. Zoller, Long-distance quantum communication with atomic ensembles and linear optics, *Nature.* **414**(6862), 413–418 (11, 2001).
11. W. Yao, R.-B. Liu, and L. J. Sham, Theory of control of the spin-photon interface for quantum networks, *Phys. Rev. Lett.* **95**, 030504 (Jul, 2005).
12. R. J. Warburton, C. Schaflein, D. Haft, F. Bickel, A. Lorke, K. Karrai, J. M. Garcia, W. Schoenfeld, and P. M. Petroff, Optical emission from a charge-tunable quantum ring, *Nature.* **405**(6789), 926–929 (06, 2000).
13. M. E. Ware, E. A. Stinaff, D. Gammon, M. F. Doty, A. S. Bracker, D. Gershoni, V. L. Korenev, i. m. c. C. Bădescu, Y. Lyanda-Geller, and T. L. Reinecke, Polarized

fine structure in the photoluminescence excitation spectrum of a negatively charged quantum dot, *Phys. Rev. Lett.* **95**, 177403 (Oct, 2005).

14. B. Alén, F. Bickel, K. Karrai, R. J. Warburton, and P. M. Petroff, Stark-shift modulation absorption spectroscopy of single quantum dots, *Applied Physics Letters.* **83** (11), 2235, (2003).

15. S. E. Economou, R.-B. Liu, L. J. Sham, and D. G. Steel, Unified theory of consequences of spontaneous emission in a λ system, *Phys. Rev. B.* **71**, 195327 (May, 2005).

16. R. H. Brown and R. Twiss, Correlation between photons in two coherent beams of light, *Nature.* **177**(4497), 27–29, (1956).

17. A. Muller, E. B. Flagg, P. Bianucci, X. Y. Wang, D. G. Deppe, W. Ma, J. Zhang, G. J. Salamo, M. Xiao, and C. K. Shih, Resonance fluorescence from a coherently driven semiconductor quantum dot in a cavity, *Phys. Rev. Lett.* **99**, 187402 (Nov, 2007).

18. J. R. Schaibley, A. P. Burgers, G. A. McCracken, D. G. Steel, A. S. Bracker, D. Gammon, and L. J. Sham, Direct detection of time-resolved rabi oscillations in a single quantum dot via resonance fluorescence, *Phys. Rev. B.* **87**, 115311 (Mar, 2013).

19. X. Xu, B. Sun, E. D. Kim, K. Smirl, P. R. Berman, D. G. Steel, A. S. Bracker, D. Gammon, and L. J. Sham, Single charged quantum dot in a strong optical field: Absorption, gain, and the ac-stark effect, *Phys. Rev. Lett.* **101**, 227401 (Nov, 2008).

20. X. Xu, B. Sun, P. R. Berman, D. G. Steel, A. S. Bracker, D. Gammon, and L. J. Sham, Coherent optical spectroscopy of a strongly driven quantum dot, *Science.* **317** (5840), 929–932, (2007).

21. A. Högele, S. Seidl, M. Kroner, K. Karrai, R. J. Warburton, B. D. Gerardot, and P. M. Petroff, Voltage-controlled optics of a quantum dot, *Phys. Rev. Lett.* **93**, 217401 (Nov, 2004).

22. T. H. Stievater, X. Li, D. G. Steel, D. Gammon, D. S. Katzer, D. Park, C. Piermarocchi, and L. J. Sham, Rabi oscillations of excitons in single quantum dots, *Phys. Rev. Lett.* **87**, 133603 (Sep, 2001).

23. B. B. Blinov, D. L. Moehring, L. M. Duan, and C. Monroe, Observation of entanglement between a single trapped atom and a single photon, *Nature.* **428**(6979), 153–157 (03, 2004).

24. J. Volz, M. Weber, D. Schlenk, W. Rosenfeld, J. Vrana, K. Saucke, C. Kurtsiefer, and H. Weinfurter, Observation of entanglement of a single photon with a trapped atom, *Phys. Rev. Lett.* **96**, 030404 (Jan, 2006).

25. T. Wilk, S. C. Webster, A. Kuhn, and G. Rempe, Single-atom single-photon quantum interface, *Science.* **317**(5837), 488–490, (2007).

26. E. Togan, Y. Chu, A. S. Trifonov, L. Jiang, J. Maze, L. Childress, M. V. G. Dutt, A. S. Sorensen, P. R. Hemmer, A. S. Zibrov, and M. D. Lukin, Quantum entanglement between an optical photon and a solid-state spin qubit, *Nature.* **466**(7307), 730–734 (08, 2010).

27. J. R. Schaibley and P. R. Berman, The effect of frequency-mismatched spontaneous emission on atom-field entanglement, *Journal of Physics B: Atomic, Molecular and Optical Physics.* **45**(12), 124020, (2012).

28. D. L. Moehring, P. Maunz, S. Olmschenk, K. C. Younge, D. N. Matsukevich, L. M. Duan, and C. Monroe, Entanglement of single-atom quantum bits at a distance, *Nature.* **449**(7158), 68–71 (09, 2007).

29. H. Bernien, B. Hensen, W. Pfaff, G. Koolstra, M. S. Blok, L. Robledo, T. H. Taminiau, M. Markham, D. J. Twitchen, L. Childress, and R. Hanson, Heralded entanglement between solid-state qubits separated by three metres, *Nature.* **497**(7447), 86–90 (05, 2013).

30. C. K. Hong, Z. Y. Ou, and L. Mandel, Measurement of subpicosecond time intervals between two photons by interference, *Phys. Rev. Lett.* **59**, 2044–2046 (Nov, 1987).

31. W. B. Gao, P. Fallahi, E. Togan, A. Delteil, Y. S. Chin, J. Miguel-Sanchez, and A. Imamoğlu, Quantum teleportation from a propagating photon to a solid-state spin qubit, *Nat Commun.* **4** (11, 2013).

III

Coherent Nanophotonics
and Plasmonics

Chapter 6

Enhancement of Single-Photon Sources with Metamaterials

M. Y. Shalaginov[1], S. Bogdanov[1], V. V. Vorobyov[2,4], A. S. Lagutchev[1],
A. V. Kildishev[1], A. V. Akimov[2,3,4], A. Boltasseva[1], and V. M. Shalaev[1,*]

[1] *School of Electrical and Computer Engineering and Birck Nanotechnology Center,*
Purdue University, West Lafayette, IN 47907, USA,
[2] *Moscow Institute of Physics and Technology, Dolgoprudny,*
Moscow Region,141700, Russia
[3] *Russian Quantum Center, Skolkovo Innovation Center,*
Moscow Region, 143025, Russia
[4] *Lebedev Physical Institute RAS, Moscow, 119991, Russia*

Scientists are looking for new, breakthrough solutions that can greatly advance computing and networking systems. These solutions will involve quantum properties of matter and light as promised by the ongoing experimental and theoretical work in the areas of quantum computation and communication. Quantum photonics is destined to play a central role in the development of such technologies due to the high transmission capacity and outstanding low-noise properties of photonic information channels.

Among the vital problems to be solved in this direction, are efficient generation and collection of single photons. One approach to tackle these problems is based on engineering emission properties of available single-photon sources using metamaterials. Metamaterials are artificially engineered structures with sub-wavelength features whose optical properties go beyond the limitations of conventional materials. As promising single-photon sources, we have chosen nitrogen-vacancy (NV) color centers in diamond, which are capable to operate stably in a single-photon regime at room temperature in a solid state environment.

In this chapter, we report both theoretical and experimental studies of the radiation from a nanodiamond single NV center placed near a hyperbolic metamaterial (HMM). In particular, we derive the reduction of excited-state lifetime and the enhancement of collected single-photon emission rate and compare them with the experimental observations. These results could be of great impact for future integrated quantum sources, especially owing to a CMOS-compatible approach to HMM synthesis.

---

*shalaev@purdue.edu

1. Introduction

The strongest driving force behind the study of quantum systems is to build quantum computing and networking technologies. These technologies would employ entirely different algorithms, unattainable for classical computers, that will reduce the complexity of a variety of computational problems. Among these problems are algorithmic searching,[1] number factorization,[2] and simulation of quantum systems.[3]

Low noise of photonic information channels and low cost of standard photonic device fabrication give photonics a central role in the development of quantum technologies. It seems probable that quantum information will be transmitted using quantum states of light and that at some degree information processing will be performed on these states.[5] The simplest and most fundamental quantum photonic states are single photons. The typical generation rates of available isolated single-photon sources are of the order of 100 kcounts per second. For fast communication and computation these rates have to be improved. In this chapter, we present our research towards the development of an efficient source of single photons.

1.1. *Available Single-photon Sources*

At present, several types of quantum systems capable of serving as truly deterministic single-photon sources have been investigated[7] (see Fig. 1), including trapped atoms[8] and ions,[9] single molecules,[10] quantum dots,[11] color centers,[12] atomic ensembles,[13] and mesoscopic quantum wells.[14] The solid state sources such as quantum dots and lattice defects are naturally interesting for future commercial quantum devices.

A nitrogen-vacancy (NV) color center in diamond, formed by a substitutional nitrogen atom and a vacancy at an adjacent lattice site, is a candidate of particular interest for the implementation of a solid state single-photon source.[15] It is resistant against photobleaching and operates at room temperature.[16] Its intrinsic quantum yield is close to unity[15] and it can be deterministically manipulated by a pick-and-place procedure.[17]

In addition, the diamond NV center gained significant attention thanks to its spin degree of freedom. Its electron spin state has coherence time in the range of milliseconds at room temperature,[21] far beyond that of, e.g., quantum dots made of III-V materials, and can be read out optically.[22] This unique spin-photon interface makes it possible to build reliable quantum-memory units[23] and perform high-fidelity single-shot projective quantum measurement.[24] Moreover, NV centers as solid-state structures are promising in sharing quantum entanglement among large number of qubits[25] and implementing scalable quantum information systems.[26] NV centers also could be used as sensors of electric[27] and magnetic[28] fields, temperature[29] and rotation[30] with high resolution and sensitivity.

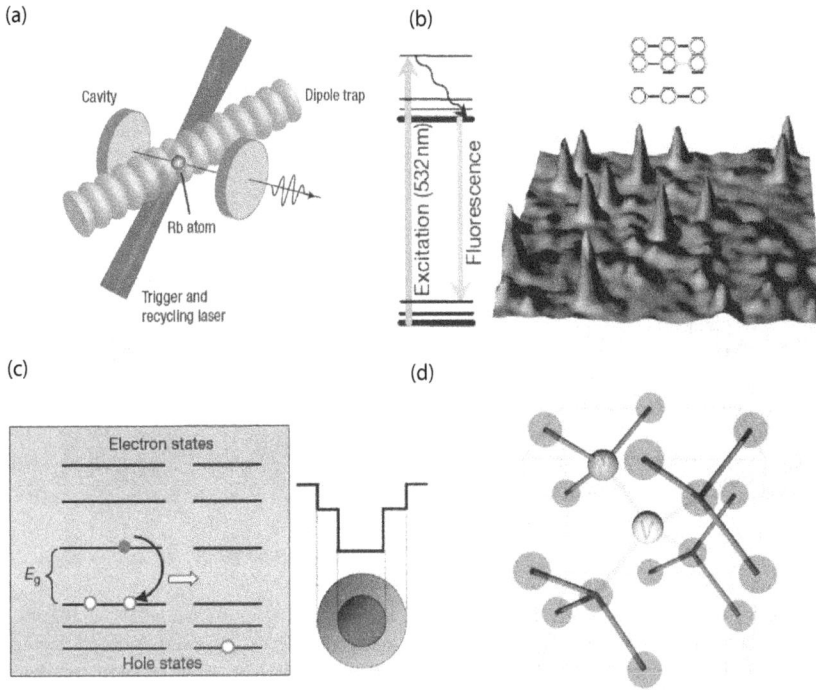

Fig. 1.   Examples of available deterministic single-photon sources: (a) ^{85}Rb atom trapped in a high-finesse optical microcavity,[8] (b) single terrylene molecules in a p-terphenyl crystal,[18] (c) CdSe/ZnSe quantum dots,[19] (d) NV color centers in diamond.[20]   Figures reproduced with permission: (a) Ref. [8] from Copyright 2007 NPG, (b) Ref. [18] from Copyright 2000 NPG, (c) Ref. [19] from Copyright 2009 NPG, (d) Ref. [20] from Copyright 2006 Wiley-VCH.

Strictly speaking, the NV center exists in two different charge states: the electrically neutral state NV^0 and the negatively charged state NV^-. In the rest of the chapter by "NV" we will refer to the NV^- state. The NV center can be identified thanks to the zero phonon line (ZPL) located at 637 nm. At room temperature the NV center emission spectrum exhibits phononic sideband that covers the range from 600 to 800 nm and the ZPL is largely obscured by this broadening. But even at liquid helium temperatures, only a few percent of the emission falls into the ZPL.[31] Therefore the NV center is intrinsically a broadband source.

1.2. *Methods of Spontaneous Emission Enhancement*

The applications using NV center as a single-photon generator and spin-photon interface could substantially benefit from enhanced emission rate and photon collection efficiency. Emission rate enhancement can be achieved by utilizing Purcell

(a)   (b)

(c)   (d)

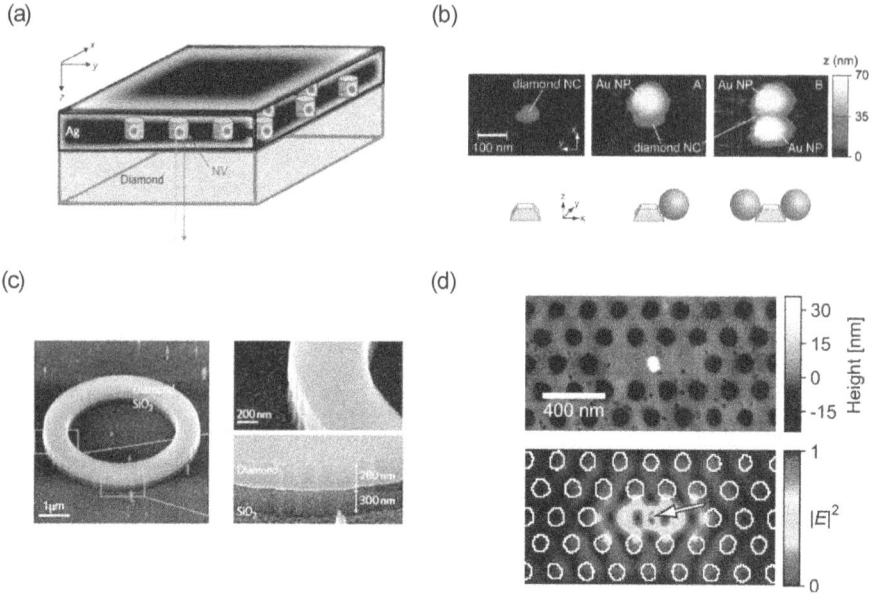

Fig. 2. Examples of resonance-based structures enhancing the emission from NV centers by coupling to (a) diamond silver apertures,[43] (b) gold nanoparticles,[44] (c) diamond microring resonators,[45] and (d) photonic crystal cavities.[34] Figures reproduced with permission: (a) Ref. [43]
from Copyright 2011 NPG, (b) Ref. [44] from Copyright 2009 ACS, (c) Ref. [45] from Copyright
2011 NPG, (d) Ref. [34] from Copyright 2010 AIP.

effect.[32] So far, this has been accomplished by using resonant photonic structures
(see Fig. 2), such as microspherical resonators,[33] photonic crystal microcavities,[34]
and microring resonators,[45] which are all bandwidth limited. Tuning them to the
exact frequency of a sharp transition is a considerable technological challenge. The
Purcell effect in these systems strongly depends on the position of the emitter with
respect to the mode power density profile. In addition, resonant structures limit the
emitter response time. In the case of NV center at room temperature the resonant
techniques would be unapplicable anyway in view of its broad emission spectrum
(600-800 nm) and the development of new broadband enhancement techniques is
being sought.

Coupling NV centers to hyperbolic metamaterials (HMM)[46–49] allows for the
spontaneous emission to be enhanced over a broad spectral range. This enhancement is a result of a broadband singularity in the photonic local density of states
(LDOS) within the HMM.[50–54]

The photonic LDOS, similar to its electronic counterpart, can be quantified
as the volume in k-space between iso-frequency surfaces. For extraordinary waves
in a uniaxial anisotropic medium with dielectric tensor $\overleftrightarrow{\varepsilon} = diag\left(\varepsilon_{\parallel}, \varepsilon_{\parallel}, \varepsilon_{\perp}\right)$, the
iso-frequency surfaces are defined by the following equation:

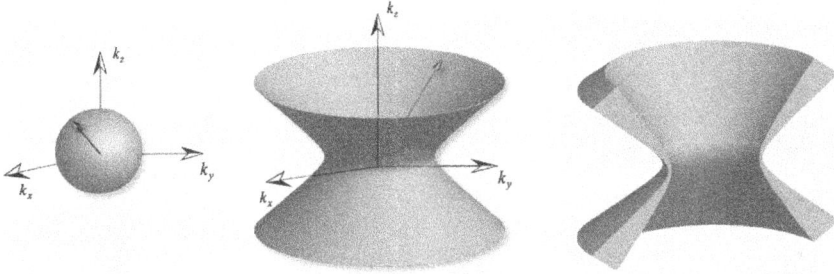

Fig. 3. Iso-frequency surfaces in the case of an isotropic dielectric (spheroid) and an extremely anisotropic medium with hyperbolic dispersion (hyperboloid).[53] Figure reproduced with permission from Copyright 2010 SSBM.

$$\omega^2/c^2 = k_\parallel^2/\varepsilon_\perp + k_\perp^2/\varepsilon_\parallel, \tag{1}$$

where subscripts "$\perp$" and "$\parallel$" indicate the directions perpendicular and parallel to the plane of anisotropy, respectively. In the case of dielectric materials with $\varepsilon_\perp, \varepsilon_\parallel > 0$, LDOS is equivalent to the volume of an infinitesimally thin ellipsoidal shell in k-space (see Fig. 3, left). However, in a medium with hyperbolic dispersion, $\varepsilon_\perp$ and $\varepsilon_\parallel$ are of opposite signs which produces a hyperboloidal shell (see Fig. 3, middle and right) whose volume is infinitely large (i.e. broadband singularity in LDOS appears). As a result, such a medium allows the propagation of high-k modes with arbitrarily large wavevectors.

The HMM is practically realized as a lamellar structure consisting of alternating subwavelength-thick layers of metal and dielectric,[51,55] or as an array of nanowires embedded into a dielectric host matrix.[50,56] Between the two approaches to implement HMMs, planar HMMs are important from a technological point of view, as they can be fabricated using existing planar processing techniques. In these practical implementations, the maximum value of the wavevector is restricted by the size of the metamaterial unit cell and hence the singularity disappears. However, with nanometer-size layers even the available k-space yields a very high photonic LDOS, providing extra radiative channels for the coupled emitter. These channels are high-k metamaterial modes which result from hybridization of surface plasmon polariton modes at the interfaces of the layers constituting the HMM.[57]

In this chapter, the radiation from a nanodiamond single NV center placed near a hyperbolic metamaterial is studied both theoretically and experimentally.[58] The chapter is organized in the following way. First, we discuss sample fabrication and characterization as well as the experimental setup for NV center optical lifetime and collected power measurement. Next, we provide theoretical estimates of these quantities as a function of experimental parameters such as separation distance between an emitter and HMM surface and presence of scattering particles in the vicinity of the emitter. Then, we proceed with the discussion of the experimentally obtained results.

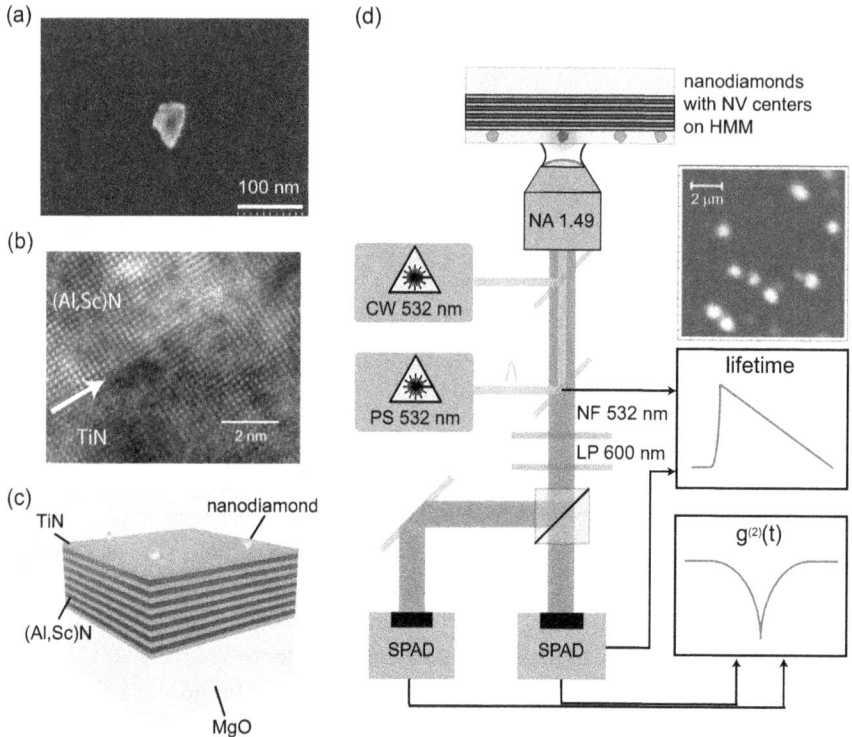

Fig. 4. (a) SEM scan of a ND on top of the HMM sample.[58] (b) HRTEM image showing the interface of TiN and (Al,Sc)N matching lattices in the superlattice.[61] (c) Sample structure and (d) schematic of the fluorescence analysis setup.[58] Figures reproduced with permission: (a), (c), (d) Ref. [58] from Copyright 2014 Wiley-VCH and (b) Ref. [61] from Copyright 2014 USNAS.

2. Experiment

Here we outline the details of the experiment aimed at studying radiation from single NV center nanodiamonds coupled to an epitaxially grown HMM superlattice composed of titanium nitride (TiN) and aluminum scandium nitride ($Al_xSc_{1-x}N$).[61]

2.1. Nanodiamond Samples

An aqueous suspension of nanodiamonds (NDs) with nominal size of 50 nm, 0.5% w/v was obtained from Microdiamant AG. The size of NDs was chosen according to a previous study, which showed that NDs with the size of 50-70 nm have the highest probability to host a single NV center.[66] However, the suspension also contained NDs smaller than the nominal size. The smallest NDs were eliminated and the suspension was cleaned by a multistep procedure. First, the suspension was centrifuged at 5000 rpm for 5 minutes. The supernatant (80% volume fraction) was then replaced with distilled water and the suspension was shaken until homogeneity.

The above procedure was repeated 5 times. Subsequently, a 20 μl droplet of the processed suspension was spin-coated onto a superlattice sample at 2000 rpm for 2 minutes. As a result, NDs with average size close to 50 nm were observed at the HMM surface as shown by the SEM scan in Fig. 4(a). Finally, the sample was covered with a 60-nm-thick layer of polyvinyl alcohol (PVA, 1.5% w/v) to guarantee the immobility of the NDs. In addition, the PVA layer separated the NDs from the immersion oil which was employed in the microscope setup for efficient collection of emitted light by an objective lens with high numerical aperture.

The reference sample consisted of NDs deposited by the same method onto a standard 150-μm-thick glass coverslip (VWR VistaVision[TM] Cover Glasses, No. 1). Since the dielectric permittivities of these glass substrates (2.34) are close to those of the immersion oil (2.31) and the PVA layer (2.25) [78], the NDs in reference samples were considered as being effectively immersed into an infinite homogeneous medium with a permittivity of 2.25.

2.2. *Fabrication and Characterization of HMM*

TiN is a new plasmonic material known for its CMOS compatibility, mechanical strength, and thermal stability at high temperatures (melting point > 2700°C).[62,63] This material can be epitaxially deposited on a variety of different substrates, such as magnesium oxide, silicon, aluminum scandium nitride, sapphire, down to thickness of a few nanometers. Conventional plasmonic materials (e.g. gold or silver) cannot form continuous layers of such thicknesses. This property is of fundamental importance for increasing LDOS, since LDOS for multilayer HMM is inversely proportional to the cube of the layer thickness.[64,65]

Our HMM consisted of a TiN/Al$_{0.7}$Sc$_{0.3}$N superlattice epitaxially grown on a [001]-oriented magnesium oxide (MgO) 0.5-mm-thick substrate using reactive DC magnetron sputterer (PVD Products) with a base pressure of 10^{-7} torr at 750°C. The superlattice was formed by a stack of 10 pairs of layers each consisting of 8.5-nm-thick film of TiN and 6.3-nm-thick film of Al$_{0.7}$Sc$_{0.3}$N similar to the structure reported in Ref. [61]. Fig. 4(b) shows the cross-sectional high-resolution TEM image of the superlattice with matching latices of TiN and (Al,Sc)N.[61] The NDs were spun on top of the HMM as shown in Fig. 4(c). Since the TiN and (Al,Sc)N layer thicknesses were much smaller than the wavelength of light, the HMM can be approximated as a uniaxial anisotropic effective medium with dielectric permittivities $\varepsilon_\parallel$ and $\varepsilon_\perp$.[68] The latter were measured using a spectroscopic ellipsometer (J. A. Woollam Co.; W-VASE).[59] The dielectric permittivities of the constituent materials (TiN and Al$_{0.7}$Sc$_{0.3}$N) as well as substrate (MgO) were retrieved from the values of Ψ and Δ measured in the wavelength range 300 – 2000 nm at angles of incidence 50° and 70°. The Drude-Lorentz model used for data fit is the following:

$$\varepsilon(\hbar\omega) = \varepsilon_1 + i\varepsilon_2 = \varepsilon_\infty + \sum_k \frac{A_k}{E_k^2 - (\hbar\omega)^2 - iB_k\hbar\omega}, \tag{2}$$

Table 1. Experiment-fitted parameters of
the Drude-Lorentz model for (Al,Sc)N and
TiN films.[58]   Table reproduced with permission from Copyright 2014 Wiley-VCH.

	TiN	(Al,Sc)N
ε_∞	4	4
A_1 (eV)^2	52.536	0
B_1(eV)	0.29197	0
A_2(eV)^2	130.93	11.162
B_2(eV)	4.9807	0.63261
E_2(eV)	5.9784	3.6857

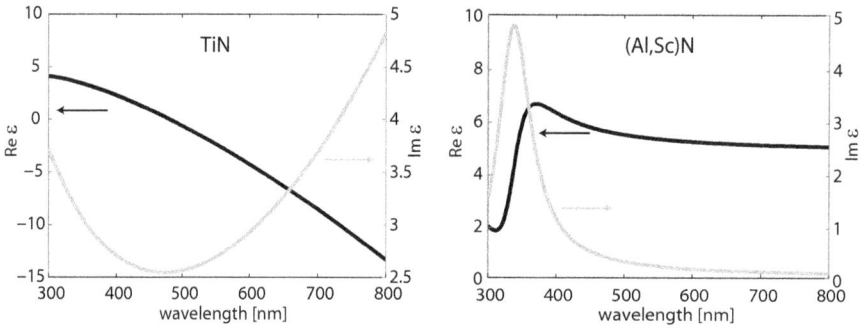

Fig. 5.   The dielectric functions of TiN and (Al,Sc)N: real and imaginary parts are shown with black and light gray lines, respectively.[58]   Figure reproduced with permission from Copyright 2014 Wiley-VCH.

where $\hbar$ is the reduced Plank's constant, A_k is the amplitude, E_k is the center energy and B_k is the broadening of each oscillator.[69] The experiment-fitted parameters are presented in the Table 1. The corresponding dependences of the TiN and (Al,Sc)N dielectric functions (Re ε, Im ε) versus wavelength are plotted in Fig. 5.

The TiN-based metamaterial exhibited hyperbolic dispersion in the wavelength range of the NV center emission (600-800 nm, Fig. 6, highlighted area): with $\varepsilon_\perp = 16.4 + i21.1$ and $\varepsilon_\parallel = -2.3 + i2.1$ at emission peak wavelength (685 nm). Thus, for the electric field parallel to the interface, the metamaterial at this wavelength behaved as a metal. However, at the excitation wavelength of 532 nm (Fig. 6, vertical dashed gray line) the TiN/(Al,Sc)N metamaterial behaved as a lossy dielectric with $\varepsilon_\parallel = 1.27 + i1.6$, allowing for transmission of the tangential component of the pump electric field. This facilitated efficient delivery of the excitation field to the NV center. The transmittance, reflectance and absorbance spectra of the HMM are shown in Fig. 7.

2.3. Experimental Setup

The increase in the spontaneous emission rate, the reduction of the excited state lifetime, and the second-order correlation function of the emission were measured using

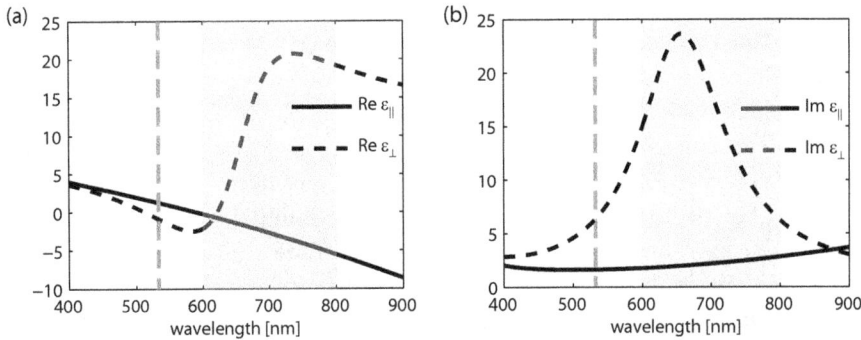

Fig. 6. (a) Real and (b) imaginary parts of the dielectric functions of the uniaxial effective medium that approximates the fabricated HMM.[58] The permittivities were obtained by spectroscopic ellipsometry. Within the range of the NV center emission (600-800 nm-highlighted), the metamaterial showed hyperbolic dispersion (Re$[\varepsilon_\perp]$ >0, Re$[\varepsilon_\parallel]$ <0). At the excitation wavelength (532 nm, vertical dashed gray line) the metamaterial behaved as a lossy dielectric. Figures reproduced with permission from Copyright 2014 Wiley-VCH.

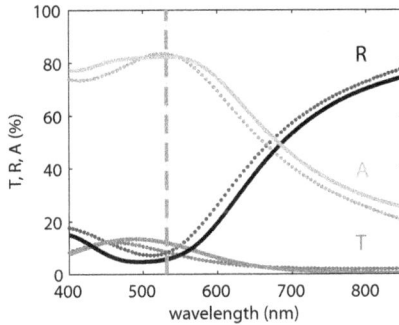

Fig. 7. Transmittance (T), reflectance (R), and absorbance (A) of the TiN HMM calculated using direct T-matrix method (solid lines)[84] and measured by spectrophotometry (dotted lines); angle of incidence for $T - 0°$, for $R - 8°$; $A = 1 - T - R$.[58] Figure reproduced with permission from Copyright 2014 Wiley-VCH.

a home-made confocal microscope, which is schematically represented in Fig. 4(d). The fluorescent radiation from an NV center was collected with a wide aperture (NA = 1.49) oil immersion objective (Nikon; CFI Apo TIRF 100X Oil) and separated from the pump with a series of optical filters. A single-photon avalanche photodiode, SPAD (PerkinElmer; SPCM-AQRH-14-FC) was used as a detector. Statistical measurement of the time delay between the excitation pulse and the first detected fluorescence photon, known as time-correlated single-photon counting (TCSPC),[70] has been employed. The lifetime was obtained from the exponential fit of the fluorescence decay histogram. The excitation of NV center for fluorescence lifetime measurements was performed by 532-nm picosecond pulsed laser

(PicoQuant; LDH-P-FA-530XL). The setup was converted as necessary to a Hanbury Brown-Twiss interferometer[71] to measure the second-order correlation function $g^{(2)}(t)$ by addition of a 50/50 beamsplitter and a second avalanche photodiode with a coaxial cable 80-ns-delay line. In this configuration, the NV centers were excited by a 532 nm continuous-wave laser (Coherent; Compass 315M-100). The measurement of $g^{(2)}(t)$ allows to identify NDs with single NV centers. $g^{(2)}(t)$ was obtained by measuring the background corrected coincidence count rate, normalized by the total counts on each detector, time-bin width and data collection time.[72] For a single-photon emitter the $g^{(2)}(t)$ curve is expected to exhibit a dip around $t = 0$ due to impossibility of simultaneous photon arrival at both detectors. The measured dip was fitted with the expression $\exp\left(-\left[\Gamma + R\right]|t|\right)$, where Γ and R are the total decay rate and pump rate, respectively. The signal from SPAD was processed by an electronic correlation module (PicoHarp 300, PicoQuant).

3. Theory

In this section, we calculate the spontaneous emission rate (represented by the Purcell factor F_P) and the normalized emission power f_{rad} collected in the far-field by an objective located above the HMM surface. Rigorous account of the spontaneous emission process can only be done in the frame of quantum electrodynamics. Since our quantum emitter is weakly coupled to the metamaterial environment, a calculation of the above parameters based on principles of classical electrodynamics is expected to be in reasonable agreement with the experiment.[73] In order to model our system, we considered the problem of dipole radiation near a planar layered medium, which is described in more detail in Refs. [74–76].

3.1. Semi-analytical Calculations of the Purcell Factor and Normalized Collected Emission Power

A single emitter was modeled as an oscillating electric dipole with dipole moment $\mathbf{p}$ and angular frequency ω. The energy dissipation rate in an inhomogeneous environment is given by[76]

$$P = \frac{\omega}{2} Im\left[p^* \cdot (E_0(r_0) + E_s(r_0))\right], \tag{3}$$

where $E_0(r_0)$ and $E_s(r_0)$ are the primary dipole field and scattered field at the dipole position (r_0), respectively. These electric fields were calculated using the dyadic Green's function formalism.[76] We analyzed the contribution of each spatial frequency mode (k-mode) using angular spectrum representation of the Green's functions.

The inset in Fig. 8(a) depicts an oscillating dipole situated at height h above the HMM uppermost layer (Al$_{0.7}$Sc$_{0.3}$N), the upper half-space (superstrate) filled with homogeneous medium with $\varepsilon_{sup} = 2.25$, planar multilayer TiN/Al$_{0.7}$Sc$_{0.3}$N

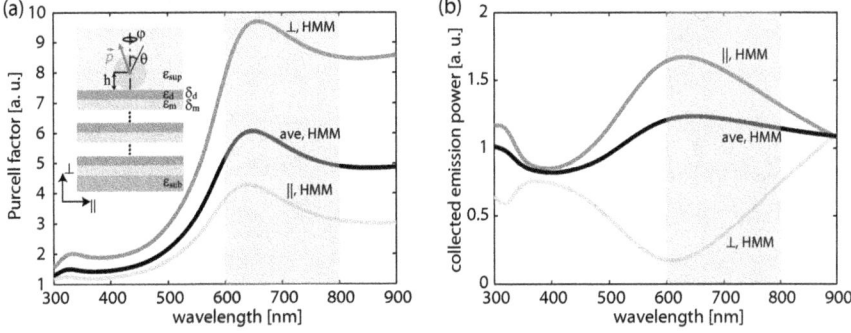

Fig. 8.   Calculated estimates of Purcell factor (a) and collected emission power (b) for a dipole located in homogeneous medium with $\varepsilon_{sup} = 2.25$ at the distance of 25 nm above the HMM surface.[58] Dark gray, light gray, and black curves correspond to the dipole orientations perpendicular ($\perp$), parallel ($\parallel$) to the HMM interface, and averaged (ave), respectively. Highlighted area indicates the emission spectral range of NV center at room temperature. Collection angle is 79.6°, which corresponds to NA 1.49. Layout of the modeled structure is shown in the inset. Figures reproduced with permission from Copyright 2014 Wiley-VCH.

superlattice ($\varepsilon_m/\varepsilon_d$) and the lower half-space (substrate) made of MgO (ε_{sub}). For this simulation we considered the HMM, used in our experiments, consisting of 10 pairs of 8.5-nm-thick TiN and 6.3-nm-thick (Al,Sc)N on MgO substrate. The Purcell factors F_P for in-plane ($\parallel$) and perpendicular ($\perp$) oriented single-dipole emitters placed at a distance h above a multilayer planar structure were calculated using the following formulae[76]

$$F_P^{\perp} = 1 + \frac{3}{2} \frac{1}{\varepsilon_{sup}^{3/2}} \int_0^{\infty} Re \left\{ \frac{s^3}{s_{\perp,sup}(s)} \tilde{r}^p(s) e^{2ik_0 s_{\perp,sup}(s)h} \right\} ds, \qquad (4)$$

$$F_P^{\parallel} = 1 + \frac{3}{4} \frac{1}{\varepsilon_{sup}^{1/2}} \int_0^{\infty} Re \left\{ \frac{s}{s_{\perp,sup}(s)} \left[\tilde{r}^s(s) - \frac{s_{\perp,sup}^2(s)}{\varepsilon_{sup}} \tilde{r}^p(s) \right] e^{2ik_0 s_{\perp,sup}(s)h} \right\} ds. \qquad (5)$$

The value of F_P for the averaged (ave) dipole orientation is given by

$$F_P^{iso} = \frac{2}{3} F_P^{\parallel} + \frac{1}{3} F_P^{\perp}. \qquad (6)$$

The normalized collected emission powers f_{rad} for the same dipole orientations are

$$f_{rad}^{\perp} = \frac{3}{4} \int_0^{\theta_{max}} \sin^3 \theta \left| e^{-i\varepsilon_{sup}^{1/2} k_0 h \cos \theta} + \tilde{r}^p(\theta) e^{i\varepsilon_{sup}^{1/2} k_0 h \cos \theta} \right|^2 d\theta, \tag{7}$$

$$f_{rad}^{\parallel} = \frac{3}{8} \int_0^{\theta_{max}} \left(\cos^2 \theta \left| e^{-i\varepsilon_{sup}^{1/2} k_0 h \cos \theta} - \tilde{r}^p(\theta) e^{i\varepsilon_{sup}^{1/2} k_0 h \cos \theta} \right|^2 \right.$$
$$\left. + \left| e^{-i\varepsilon_{sup}^{1/2} k_0 h \cos \theta} + \tilde{r}^s(\theta) e^{i\varepsilon_{sup}^{1/2} k_0 h \cos \theta} \right|^2 \right) \sin \theta d\theta, \tag{8}$$

$$f_{rad}^{iso} = \frac{2}{3} f_{rad}^{\parallel} + \frac{1}{3} f_{rad}^{\perp}. \tag{9}$$

In equations (4) - (9), $s = k_{\parallel}/k_0$, $s_{\perp,sup}(s) = k_{\perp,sup}(s)/k_0 = \left(\varepsilon_{sup} - s^2 \right)^{1/2}$, $k_0 = \omega/c$, θ is a polar angle measured from the $\perp$ direction, the collection angle $\theta_{max} = 79.6°$. $k_{\parallel}$ is the in-plane component of the k-vector varying from 0 to infinity. The generalized superlattice's Fresnel reflection coefficients for p- and s-polarized light $\tilde{r}^p$ and $\tilde{r}^s$ were calculated utilizing the recursive imbedding method (see Appendix), which is more precise and efficient than the direct transfer matrix approach.[77] The integrals were numerically evaluated by using an adaptive Gauss–Kronrod quadrature method.[78] In the formulae, we assumed that intrinsic quantum yield of NV centers is close to unity.[79] F_P and f_{rad} were normalized by the total radiation power and the power emitted into the collection angle, respectively. Both powers corresponded to the case of the emitter immersed into homogeneous medium with dielectric permittivity ε_{sup}, which models well the reference sample in the experiment.

In Fig. 8 we represent the calculated F_P and f_{rad} for in-plane (light gray), perpendicular (dark gray) and averaged (black) dipole orientations as a function of emission wavelength. The values of F_P and f_{rad} for the averaged orientation are defined as statistical averages of F_P and f_{rad} over all possible orientations. Assuming that the NV center is located at the crystal center, the expected Purcell factor for the NDs with a mean diameter of 50 nm should be on average around 4.5. The detected count rates corresponding to the normalized collected emission power for the same type of NDs were anticipated to increase on average by about 20%. The estimates for F_P and f_{rad} are modest in our experimental configuration. In the next two subsections we will show that the experimental conditions can be optimized to obtain higher values of both quantities.

3.2. Dependence of F_P and f_{rad} on the Distance between an Emitter and HMM Surface

The Purcell factor F_P increases as the separation distance h between the emitter and the surface is reduced. This is a common feature for dipoles located near metallic surfaces[4] and is expected to be observed as well in the case of an NV center on

HMM. In our case, the smallest separation distance is dictated by the ND size. The Purcell factor dependence on h is explained by the fact that at short distances h the evanescent fields created by the emitter are better coupled to the metamaterial modes.[85]

We have calculated the Purcell factor as a function of h for different orientations of a dipole located in the homogeneous medium above the HMM structure previously described (see Fig. 9(a)). At the distance of 25 nm, corresponding to the size of our NDs, the Purcell enhancement is on the order of 10, as shown in Fig. 8. However, at distances h of a few nanometers the Purcell factor reaches two orders of magnitude.

In Fig. 9(b) we represent the dependence of f_{rad} on h. The averaged f_{rad} improvement (black curve) from different dipole orientations drops down from 23% to 0% as h decreases from 25 nm to zero. However, for the in-plane and out-of-plane dipole orientations considered separately, the trends are opposite, which can be understood as an interference effect. The collective behavior of free electrons in a metal can be represented as an oscillation of an image dipole. The interference happens between the real and image dipole sources whose relative phase and amplitudes are dictated by the complex generalized reflection coefficients $\tilde{r}^p$ and $\tilde{r}^s$ calculated in Appendix.[4]

Thus, bringing the dipole closer to the metamaterial allows for better energy coupling to metamaterial modes. However, this increase in the Purcell factor does not translate into higher collected emission power, which encourages to look for techniques of metamaterial mode outcoupling.

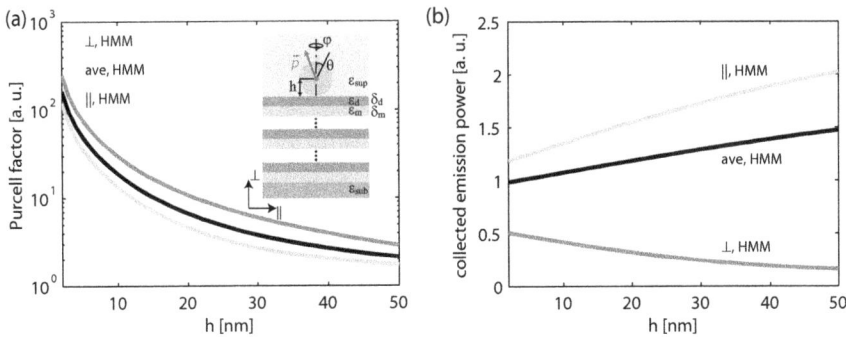

Fig. 9. Dependencies of the Purcell factor F_P (a) and the collected emission power f_{rad} (b) on a dipole position h above the HMM surface.[58] Dark gray, light gray, and black curves correspond to the dipole orientations perpendicular ($\perp$), parallel ($\parallel$) to the HMM interface and averaged (ave), respectively. Figures reproduced with permission from Copyright 2014 Wiley-VCH.

3.3. *Influence of Neighboring NDs and Structural Surface Defects of HMM on Collected Emission Power*

The coupled plasmonic modes are collective oscillations of electromagnetic field and electron plasma in the metallic layers of the HMM. One can expect the coupled plasmonic modes (corresponding to $k_{\parallel} > \sqrt{\epsilon_{sup}}k_0$) to be efficiently outcoupled to low-k superstrate modes via small size structural irregularities. Examples of such irregularities (adacent nanodiamonds and superlattice defects) are demonstrated in Fig. 10(a). To qualitatively estimate the influence of these scattering objects, we have performed finite-element method (FEM) analysis using a commercial numerical simulation tool (COMSOL Multiphysics). Beforehand, we have checked the consistency of the FEM with the semi-analytical method in the case of planar HMM without any scatterers and obtained a reasonable agreement between the two methods (Fig. 11). Then, we have simulated the power emitted by the dipole

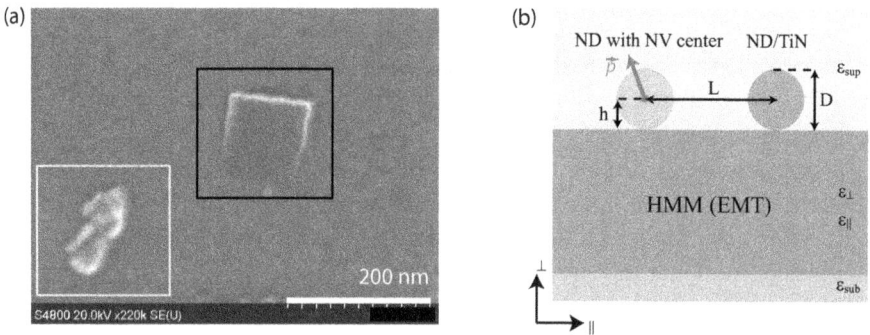

Fig. 10.   (a) SEM scan showing a superlattice defect (inside black frame) and agglomerated NDs (inside white frame). (b) Schematic of the simulated structure.[58] Figures reproduced with permission from Copyright 2014 Wiley-VCH.

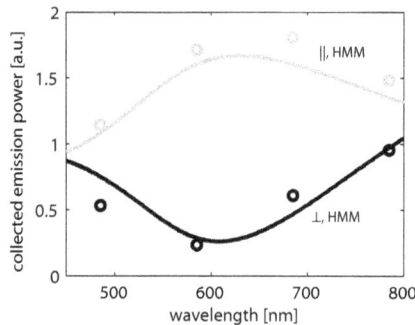

Fig. 11.   Agreement between semi-analytical (solid lines) and FEM calculations (circles). The calculations are performed for the HMM without any scatterers. Black and light gray curves correspond to the dipole orientations perpendicular ($\perp$) and parallel ($\parallel$) to the HMM interface, respectively.[58] Figure reproduced with permission from Copyright 2014 Wiley-VCH.

Table 2.   Influence of NDs lacking NV centers and structural surface defects (TiN particles) on normalized collected emission power f_{rad}.   Demonstrated values are the enhancement factors of f_{rad}.[58]   Table reproduced with permission from Copyright 2014 Wiley-VCH.

	on HMM	on HMM with ND			on HMM with TiN particle		
		10 nm	69 nm	343 nm	10 nm	69 nm	343 nm
in-plane (∥)	1.8	2.3	1.8	1.8	3.9	1.9	1.8
perpendicular (⊥)	0.6	0.6	0.6	0.6	0.2	0.4	0.6
isotropic (iso)	1.5	1.9	1.5	1.5	3.0	1.5	1.5

source on top of HMM in the presence of a spherical scatterer. The schematic of the simulated structure is shown in Fig. 10(b). The HMM was considered as an effective anisotropic medium, whereas scatterers are modelled as either diamond or TiN spheres of 50 nm in diameter. The emitter was emulated as an electric point dipole embedded into a 50-nm-diameter diamond sphere. The superstrate and substrate are respectively the immersion oil and MgO. Three values of emitter-scatterer distance of 10, 69 and 343 nm have been considered. The emission wavelength was fixed to 685 nm.

We retrieved the total NV centers lifetimes from the exponential fitting of the fluorescence decays such as the ones shown in Fig. 13(a). The average values measured were 17.1 ns and 4.3 ns on glass and HMM, respectively (see Fig. 13(b)). Hence, the NV centers on HMM exhibit an average decrease in lifetime by a factor of 4, which is consistent with the above Purcell effect calculations. The shortest recorded lifetime for a single NV center on top of HMM is 1.5 ns, which corresponds to a Purcell factor of 11.4. The spreads in the lifetime statistics are likely due to the variation in NV center distance from the HMM surface and its dipole orientation.

Results of the FEM simulations, the enhancement factors of f_{rad} values, are summarized in Table 2. The noticeable influence is observed at small separation distances. For the averaged dipole orientation, an adjacent (at a distance of 10 nm) ND or TiN scatterer can increase f_{rad} by 30% or 200%, respectively. For an in-plane oriented emitter, the increase in emitted power from a nearby surface defect is up to 220%, giving an overall emission enhancement factor of 4 compared to the emitter on coverslip.

4. Results and Discussions

The NDs with single NV centers were selected by measuring the second-order correlation function $g^{(2)}(t)$ of the radiation from the detected fluorescence spots. Only NV centers with $g^{(2)}(0) < 0.5$ were considered for further experiments. This is a commonly accepted criterion for antibunched emission that is characteristic of a single photon source.[86] Typical photon antibunching effect is shown for a single NV center on glass coverslip and on HMM in Fig. 12.

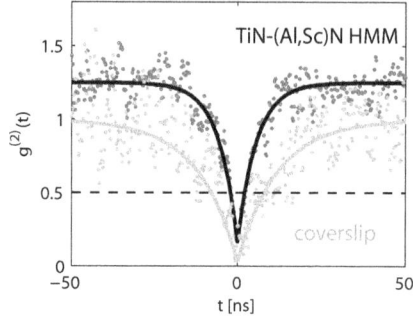

Fig. 12.   Second-order correlation function $g^{(2)}(t)$ of the emission from a representative ND with single NV center on top of coverslip and HMM.[58] Figure reproduced with permission from Copyright 2014 Wiley-VCH.

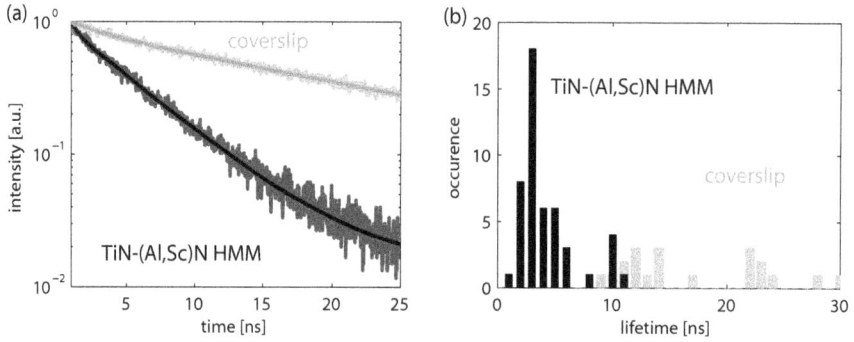

Fig. 13.   (a) Representative normalized fluorescence decays and (b) histograms of lifetimes for the NV centers located on glass coverslip (reference sample) and on HMM.[58] The average and the lowest recorded lifetimes on HMM are 4 and 11.4 times shorter, respectively, than the average lifetime recorded on coverslip. Figures reproduced with permission from Copyright 2014 Wiley-VCH.

The single-photon emission rate dependences versus the excitation power are shown in Fig. 14(a), for the brightest NV centers on both the HMM and the glass coverslip. The data have been corrected for the background emission measured at a ND-free location. The emission count rate saturates around 1 mW of pump power. By fitting the experimentally measured saturation curves using the expression $I(P) = I_0/(1 + P_{sat}/P)$[87] we obtain the saturated single-photon count rates for each individual NV center. The histograms of these saturated single-photon count rates are shown in Fig. 14(b) for the NV centers both on glass and on HMM. The histogram corresponding to the HMM presents multiple groups. The NV centers in the group around 200 kcounts/s present an average enhancement against the coverslip of 1.8±1.1. This enhancement is consistent with the predicted value of 1.2, within the error margin. This large error is due to the enhancement value being the ratio of two measured parameters both having significant uncertainties

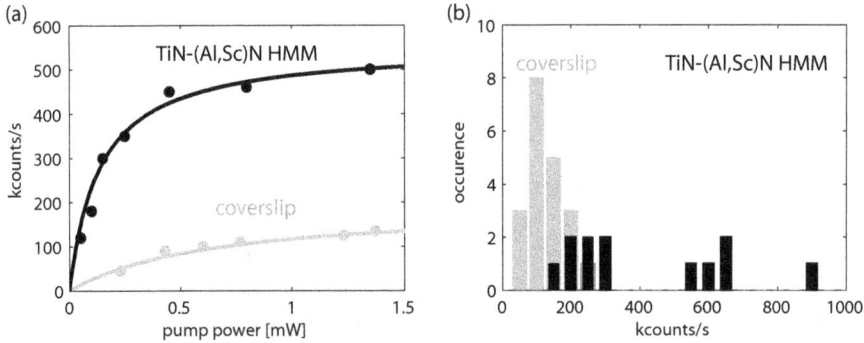

Fig. 14. Collected single-photon count rates (corrected for background emission) from NV centers in 50 nm NDs.[58] (a) Typical saturation curves and (b) histograms of count rates for NDs on glass coverslip (light grey) and HMM (black). The average count rate enhancements for the first and second group are 1.8 and 4.7, respectively. Figures reproduced with permission from Copyright 2014 Wiley-VCH.

due to random dipole orientation. The effect of h on collected power uncertainty is negligible as it follows from Fig. 9. The NV centers in the next group in the HMM histogram present an average count rate enhancement of 4.7±2.2, which is beyond the theory prediction. Finally there is one diamond showing even higher count rate enhancement. Excessive emission rates from certain NDs potentially could arise from the influence of either neighboring superlattice defects or adjacent "dark" NDs lacking NV centers.

The results of section 3.3 suggest that a substantial portion of the photons emitted by the dipole in the vicinity of HMM surface gets converted into high-k modes.[64] Once excited, these modes propagate through the HMM never leaving its bulk where they are eventually absorbed and do not provide any contribution to the collected photon flux. However, the presence of surface scatterers may result in outcoupling of these modes into the far field and therefore provide additional contribution to the emitted signal. We have shown above that taking into account energy outcoupling from defects surrounding the NV center can account for a further twofold increase in collected power and constitute a plausible cause for the observed count rate enhancement.

5. Summary

We have experimentally demonstrated an improvement in the emission rate of single NV centers in NDs placed on top of an epitaxial, TiN/(Al,Sc)N hyperbolic metamaterial compared to an isotropic environment with a uniform permittivity. The observed average lifetime decreased from 17 ns to 4.3 ns, corresponding to a Purcell factor of 4. This result is in good agreement with the theoretical prediction for an emitter located 25 nm away from the HMM surface. At the same time the collected emission power for NV centers near the hyperbolic metamaterial was increased by a factor of 1.8 on average compared to NV centers on coverslip, although an enhancement of 4.7 was detected for several NV centers. The enhancement effect does not depend on in-plane location and does not require deterministic ND placement. At the same time, the emission rate of the ND-based single-photon source could be further enhanced by placing it closer to the surface of a TiN/(Al,Sc)N hyperbolic metamaterial. Most of the dipole emission is coupled to the hyperbolic metamaterial modes that are not directly detected in the far field. However, a significant portion of this power can be recovered in the far field by the presence of an adjacent growth defect or specifically engineered nanostructure. The results of this research can be used to build a highly efficient room temperature CMOS-compatible single-photon source.

Acknowledgments

The authors would like to thank R. Chandrasekar, J. Liu, M. Ferrera, J. Irudayaraj, V. V. Klimov, and A. N. Smolyaninov for their help with chapter preparation. This work was partially supported by the AFOSR-MURI grant (FA9550-10-1-0264) and NSF-MRSEC grant (DMR-1120923).

References

1. L. K. Grover, A fast quantum mechanical algorithm for database search, *Proceedings of the twenty-eighth annual ACM symposium on theory of computing*, 212-219, ACM,(1996).
2. P. W. Shor, Polynomial-time algorithms for prime factorization and discrete logarithms on a quantum computer, *SIAM J. Comput.* **26**(5), 1484-1509, (1997).
3. R. P. Feynman, Simulating physics with computers, *Int. J. Theor. Phys.* **21**(6), 467-488, (1982)
4. W. L. Barnes, Fluorescence near interfaces: the role of photonic mode density, *J. Mod. Opt.* **45**(4), 661-699, (1998).
5. J. L. O'Brien, A. Furusawa, and J. Vukovi, Photonic quantum technologies, *Nat. Photon.* **3**(12), 687-695, (2009).
6. J. Mower, N. C. Harris, G. R. Steinbrecher, Y. Lahini, and D. Englund, High-fidelity quantum photonics on a programmable integrated circuit, *arXiv preprint*, arXiv:1406.3255 (2014).
7. M. D. Eisaman, J. Fan, A. Migdall, and S. V Polyakov, Invited review article: Single-photon sources and detectors, *Rev. Sci. Instrum.* **82**(7), 071101, (2011).
8. M. Hijlkema, B. Weber, H. P. Specht, S. C. Webster, A. Kuhn, and G. Rempe, A single-photon server with just one atom, *Nat. Phys.* **3**(4), 253-255, (2007).
9. C. Maurer, C. Becher, C. Russo, J. Eschner, and R. Blatt, A single-photon source based on a single Ca^+ ion, *New J. Phys.* **6**, 94–94, (2004).
10. M. Steiner, A. Hartschuh, R. Korlacki, and A. J. Meixner, Highly efficient, tunable single-photon source based on single molecules, *Appl. Phys. Lett.* **90**(18), 183122, (2007).
11. A. V. Akimov, A. Mukherjee, C. L. Yu, D. E. Chang, A. S. Zibrov, P. R. Hemmer, H. Park, and M. D. Lukin, Generation of single optical plasmons in metallic nanowires coupled to quantum dots., *Nature* **450**(7168), 402–6, (2007).

12. E. Wu, J. R. Rabeau, G. Roger, F. Treussart, H. Zeng, P. Grangier, S. Prawer, and J.-F. Roch, Room temperature triggered single-photon source in the near infrared, *New J. Phys.* **9**(12), 434–434, (2007).

13. S. Chen, Y.-A. Chen, T. Strassel, Z.-S. Yuan, B. Zhao, J. Schmiedmayer, and J.-W. Pan, Deterministic and Storable Single-Photon Source Based on a Quantum Memory, *Phys. Rev. Lett.* **97**(17), 173004, (2006).

14. J. Kim, O. Benson, H. Kan, and Y. Yamamoto, A single-photon turnstile device, *Nature* **397**, 500–503, (1999).

15. A. Gruber, A. Drbenstedt, C. Tietz, L. Fleury, J. Wrachtrup, and C. Von Borczyskowski, Scanning Confocal Optical Microscopy and Magnetic Resonance on Single Defect Centers, *Science* **276**(5321), 2012–2014, (1997).

16. C. Kurtsiefer, S. Mayer, P. Zarda, and H. Weinfurter, Stable solid-state source of single photons, *Phys. Rev. Lett.* **85**(2), 290–3, (2000).

17. A. W. Schell, G. Kewes, T. Schrder, J. Wolters, T. Aichele, and O. Benson, A scanning probe-based pick-and-place procedure for assembly of integrated quantum optical hybrid devices, *Rev. Sci. Instrum.* **82**(7), 073709, (2011).

18. B. Lounis and W. Moerner, Single photons on demand from a single molecule at room temperature, *Nature* **407** (6803), 491–3, (2000).

19. X. Wang, X. Ren, K. Kahen, M. a Hahn, M. Rajeswaran, S. Maccagnano-Zacher, J. Silcox, G. E. Cragg, A. L. Efros, and T. D. Krauss, Non-blinking semiconductor nanocrystals, *Nature* **459**(7247), 686–9, (2009).

20. F. Jelezko and J. Wrachtrup, Single defect centres in diamond: A review, *Phys. Status Solidi*, **203**(13), 3207–3225, (2006).

21. T. A. Kennedy, J. S. Colton, J. E. Butler, R. C. Linares, and P. J. Doering, Long coherence times at 300 K for nitrogen-vacancy center spins in diamond grown by chemical vapor deposition, *Appl. Phys. Lett.* **83**(20),4190, (2003).

22. P. Neumann, J. Beck, M. Steiner, F. Rempp, H. Fedder, P. R. Hemmer, J. Wrachtrup, and F. Jelezko, Single-shot readout of a single nuclear spin, *Science*, **329**(5991), 542–4, (2010).

23. P. C. Maurer, G. Kucsko, C. Latta, L. Jiang, N. Y. Yao, S. D. Bennett, F. Pastawski, D. Hunger, N. Chisholm, M. Markham, D. J. Twitchen, J. I. Cirac, and M. D. Lukin, Room-temperature quantum bit memory exceeding one second, *Science*, **336**(6086), 1283–6, (2012).

24. L. Robledo, L. Childress, H. Bernien, B. Hensen, P. F. A. Alkemade, and R. Hanson, High-fidelity projective read-out of a solid-state spin quantum register, *Nature* **477**(7366), 574–8, (2011).

25. F. Dolde, I. Jakobi, B. Naydenov, N. Zhao, S. Pezzagna, C. Trautmann, J. Meijer, P. Neumann, F. Jelezko, and J. Wrachtrup, Room-temperature entanglement between single defect spins in diamond, *Nat. Phys.***9**(3), 139–143, (2013).

26. M. V. G. Dutt, L. Childress, L. Jiang, E. Togan, J. Maze, F. Jelezko, a S. Zibrov, P. R. Hemmer, and M. D. Lukin, Quantum register based on individual electronic and nuclear spin qubits in diamond, *Science* **316**(5829), 1312–6, (2007).

27. F. Dolde, H. Fedder, M. W. Doherty, T. Nbauer, F. Rempp, G. Balasubramanian, T. Wolf, F. Reinhard, L. C. L. Hollenberg, F. Jelezko, and J. Wrachtrup, Electric-field sensing using single diamond spins, *Nat. Phys.* **7**(6), 459–463, (2011).

28. J. R. Maze, P. L. Stanwix, J. S. Hodges, S. Hong, J. M. Taylor, P. Cappellaro, L. Jiang, M. V. G. Dutt, E. Togan, S. Zibrov, A. Yacoby, R. L. Walsworth, and M. D. Lukin, Nanoscale magnetic sensing with an individual electronic spin in diamond, *Nature* **455**(7213), 644–7, (2008).

29. G. Kucsko, P. C. Maurer, N. Y. Yao, M. Kubo, H. J. Noh, P. K. Lo, H. Park, and M. D. Lukin, Nanometre-scale thermometry in a living cell, *Nature* **500**(7460), 54–8, (2013).
30. A. Ajoy and P. Cappellaro, Stable three-axis nuclear-spin gyroscope in diamond, *Phys. Rev. A* **86**(6), 062104, (2012).
31. C. Santori, P. E. Barclay, K. C. Fu, R. G. Beausoleil, S. Spillane, and M. Fisch, Nanophotonics for quantum optics using nitrogen-vacancy centers in diamond, *Nanotechnology* **21**(27), 274008, (2010).
32. E. M. Purcell, Spontaneous emission probabilities at radio frequencies, *Phys. Rev.* **69**, (1946).
33. S. Schietinger and O. Benson, Coupling single NV-centres to high-Q whispering gallery modes of a preselected frequency-matched microresonator, *J. Phys. B At. Mol. Opt. Phys.* **42**(11), 114001, (2009).
34. J. Wolters, A. W. Schell, G. Kewes, N. Nsse, M. Schoengen, H. Dscher, T. Hannappel, B. Lchel, M. Barth, and O. Benson, Enhancement of the zero phonon line emission from a single nitrogen vacancy center in a nanodiamond via coupling to a photonic crystal cavity, *Appl. Phys. Lett.* **97**(14), 141108, (2010).
35. J. Riedrich-Mller, L. Kipfstuhl, C. Hepp, E. Neu, C. Pauly, F. Mcklich, A. Baur, M. Wandt, S. Wolff, M. Fischer, S. Gsell, M. Schreck, and C. Becher, One- and two-dimensional photonic crystal microcavities in single crystal diamond, *Nat. Nanotechnol.* **7**(1), 69–74, (2012).
36. A. Faraon, C. Santori, Z. Huang, V. M. Acosta, and R. G. Beausoleil, Coupling of Nitrogen-Vacancy Centers to Photonic Crystal Cavities in Monocrystalline Diamond, *Phys. Rev. Lett.* **109**(3), 033604, (2012).
37. B. J. M. Hausmann, B. J. Shields, Q. Quan, Y. Chu, N. P. de Leon, R. Evans, M. J. Burek, A. S. Zibrov, M. Markham, D. J. Twitchen, H. Park, M. D. Lukin, and M. Lonc R, Coupling of NV centers to photonic crystal nanobeams in diamond, *Nano Lett.* **13**(12), 5791–6, (2013).
38. T. Schrder, F. Gdeke, M. J. Banholzer, and O. Benson, Ultrabright and efficient single-photon generation based on nitrogen-vacancy centres in nanodiamonds on a solid immersion lens, *New J. Phys.* **13** (5), 055017, (2011).
39. J. P. Hadden, J. P. Harrison, C. Stanley-Clarke, L. Marseglia, Y.-L. D. Ho, B. R. Patton, J. L. O'Brien, and J. G. Rarity, Strongly enhanced photon collection from diamond defect centers under microfabricated integrated solid immersion lenses, *Appl. Phys. Lett.* **97**(24), 241901, (2010).
40. T. M. Babinec, B. J. M. Hausmann, M. Khan, Y. Zhang, J. R. Maze, P. R. Hemmer, and M. Loncar, A diamond nanowire single-photon source, *Nat. Nanotechnol.* **5**(3), 195–9, (2010).
41. J. T. Choy, I. Bulu, B. J. M. Hausmann, E. Janitz, I.-C. Huang, and M. Loncar, Spontaneous emission and collection efficiency enhancement of single emitters in diamond via plasmonic cavities and gratings, *Appl. Phys. Lett.* **103**(16), 161101, (2013).
42. T. Galfsky, H. N. S. Krishnamoorthy, W. D. Newman, and E. Narimanov, Directional out-coupling from active hyperbolic metamaterials, *arXiv Prepr.* arXiv1404.1535 (2014).
43. J. T. Choy, B. J. M. Hausmann, T. M. Babinec, I. Bulu, M. Khan, P. Maletinsky, A. Yacoby, and M. Lonar, Enhanced single-photon emission from a diamond–silver aperture, *Nat. Photon.* **5**(12), 738–743, (2011).
44. S. Schietinger, M. Barth, T. Aichele, and O. Benson, Plasmon-enhanced single photon emission from a nanoassembled metal-diamond hybrid structure at room temperature, *Nano Lett.* **9**(4), 1694–8, (2009).

45. A. Faraon, P. Barclay, and C. Santori, Resonant enhancement of the zero-phonon emission from a colour centre in a diamond cavity, *Nat. Photon.* **5**, 301–305, (2011).
46. D. Smith and D. Schurig, Electromagnetic Wave Propagation in Media with Indefinite Permittivity and Permeability Tensors, *Phys. Rev. Lett.* **90**(7), 077405, (2003).
47. Z. Jacob, L. V Alekseyev, and E. Narimanov, Optical Hyperlens: Far-field imaging beyond the diffraction limit, *Opt. Express* **14**(18), 8247–56, (2006).
48. A. Poddubny, I. Iorsh, P. Belov, and Y. Kivshar, Hyperbolic metamaterials, *Nat. Photon.* **7**(12), 948–957, (2013).
49. Z. Liu, H. Lee, Y. Xiong, C. Sun, and X. Zhang, Far-field optical hyperlens magnifying sub-diffraction-limited objects, *Science* **315** (5819), 1686, (2007).
50. M. A. Noginov, H. Li, Y. A. Barnakov, D. Dryden, G. Nataraj, G. Zhu, C. E. Bonner, M. Mayy, Z. Jacob, and E. E. Narimanov, Controlling spontaneous emission with metamaterials, *Opt. Lett.* **35**(11), 1863–5, (2010).
51. J.-Y. Kim, V. P. Drachev, Z. Jacob, G. V. Naik, A. Boltasseva, E. Narimanov, and V. M. Shalaev, Improving the radiative decay rate for dye molecules with hyperbolic metamaterials, *Opt. Express* **20**(7),8100–8116, (2012).
52. D. Lu, J. J. Kan, E. E. Fullerton, and Z. Liu, Enhancing spontaneous emission rates of molecules using nanopatterned multilayer hyperbolic metamaterials, *Nat. Nanotechnol.* **9**(1), 48–53, (2014).
53. Z. Jacob, J.-Y. Kim, G. V. Naik, a. Boltasseva, E. E. Narimanov, and V. M. Shalaev, Engineering photonic density of states using metamaterials, *Appl. Phys. B* **100**(1), 215–218, (2010).
54. H. N. S. Krishnamoorthy, Z. Jacob, E. Narimanov, I. Kretzschmar, and V. M. Menon, Topological transitions in metamaterials, *Science* **336**(6078), 205–9, (2012).
55. O. Kidwai, S. V. Zhukovsky, and J. E. Sipe, Effective-medium approach to planar multilayer hyperbolic metamaterials: Strengths and limitations, *Phys. Rev. A* **85**(5), 053842, (2012).
56. J. Elser, R. Wangberg, V. A. Podolskiy, and E. E. Narimanov, Nanowire metamaterials with extreme optical anisotropy, *Appl. Phys. Lett.*, **89**(26), 261102, (2006).
57. S. V. Zhukovsky, O. Kidwai, and J. E. Sipe, Physical nature of volume plasmon polaritons in hyperbolic metamaterials, *Opt. Express* **21**(12), 14982-14987, (2013).
58. M. Y. Shalaginov, V. V. Vorobyov, J. Liu, M. Ferrera, A. V. Akimov, A. Lagutchev, A. N. Smolyaninov, V. V. Klimov, J. Irudayaraj, A. V. Kildishev, A. Boltasseva, and V. M. Shalaev, Enhancement of single-photon emission from nitrogen-vacancy centers with TiN/(Al,Sc)N hyperbolic metamaterial *Laser & Photon. Rev.* **9**(1), 120–7, (2015).
59. M. A. Azzam and N. M. Bashara, *Ellipsometry and polarized light.* (North Holland, 1987).
60. M. Y. Shalaginov, S. Ishii, J. Liu, J. Irudayaraj, A. Lagutchev, A. V. Kildishev, and V. M. Shalaev, Broadband enhancement of spontaneous emission from nitrogen-vacancy centers in nanodiamonds by hyperbolic metamaterials, *Appl. Phys. Lett.* **102**(17), p. 173114, (2013).
61. G. V Naik, B. Saha, J. Liu, S. M. Saber, E. Stach, J. M. K. Irudayaraj, T. D. Sands, V. M. Shalaev, and A. Boltasseva, Epitaxial superlattices with titanium nitride as a plasmonic component for optical hyperbolic metamaterials, *Proc. Natl. Acad. Sci. U. S. A.* **111**(21), 7546-7551, (2014).
62. G. V. Naik, J. L. Schroeder, X. Ni, A. V. Kildishev, and T. D. Sands, Titanium nitride as a plasmonic material for visible and near-infrared wavelengths, *Opt. Mater.* **2**(4), 1–13, (2012).

63. U. Guler, A. Boltasseva, and V. M. Shalaev, Applied physics. Refractory plasmonics, *Science* **344**(6181), 263–4, (2014).
64. Z. Jacob, I. I. Smolyaninov, and E. E. Narimanov, Broadband Purcell effect: Radiative decay engineering with metamaterials, *Appl. Phys. Lett.* **100**(18), 181105, (2012).
65. O. Kidwai, S. V Zhukovsky, and J. E. Sipe, Dipole radiation near hyperbolic metamaterials: applicability of effective-medium approximation, *Opt. Lett.* **36**(13), 2530–2, (2011).
66. C. Bradac, T. Gaebel, N. Naidoo, J. R. Rabeau, and A. S. Barnard, Prediction and measurement of the size-dependent stability of fluorescence in diamond over the entire nanoscale, *Nano Lett.* **9**(10), 3555–64, (2009).
67. S. Mahendia, a K. Tomar, R. P. Chahal, P. Goyal, and S. Kumar, Optical and structural properties of poly(vinyl alcohol) films embedded with citrate-stabilized gold nanoparticles, *J. Phys. D. Appl. Phys.* **44**(20), 205105, (2011).
68. J. Elser, V. Podolskiy, I. Salakhutdinov, and I. Avrutsky, Nonlocal effects in effective-medium response of nanolayered metamaterials, *Appl. Phys. Lett.* **90**(19), 191109, (2007).
69. *Guide to Using WVASE 32: Spectroscopic Ellipsometry Data Acquisition and Analysis Software*, J. A. Woollam Company, Inc., (2008).
70. D. O'Connor and D. Phillips, *Time-correlated Single Photon Counting.* (Academic Press, London, 1984).
71. R. Brown and R. Twiss, Correlation between photons in two coherent beams of light, *Nature* **177**(4497), 27–29, (1956).
72. A. Beveratos, R. Brouri, T. Gacoin, J. P. Poizat, and P. Grangier, Nonclassical radiation from diamond nanocrystals, *Phys. Rev. A* **64**(6), 061802, (2001).
73. M. S. Yeung and T. K. Gustafson, Spontaneous emission near an absorbing dielectric surface, *Phys. Rev. A* **54**(6), 5227–5242, (1996).
74. J. Sipe, The dipole antenna problem in surface physics: a new approach, *Surf. Sci.* **5**, 489–504, (1981).
75. G. Ford and W. Weber, Electromagnetic interactions of molecules with metal surfaces, *Phys. Rep.* **113**(4), 195-287, (1984).
76. L. Novotny and B. Hecht, *Principles of Nano-Optics.* (Cambridge University Press, Cambridge, England 2006).
77. C. L. Cortes, W. Newman, S. Molesky, and Z. Jacob, Quantum nanophotonics using hyperbolic metamaterials, *J. Opt.* **14** (6), 063001, (2012).
78. L. F. Shampine, Vectorized adaptive quadrature in MATLAB, *J. Comput. Appl. Math.* **211**(2), 131–140, (2008).
79. M. Frimmer, A. Mohtashami, and A. F. Koenderink, Nanomechanical method to gauge emission quantum yield applied to nitrogen-vacancy centers in nanodiamond, *Appl. Phys. Lett.* **102**(12), 121105, (2013).
80. W. Chew and S. Chen, Response of a point source embedded in a layered medium, *Antennas Wirel. Propag. Lett.* **2**(1), 1–9, (2003).
81. P. Drude, Ueber Oberflachenschichten, *Ann. Phys.* **272**(4), 865-897, (1889).
82. K. Visscher, G. Brakenhoff, and T. Visser, Fluorescence saturation in confocal microscopy, *J. Microsc.* **175**(2), 162–165, (1994).
83. S. V Zhukovsky, O. Kidwai, and J. E. Sipe, Physical nature of volume plasmon polaritons in hyperbolic metamaterials, *Opt. Express* **21**(12), 14982–14987, (2013).
84. P. Yeh, Optical waves in layered media. New York: Wiley, 1988.
85. Z. Jacob, I. I. Smolyaninov, and E. E. Narimanov, Broadband Purcell effect: Radiative decay engineering with metamaterials, *Appl. Phys. Lett.* **100**(18), 181105 (2012).

86. H. J. Kimble, M. Dagenais, and L. Mandel, Photon antibunching in resonance fluorescence, *Phys. Rev. Lett.* **39**(11), 691, (1977).
87. K. Visscher, G. J. Brakenhoff, and T. D. Visser, *J. Microsc.* **175**(2), 162-165, (1994).
88. P. B. Johnson and R. W. Christy, Optical Constants of the Noble Metals, *Phys. Rev. B* **6**, 4370-4379, (1972).

Appendix: Calculation of Generalized Reflection and Transmission Coefficients of a Planar Multilayer Slab

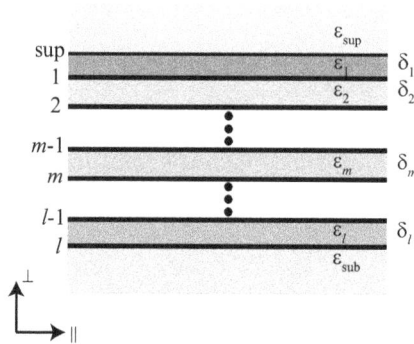

Fig. 15. Schematic of an arbitrary planar lamellar slab consisting of l layers.[58] Figure reproduced with permission from Copyright 2014 Wiley-VCH.

The generalized reflection coefficients $\tilde{r}^{(p,s)}(\omega, s)$ incorporate the reflection not only from the first interface but all the subsequent reflections from the interfaces positioned below. Therefore, $\tilde{r}^{(p,s)}$ represent the reflection from the whole multilayer structure, including the substrate. The schematic of the structure is shown in Fig. 15. Each layer is characterized by relative permittivity ε_m and thickness d_m. Above and below the multilayer slab, there are superstrate (ε_{sup}) and substrate (ε_{sub}) half-spaces.

The coefficients $\tilde{r}^{(p,s)}(\omega, s)$ are usually calculated using the transfer matrices (T-matrices). However, direct multiplication of T-matrices corresponding to the different layers result in large calculation errors and even in some cases fail to produce the output. Fig. 16 shows the relative errors when calculating the k-dependent

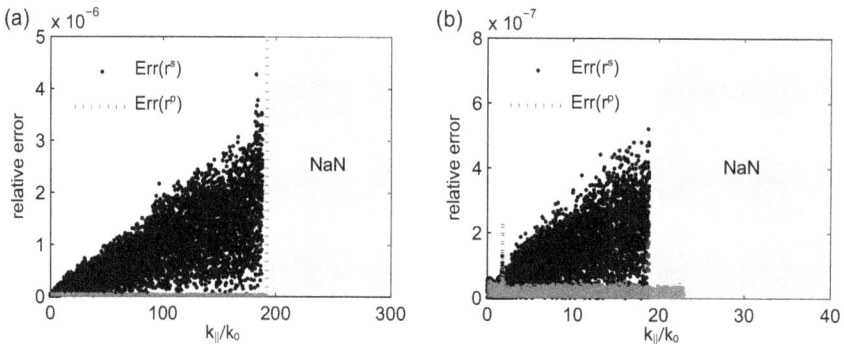

Fig. 16. Relative errors when calculating the k-dependent reflection coefficients $\tilde{r}^p$, $\tilde{r}^s$ for the binary lamellar structure made of (a) 10 and (b) 100 metal-dielectric (gold-alumina) layers.

reflection coefficients $\tilde{r}^p$, $\tilde{r}^s$ for a binary lamellar structure using direct T-matrices multiplication. The example structures in Figs. 16 (a) and (b) are made of 10 and 100 metal-dielectric pairs, respectively. The considered wavelength is 503.3 nm. The dielectric material is alumina ($\varepsilon_{diel} = 3.15$),[88] the metal is gold ($\varepsilon_{met} = -2.69 + i3.45$),[88] substrate and superstrate are glass ($\varepsilon_{sub} = 2.25$) and air($\varepsilon_{sup} = 1$), correspondingly. The thickness of each layer is 15 nm, except for the uppermost layer (alumina), which is 17-nm-thick. Above a certain value of $k_{\parallel}$, calculations using direct multiplication of T-matrices fail (NaN is returned). On one hand, for a fixed value of k, as the number of layers in the structure increases, the errors produced at different T-matrix multiplication steps are accumulated. On the other hand, for a fixed number of layers, larger k-vectors will negatively affect the errors during each T-matrix multiplication, because the numbers being multiplied and added become increasingly mismatched. Therefore, this calculation method restricts both the number of layers in the structure and the maximum value of the k-vector.

In order to avoid the issues of applying the direct T-matrix approach, we use the recursive invariant imbedding method.[80] In this method, in order to obtain $\tilde{r}^{(p,s)}$ instead of directly multiplying the T-matrices, the former are calculated recursively starting from the Fresnel coefficient of the last metamaterial layer. Since the recursive formula is explicit, we bypass the T-matrix multiplication and avoid the errors associated with it. As a result, our calculations converge for arbitrary numbers of layers and k-vector values.

The iterative formula to calculate $\tilde{r}^{(p,s)}(s)$ is the following

$$\tilde{r}_m^{(p,s)}(s) = \frac{r_m^{(p,s)}(s) + \tilde{r}_{m+1}^{(p,s)}(s)e^{i2k_0 s_{\perp,m}(s)\delta_m}}{1 + r_m^{(p,s)}(s)\tilde{r}_{m+1}^{(p,s)}(s)e^{i2k_0 s_{\perp,m}(s)\delta_m}}, \quad (m \in \text{sup}, \overline{1,l}) \tag{10}$$

where $s_{\perp,m} = k_{\perp,m}/k_0 = \left(\varepsilon_m - s^2\right)^{1/2}$, $r_m^{(p,s)}$ are the conventional single-interface Fresnel coefficients

$$r_m^p(s) = \frac{\varepsilon_m s_{\perp,m-1}(s) - \varepsilon_{m-1} s_{\perp,m}(s)}{\varepsilon_m s_{\perp,m-1}(s) + \varepsilon_{m-1} s_{\perp,m}(s)}, r_m^s(s) = \frac{s_{\perp,m-1}(s) - s_{\perp,m}(s)}{s_{\perp,m-1}(s) + s_{\perp,m}(s)}. \tag{11}$$

The iteration starts from the substrate ($\tilde{r}_{l+1}^{(p,s)}(s) = 0$), then we come up with the seeding iteration ($\tilde{r}_l^{(p,s)}(s) = r_l^{(p,s)}(s)$) and continue with the next step:

$$\tilde{r}_{l-1}^{(p,s)}(s) = \frac{r_{l-1}^{(p,s)}(s) + r_l^{(p,s)}(s)e^{i2k_0 s_{\perp,l-1}(s)\delta_{l-1}}}{1 + r_{l-1}^{(p,s)}(s)r_l^{(p,s)}(s)e^{i2k_0 s_{\perp,l-1}(s)\delta_{l-1}}} \tag{12}$$

It should be noted that (12) can be turned into the familiar Drude formula for the reflection coefficient for a single layer of thickness δ [89]:

$$r^{(p,s)}(s) = \frac{r_{1,2}^{(p,s)}(s) + r_{2,3}^{(p,s)}(s)e^{i2k_0 s_\perp(s)\delta}}{1 + r_{1,2}^{(p,s)}(s)r_{2,3}^{(p,s)}(s)e^{i2k_0 s_\perp(s)\delta}} \tag{13}$$

In the formulas (7) and 8, $\tilde{r}^p$ and $\tilde{r}^s$ are expressed as functions of a polar angle θ instead of normalized in-plane wavevector s. The relation between s and θ for the modes propagating in the superstrate medium is $s(\theta) = \varepsilon_{\text{sup}}^{1/2} \sin(\theta)$.

Chapter 7

Linear Optical Properties of Periodic Hybrid Materials at Oblique Incidence: A Numerical Approach

Adam Blake and Maxim Sukharev*

Science and Mathematics Faculty, School of Letters and Sciences, Arizona State University, Mesa, AZ 85212, USA

We discuss optics of periodic hybrid systems — materials comprised of plasmon sustaining structures and molecular aggregates. The emphasis is on the linear regime under strong coupling conditions between molecules and surface plasmon-polaritons. The numerical algorithm used in this work relies on a utilization of two sub-cells to account for propagating numerical errors. With the developed numerical procedure we examine transmission properties of one-dimensional periodic diffraction gratings with sinusoidal spatial modulation. It is shown that such systems exhibit three surface plasmon-polariton resonances that split when the incident radiation is not at the normal incidence. By adding resonant molecular aggregates to such a grating and tilting the incident angle we are able to demonstrate strong coupling between molecules and surface plasmons.

1. Surface Plasmon Polaritons

Surface plasmon-polaritons (SPPs) are oscillations of charge on a metal-dielectric boundary that are able to produce intense, highly localized evanescent fields. SPPs can be described classically, but the name derives from the more accurate quantum mechanical description. A quantum of surface charge oscillation (a plasmon) couples to a quantum of light (a photon) leading to a system that is a hybrid of the two independent entities: a surface plasmon polariton.

Maxwell's equations are applied to the interface between a dielectric and a metal in (Raether), leading to a dispersion relation for SPPs

$$k_x = \frac{\omega}{c}\sqrt{\frac{\varepsilon_1 \varepsilon_2}{\varepsilon_1 + \varepsilon_2}} \qquad (1)$$

ε_2 is the permittivity of the dielectric, ε_1 is the permittivity of the metal, k_x is the in-plane wave vector of the SPP (the x-direction lies in the plane of the interface; see Fig. 1a), ω is the angular frequency of the SPP, and c is the speed of light in vacuum. The limiting value of ω given $\varepsilon_2 = 1.0$ is $\frac{\omega_p}{\sqrt{2}}$, where ω_p is the plasma frequency (the natural frequency of free electron plasma oscillations). The response

---

*maxim.sukharev@asu.edu

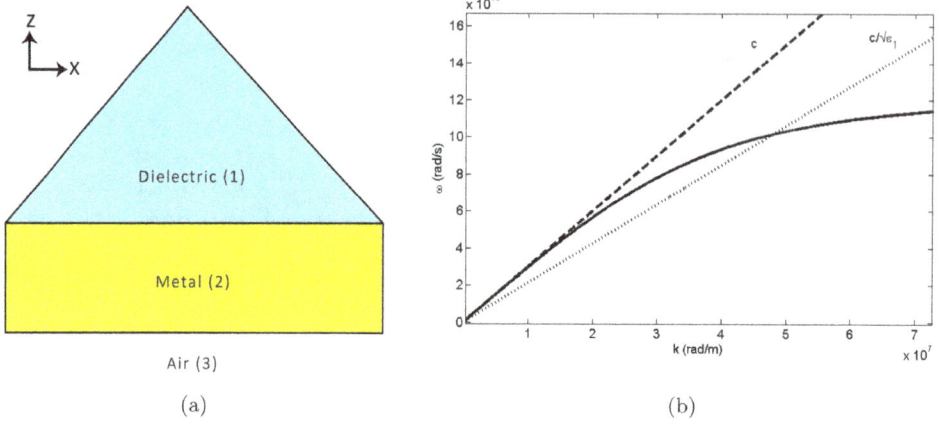

Fig. 1.   Illustration of propagating surface plasmon-polaritons and associated setup. Panel (a): A dielectric prism (medium 1) on top of a metal film (medium 2) with air underneath (medium 3). Panel (b): Dispersion relation for SPPs formed on an air/metal (2/3) interface. The curve follows the light line for air (with slope c) for low values of k but bends downward and asymptotically approaches $\frac{\omega_p}{\sqrt{2}}$ for large values of k. The dispersion line for light in a dielectric (medium 1, slope $\frac{c}{\sqrt{\varepsilon_1}}$) is shown for reference. Not shown is the dispersion relation for SPPs on the metal/dielectric (1/2) interface: this curve lies entirely to the right of the dispersion line for light in the dielectric.

of the metal, obtained from the Drude model (to be discussed in Section 2) is substituted into the above dispersion relation and the curve shown in Fig. 1b is obtained.

SPPs offer means around the diffraction limit because as can be seen from the dispersion curve, the wavelength corresponding to the SPP can be made much smaller than that of light for a given frequency of excitation.

One way to excite and observe SPPs is via the attenuated total reflection (ATR) method discussed extensively in (Raether). In order to excite SPPs, the in-plane incident wavevector must match that of the SPP. This can be achieved by placing a prism in contact with the top of the metal film (Kretschman-Raether configuration) and having air or vacuum in contact with the bottom of the film: see Fig. 1a. The dielectric (medium 1) has $\varepsilon_1 > 1$, the metal (medium 2) has ε_2 represented by the Drude model, and air (medium 3) has $\varepsilon_3 = 1$. According to the dispersion relation for SPPs on the 1/2 interface (not shown in Fig. 1b), light passing from the prism to the metal film cannot have the same in-plane wave vector k_x as an SPP (on the 1/2 interface) of the same frequency because the SPP dispersion curve for the 1/2 interface lies completely to the right of the light line in the dielectric (whose slope is $\frac{c}{\sqrt{\varepsilon_1}}$). Light that is incident upon the 1/2 interface can be made to undergo total internal reflection (TIR), and an evanescent wave whose in-plane wave vector equals $\frac{\omega}{c\sin\theta_0\sqrt{\varepsilon_1}}$ extends into the metal to the 2/3 boundary. This line corresponding to the dispersion relation of the evanescent wave intersects the SPP dispersion relation for the 2/3 boundary: in other words, the in-plane wavevector of the light is increased to match the in-plane wave vector k_x for an SPP on the 2/3 boundary. SPPs are not excited on the 1/2 interface; the k_x is set at the 1/2 interface and that k_x "transfers over" to the dispersion relation at the 2/3 interface. One finds that over a certain range of angles beyond the critical angle, reflection is attenuated. This corresponds to the evanescent wave coupling to the SPP on the airside of the film, which absorbs the radiation and is therefore responsible for the decreased reflection (hence attenuated total reflection).

Besides the ATR method, another means of matching in-plane wavevectors for the purpose of exciting SPPs involves the use of a periodic grating (Abajo, 2007). Periodic arrays of slits (Salomon, Gordon, Prior, Seideman, & Sukharev, 2012) can be used as well as a periodic sinusoidal grating (Sukharev, Sievert, Seideman, & Ketterson, 2009). The grating modifies the in-plane wavevector of the incident light such that it increases to match that of an SPP mode supported by the grating.

2. Coupling of SPPs and Quantum Emitters

Quantum emitters, as far as this discussion is concerned, refers to a material that has one well-defined optical transition between two energy levels. While a given material of course has numerous other possible energy transitions, these are ignored. This assumption is widely used in order to analytically elucidate the time dynamics of light-matter interaction (Allen & Eberly, 1975).

SPPs and quantum emitters with a distinct optical transition are each oscillators in their own right. These oscillators can couple together given that each emits fields that are capable of driving the other. The analytical machinery of coupled harmonic oscillators can therefore be used to analyze the resulting hybrid system. When the detuning is large (i.e. the resonances do not overlap), the response of the system resembles that of the uncoupled oscillators. When the resonance of the SPP is tuned such that it begins to overlap with the profile of the uncoupled emitter resonance, normal mode splitting occurs. The minimum separation of the modes is determined by the coupling strength: the greater the coupling strength, the larger the minimum separation of the modes. The resulting system is neither that of an SPP nor that of an emitter; it now demonstrates hybrid modes that are a mix of the individual uncoupled modes.

This leads us to recognize two regimes: *weak coupling* and *strong coupling*. Two ways of defining these regimes will be discussed here. A very general definition is given in (P. Torma, 2014): the strong coupling regime is achieved if the normal mode splitting is great enough (i.e. is greater than the individual oscillators' linewidths) to be observed experimentally. Avoided crossing, in which the original SPP resonance is split into an upper polariton and a lower polariton, is therefore a signature of the strong coupling regime. In the *weak coupling* regime, the separation of the modes is less than the linewidths of the emitter and SPP and therefore is not experimentally observable.

The coupling strength can be influenced by the dipole moment of the emitters, the density of emitters, and the effective volume (originally associated with the volume of a reflecting cavity in which emitters were placed) of the fields. A specific description of the coupling strength between SPP modes and emitters is given in (Parinda Vasa, 2013)

$$\hbar\Omega_R = \sqrt{N_x}\boldsymbol{\mu}_x \cdot \mathbf{E}_v \tag{2}$$

where $\boldsymbol{E}_v$ is the SPP vacuum field, Ω_R is the Rabi splitting, μ_x is the transition dipole moment and N_x is the number of excitonic (emitter) excitations. A calculation in (P. Torma, 2014), which analyzes the SPP dispersion relation and solves for the normal mode frequencies, shows that the splitting of normal mode frequencies is proportional to $\sqrt{N_x}$. The strong coupling regime is defined in (P. Torma, 2014) as when the normal mode splitting ($\Omega_{NMS} \approx \Omega_R$) exceeds all of the damping rates (and therefore linewidths) in the system.

Prior to coupling emitters to SPPs, emitters were coupled to cavities. Bellessa *et al.* (J. Bellessa, 2004) coupled J-aggregates to a cavity and observed a Rabi splitting of about 300 meV. Pockrand *et al.* (I. Pockrand, 1982) demonstrated a splitting of the ATR minimum due to coupling between dye molecules and SPPs on a metal film and Bellessa *et al.* were able to achieve Rabi splitting of the order of that achieved in cavities. Because SPPs have very localized fields corresponding to small effective volumes, strong coupling corresponds to the values observed in small microcavities. Coupling emitters to SPPs is not a perfect scheme, however as SPP modes are dissipative (P. Torma, 2014).

When two-level emitters are added to a metal grating, the emitters can couple via electromagnetic fields to the SPPs of the grating. When the emitter resonance overlaps with the SPP resonance, the original SPP resonance is observed to undergo Rabi splitting as discussed previously. This has been observed in the case of slits with emitters deposited on top of a slit array. The SPP resonance splits into an upper and lower polariton, the energies of which are given by (Salomon, Gordon, Prior, Seideman, & Sukharev, 2012)

$$E_{u,l}(k) = \frac{1}{2}\{[E_{pl}(k) + E_m] \pm \sqrt{4\Delta^2 + (E_{pl}(k) + E_m)^2}\} \tag{3}$$

$E_{u,l}$ is the energy of either the upper or lower polariton, E_{pl} is the energy of the SPP mode, E_m is the energy of the emitter resonance, and Δ is the Rabi splitting value (which indicates the coupling strength). The above expression for $E_{u,l}(k)$ is derived by diagonalizing a 2×2 Hamiltonian with coupling energy Δ

$$H = \begin{bmatrix} E_m & \Delta \\ \Delta & E_{pl}(k) \end{bmatrix} \tag{4}$$

It must be emphasized that this is not valid at higher densities due to stronger interactions between emitters (Salomon, Gordon, Prior, Seideman, & Sukharev, 2012).

3. Modeling Plasmonic and Hybrid Materials

In this section we discuss numerical methods used to investigate the interaction of light with nanoscale plasmonic structures. The structures studied here are either silver or hybrid (silver plus quantum emitters). Silver is desirable for plasmonic applications because its parameters are well suited to SPP features in the visible range, e.g. no interband transitions in this region (Le Ru & Etchegoin, 2009). Maxwell's equations are used to describe the propagation of electromagnetic waves

$$\mu_0 \frac{\partial \vec{H}}{\partial t} = -\nabla \times \vec{E} \tag{5a}$$

$$\varepsilon_0 \frac{\partial \vec{E}}{\partial t} = \nabla \times \vec{H} - \vec{J} \tag{5b}$$

μ_0 is the permeability of free space, ε_0 is the permittivity of free space, $\vec{E}$ is the electric field, $\vec{H}$ is the magnetic field and $\vec{J}$ is the current density. The Drude model (Le Ru & Etchegoin, 2009) is used to describe the dispersion of the metal. An electron that is part of the free electron plasma is treated as a damped harmonic oscillator with resonant frequency ω_0. At $\omega_0 = 0$ we obtain the frequency-dependent dielectric constant

$$\varepsilon(\omega) = \varepsilon_r - \frac{\omega_p^2}{\omega^2 - i\gamma\omega} \tag{6}$$

ω_p is the plasma frequency (i.e. the natural frequency of the free electron plasma), γ is the damping that corresponds to collisions between the free electrons and the crystal (Le Ru & Etchegoin, 2009), and ε_r is the high-frequency limit of the dielectric function (Sukharev M., 2012). For silver, these values are 1.76×10^{16} rad/s and 3.08×10^{14} rad/s, and 8.26 respectively (Gray & Kupka, 2003). In metal, the field values (obtained from the Maxwell equations) are used to update the currents (Gray & Kupka, 2003)

$$\frac{\partial \vec{J}}{\partial t} = -\gamma\vec{J} + \varepsilon_0\omega_p^2\vec{E} \tag{7}$$

The Liouville-von Neumann equation is used to describe the time dependence of an atom interacting with an electromagnetic field

$$i\hbar\frac{d\hat{\rho}}{dt} = [\hat{H}, \hat{\rho}] - i\hbar\hat{\Gamma}\hat{\rho} \tag{8}$$

$\hat{\rho}$ is the single-atom density matrix, $\hat{\Gamma}$ describes relaxation processes such as spontaneous emission and radiationless decay, and $\hat{H}$ describes an atom with dipole

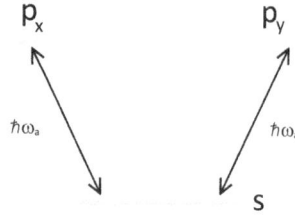

Fig. 2. Energy level diagram for emitters. The ground state is an s-type state and the excited state consists of two p-type states.

moment $\hat{\mu}$ interacting with an electric field

$$\hat{H} = \hat{H}_0 - \hat{\mu} \cdot \vec{E}(t) \tag{9}$$

The diagonal elements of the density matrix describe the population of emitters in each state and the off-diagonal elements describe how the induced transition dipole moment depends on time.

In regions containing emitters, the fields are used to update the density matrix elements via the Liouville-von Neumann equation. The expectation value of the dipole moment operator is obtained by taking using

$$\langle \hat{\mu} \rangle = Tr(\hat{\rho}\hat{\mu}). \tag{10}$$

The updated expectation value of the dipole moment is used to calculate the volume polarization and then the polarization current of the emitters, which is subsequently inserted into the Ampere law

$$\vec{P} = n_a \langle \hat{\mu} \rangle \tag{11a}$$

$$\vec{J}_p = \frac{\partial \vec{P}}{\partial t} \tag{11b}$$

n_a is the volume density of emitters, $\vec{P}$ is the polarization per unit volume, and $\vec{J}_p$ is the polarization current.

All of the results presented herein use a two-dimensional grid in which the fields E_x, E_y and H_z are evaluated in the x-y plane and the structure is taken to be infinitely long in the z-direction. The quantum emitters have three energy levels consisting of an s-type ground state and two degenerate excited p-type states (p_x and p_y) in order to account for polarization of the emitters along both x- and y-directions; see Fig. 2.

4. The Yee Cell and Introduction of the Incident Wave

Maxwell's equations are of course continuous partial differential equations with a limited number of analytical solutions. One cannot hope to find analytical solutions that account for arbitrary geometries with dispersive structures. We therefore discretize the system by evaluating derivatives at finite distances and time steps apart. For instance, a spatial derivative of the y-component of the E-field becomes

$$\frac{\partial E_y}{\partial x} \rightarrow \frac{E_y(i+1,j) - E_y(i,j)}{dx} \tag{12}$$

The indices i and j indicate grid points at which the fields are evaluated. Given that this is an approximation to the correct physics, one must take steps to ensure the accuracy of the results. The most reassuring procedure is to decrease the spatial resolution incrementally and verify convergence. When the results no longer change with decreasing spatial step size, the simulation is converged.

Several methods exist for discretizing and propagating Maxwell's equations. The method used in this research is the *finite difference time domain* (FDTD) method (Taflove & Hagness, 2005), introduced by Yee in 1966 (Yee, 1966). The mesh consists of two shifted grids: one for the electric field and one for the magnetic field. In three dimensions, each E-field vector is surrounded by four H-field vectors and vice versa; this makes evaluation of $\nabla \times \vec{E}$ and $\nabla \times \vec{H}$ very straightforward. In two-dimensional geometry considered here each H_z vector is calculated using four surrounding E-field vectors (two E_x and two E_y vectors) whereas each E-field vector is calculated using only *two* adjacent H_z values as the H_x and H_y values are zero everywhere, which must be so because there is no variation of any of the fields along z. The H-field vectors point along z whereas the E-field vector has x and y components. This corresponds to p-polarization (E-field in the plane of incidence); as s-polarization (E-field perpendicular to the plane of incidence) cannot excite SPPs in the periodic system depicted in Fig. 2.

A total-field/scattered-field (TFSF) formulation is used to introduce a plane wave into the grid. This is consistent with the illumination of a small sample with a beam that covers all of it as well as the periodic nature of the structure. A lookup table (Taflove & Hagness, 2005) is used to project the values of E and H onto the TFSF boundary generating an oblique wave. A special periodic boundary is applied to the left and right of the grid and an absorbing boundary called a *convolutional perfectly matched layers* (CPML) terminates the top and bottom of the grid.

5. Oblique Incidence

It is very useful to be able to tilt the incoming wave. Infinite, repeating structures are modeled and therefore periodic boundary conditions (PBCs) are imposed on a single unit cell. When PBCs are combined with obliquely incident waves, the situation becomes much more involved than one might expect.

For normal incidence, the PBC can be applied by simply using fields from one boundary to calculate fields at the other. For oblique incidence, however, one can see from Fig. 3 that the fields on the right boundary are equal to those on the left boundary *at a time t_{delay} in the past*. Applying the PBC at the right

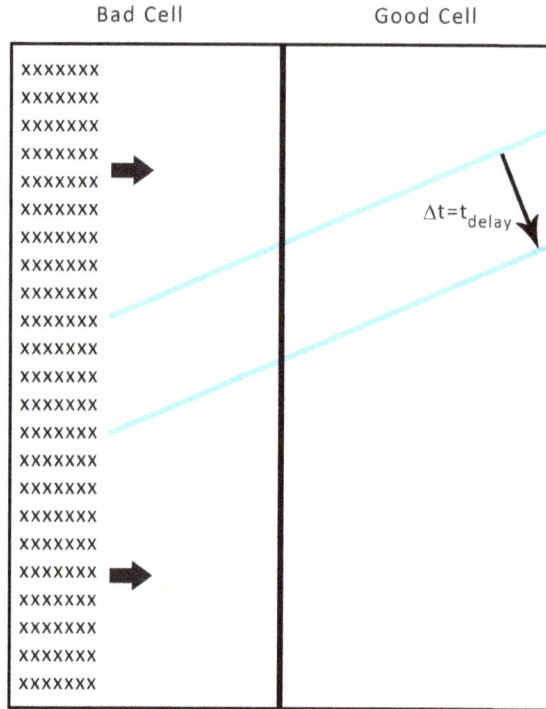

Fig. 3.   Error propagation and delay time in applying periodic boundary conditions. The error propagates from the left boundary of the bad cell at the speed of light in that medium. The periodic boundary condition is enforced on the left and right boundaries of the good cell.

boundary is straightforward: the field values at the left boundary are recorded and applied at a time t_{delay} later. Matching the left boundary to the right, however, requires that we know the fields at a time t_{delay} in *the future*, which, of course, is not possible. Several methods (spectral FDTD (Aminian & Rahmat-Samii, 2006), sin/cos (Harms, Mittra, & Ko, 1994), multiple unit cells (Taflove & Hagness, 2005), etc.) have been developed to reconcile this issue, though many were cumbersome (sin/cos) or came with excessive computing times (multiple unit cell).

A method developed by Todd Lee is both robust and straightforward (Lee & Smith, 2005). The grid is broken up into two unit cells as shown in Fig. 3. The fields in the good cell are always accurate whereas the fields in the bad cell are subject to an error that fills the cell over time. Rather than enforce a PBC or an absorbing boundary on the left side of the bad cell, it is simply left to propagate with a value of zero to the left of it. This causes an error that propagates from the boundary at the speed of light, but the critical point is that *the wave reaches the right unit cell before the error does for* $\theta_{inc} < 90°$. The wave propagates into the rightmost unit cell and is allowed to continue propagating until the error from the left boundary reaches the rightmost unit cell. At that point, the values in the right unit cell (which are still valid) are copied into the left unit cell, the right unit cell gets values pasted into it from a previous time step, and the process is allowed to run until the error again reaches the right unit cell.

We tested this method thoroughly. The electric field updates properly and is absorbed by the CPML boundaries, as expected. The simulation has been stable for every geometry modeled and every angle of wave (up to and excluding 90°).

To test the method, we implement a dispersionless dielectric ($n = 1.414$) on the lower half of the grid and vacuum ($n = 1.0$) on the upper half. The reflection (R) and transmission (T) values, which are derived from the Fresnel equations, were compared with analytical values given by Griffiths in (Griffiths, 1999)

$$\alpha = \frac{\sqrt{1 - \left[\left(\frac{n_1}{n_2}\right)\sin\theta_i\right]^2}}{\cos\theta_i} \tag{13a}$$

$$\beta = \frac{n_2}{n_1} \tag{13b}$$

$$T = \alpha\beta \left(\frac{2}{\alpha + \beta}\right)^2 \tag{13c}$$

$$R = \left(\frac{\alpha - \beta}{\alpha + \beta}\right)^2 \tag{13d}$$

The results are shown in Fig. 4. The simulations were run with a spatial step of 1.0 nm and for 300,000 time steps, which corresponds to a total propagation time of 500 fs. The agreement between theory and the simulations is excellent.

The transmittance and reflectance were found to be identical at different frequencies, which is expected given that this setup is dispersionless. Also, these

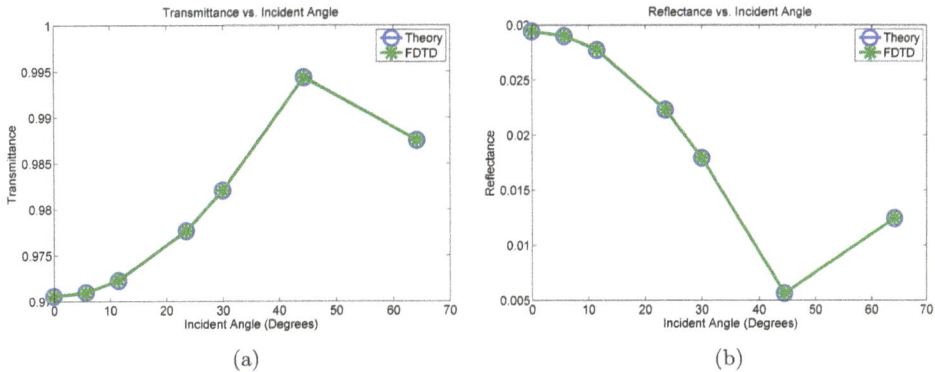

(a)

(b)

Fig. 4. Comparison of analytical transmittance and reflectance showing accurate results obtained from FDTD. Panel (a): Transmittance. Panel (b): Reflectance.

values remained constant as the grid was widened and/or lengthened. Additionally, the angles of reflection and refraction were found to behave as expected.

These comparisons with well-established but non-trivial theory give confidence that the code is working correctly.

6. Problems with Critical Angle and Total Internal Reflection

Unfortunately, problems arise in using this copy/shift routine when one tries to simulate total internal reflection (TIR). For this discussion, the top half of the grid (region 1) will have n_1 and the bottom half (region 2) will have n_2 where $n_1 > n_2$. As stated previously, the copy/shift routine assumes that an error travels from the left boundary of the bad cell to the left boundary of the good cell in a time t_{error}, which must be greater than the time t_{delay} it takes for the periodic time to be reached. As will be shown here, when the wave is launched from n_1 to n_2 at the critical angle or greater, t_{delay} is equal to or greater than t_{error}. In other words, once the time t_{delay} is reached, the error has already propagated into the good unit cell. The simulation must run for at least t_{delay} for the copying and shifting to take place and will therefore not work under these conditions.

Specifically, t_{error} is the *least* amount of time it takes for the error to propagate from the edge of the bad cell up to the good cell. This time can vary if different regions of the grid have different indices of refraction. The entire grid is copied and shifted at the same time, and this happens when the error reaches the boundary of the good cell. In the region with the lower index of refraction, we have $v_2 = c/n_2$ which is greater than $v_1 = c/n_1$. Because the error travels faster in region 2, the time that it takes for the error to reach the boundary in region 2 is used to calculate t_{error}:

$$t_{error} = \frac{unit\ cell\ width}{v_2} = \frac{unit\ cell\ width}{\left(\frac{c}{n_2}\right)} = \frac{n_2 \times unit\ cell\ width}{c} \tag{14}$$

The simulation must run for at least the region 1 delay time t_{delay} so that the copy/shift routine can take place. First, it is shown that the delay time in region 1 is equal to that of region 2. The delay time in region 1 is

$$t_{delay\ 1} = \frac{unit\ cell\ width \times sin\theta_1}{v_1} = \frac{n_1 \times (unit\ cell\ width \times sin\theta_1)}{c} \tag{15}$$

And the delay time in region 2 is

$$t_{delay\ 2} = \frac{unit\ cell\ width \times sin\theta_2}{v_2} = \frac{n_2 \times (unit\ cell\ width \times sin\theta_2)}{c} \tag{16}$$

By Snell's Law, $n_1 sin\theta_1 = n_2 sin\theta_2$ so the delay times are equal. If they were not equal, this scheme would not work!

If the incident wave is launched from region 1 at the critical angle θ_c (so that by Snell's Law, $\sin\theta_c = n_2/n_1$), then the delay time becomes:

$$t_{delay\,1} = \frac{n_1 \times (unit\ cell\ width \times \sin\theta_c)}{c} = \frac{n_1 \times (unit\ cell\ width \times \frac{n_2}{n_1})}{c}$$

$$= \frac{n_2 \times unit\ cell\ width}{c} = t_{error} \qquad (17)$$

And of course, any angle greater than the critical angle gives an even longer delay time. So any angle below the critical angle works fine and any angle equal to or greater than the critical angle will not work.

If one wishes to simulate a flat film, the film can simply be terminated instead of having the PBC applied to it. Unwanted features that depend on the length of the film will be present and these can be minimized (and moved to longer wavelengths) by increasing the length of the film, leaving accurate results in the frequency region of interest.

7. Sinusoidal Periodic Gratings with and without Emitters

Here we study optical properties of spatially modulated thin metal films. A pulse of the following form is launched at the film

$$E(t) = E_0 \sin^2\left(\frac{\pi t}{\tau}\right)\cos(\omega t) \qquad (18)$$

This form makes its value zero at t = 0 and t = τ. Additionally, its derivative is zero at those times which is important for numerical convergence.

In order to obtain the spectral response of the grating, it is of course possible to run a series of simulations with a continuous (CW) incident wave, each at a different energy. It is, however, much more efficient to launch a short pulse and analyze the impulse response of the grating.

Convergence is checked by lengthening the grid and ensuring that the results do not change. The features are unchanged on lengthening the grid from 1800 nm to 2000 nm, which suggests that they are converged. Next, using the 1800 nm grid, the number of time steps is multiplied by 10. Again the features are unchanged.

In these runs, digital Fourier transform (DFT) (Taflove & Hagness, 2005) is used to integrate the Poynting vector across the unit cell. Previously, single point detection was used and the entire region surrounding the Wood's anomaly was unconverged, possibly due to reflections from CPML. Switching from single point detection to DFT eliminated the numerical disobedience of the Wood's anomaly.

Avrutsky *et al.* examine plasmon-assisted tunneling of light through sinusoidal (corrugated) gratings in (Avrutsky, Zhao, & Kochergin, 2000). The phenomenon of extraordinary optical transmission (EOT) is well known in quantum optics and was first seen by shining light through a thin film with a periodic array of holes and observing that the fraction of light transmitted was greater than the fraction of

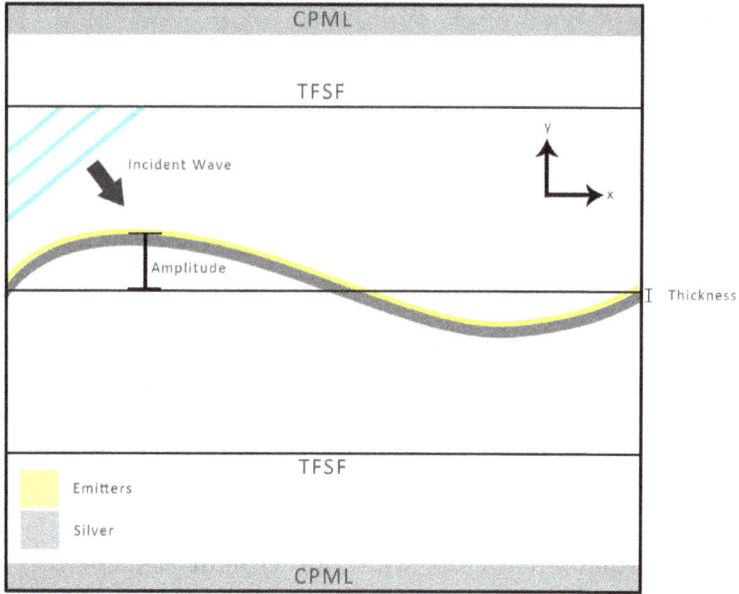

Fig. 5.   Schematic of silver/emitter hybrid system.

the film covered in holes. It is shown in (Avrutsky, Zhao, & Kochergin, 2000) that a sinusoidal grating is not only capable of producing this effect, but able to do so with a transmission peak that is stronger and more narrow than that of a periodic array of holes.

A sinusoidal grating of the form shown in Fig. 5 is studied. Such sinusoidal gratings are investigated by Sukharev *et al.* in (Sukharev, Sievert, Seideman, & Ketterson, 2009). An in-phase grating with single-ended excitation, a period of 410 nm, an amplitude of 28.8 nm and thickness 20 nm is simulated in (Sukharev, Sievert, Seideman, & Ketterson, 2009), where it is shown that the symmetry of the grating is such that the quantity T+R (rather than T or R separately) can be minimized for a single adjustable parameter, namely the grating amplitude. A graph of T + R from (Sukharev, Sievert, Seideman, & Ketterson, 2009) reveals a gently curved region at low wavelengths and SPP modes near 540 nm and 620 nm. Additionally, a Wood's anomaly is present at 532 nm (corresponding to a grating period of 400 nm and a refractive index of 1.33). This is a physical phenomenon that occurs when incoming and outgoing waves interact with waves that diffract tangent to the grating (Hessel & Oliner, 1965).

A sinusoidal grating with period 400 nm, amplitude 70 nm, and thickness 20 nm was run and the results are shown in Fig. 6. This spectrum, which compares well to that in (Sukharev, Sievert, Seideman, & Ketterson, 2009) is taken as a starting point for analyzing sinusoidal gratings. In addition to the two SPP modes seen in Fig. 6, a Wood's Anomaly is present at 3.1 eV. The energy of the Wood's anomaly

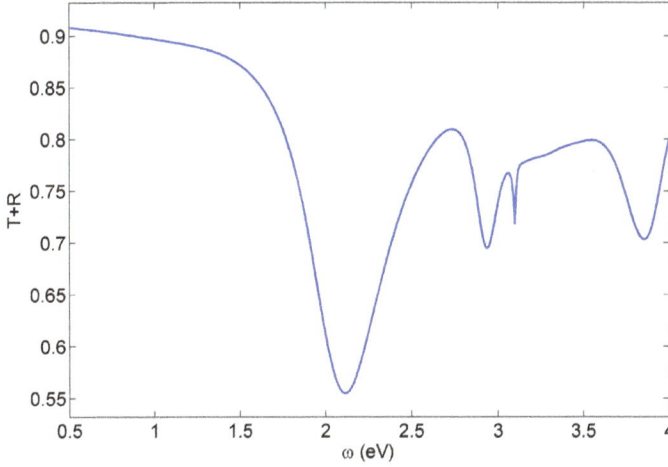

Fig. 6. The sum of transmission and reflection as a function of incident frequency for a sinusoidal grating. Several features are seen: a gently curving region at low energies, a Wood's anomaly at 3.1 eV corresponding to n = 1.0 and a period of 400 nm, a broad resonance at lower energy and a narrow resonance at a higher energy. The simulation was run for 1×10^6 time steps (corresponding to a total time of 1.67×10^{-12} s. The spatial step size is 1.0 nm.

is as expected for a 400 nm grating with a refractive index of 1.0, and it shifts to higher energies when the grating period is shortened.

The period, amplitude and thickness of the grating were adjusted. It was found that the previously mentioned period of 400 nm, amplitude of 70 nm and thickness of 20 nm give well-defined features and these parameters were therefore used for the simulations that followed.

As the incident wave is tilted it is expected that excitation of higher order Brillouin zones will cause both SPP features to split and that is precisely what is seen. A progression of T + R from 0 degrees to 10 degrees is shown in Fig. 7a and a contour plot from 0 to 40 degrees is shown in Fig. 7b.

When oscillators are coupled (in this case, the oscillators are the SPPs and the emitters), avoided crossing is expected in which the energies of the upper and lower polaritons approach each other and reach a minimum separation value (P. Torma, 2014). In this case, the emitter resonance is placed at 2.7 eV and the angle is swept from 0 to 10 degrees. Tilting the incident wave causes the SPP resonance to split. This split resonance shifts as the incident wave is tilted, effectively sweeping it across the emitter resonance. Coupling between the split SPP resonance and the emitter resonance causes Rabi splitting into upper and lower polaritons and avoided crossing is clearly demonstrated in Fig. 8.

In other simulations, the value of the emitter resonance is simply changed and the simulation is run again. It must be emphasized that this is a different way of observing avoided crossing: the incident wave is tilted such that an SPP resonance

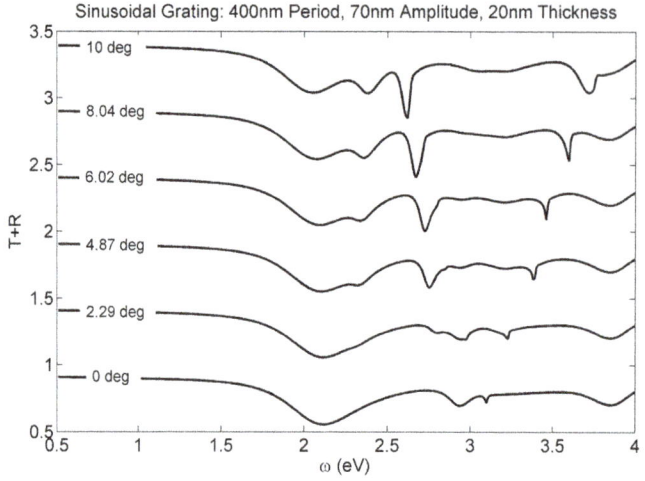

(a)

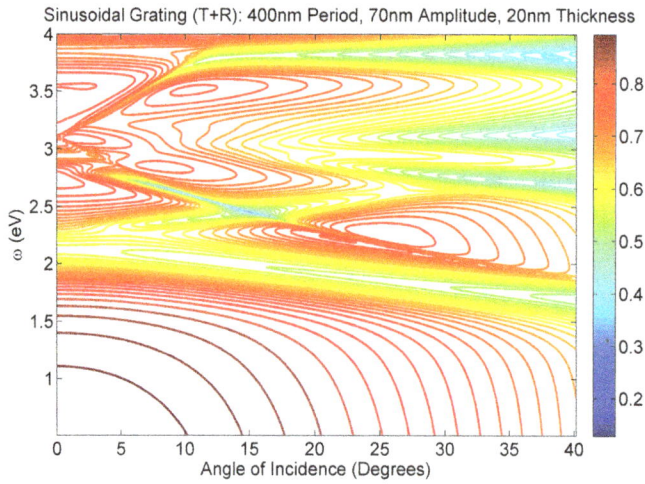

(b)

Fig. 7. Spectra of the sinusoidal grating as the incident angle is varied. Amplitude is 70 nm, thickness is 20 nm and the period is 400 nm. Panel (a): The lower energy resonance (near 2.1 eV) begins to split at 2.29 degrees and this splitting clearly increases as the angle is increased. The higher energy resonance (near 2.9 eV) appears to have split at 2.29 degrees and the splitting becomes more pronounced as the incident angle is increased. At 10 degree incidence the higher energy resonance is beginning to overlap with the lower energy resonance. The Wood's anomaly moves to higher energies as expected. Panel (b): Contour plot showing the variation in the spectrum as incident angle is varied. The higher energy SPP mode crosses the (split) lower energy mode near an incident angle of 15 degrees. An observable avoided crossing might be expected if the coupling between these modes were to be increased.

Sinusoidal Grating: 400nm Period, 70nm Amplitude, 20nm Thickness

(a)

(b)

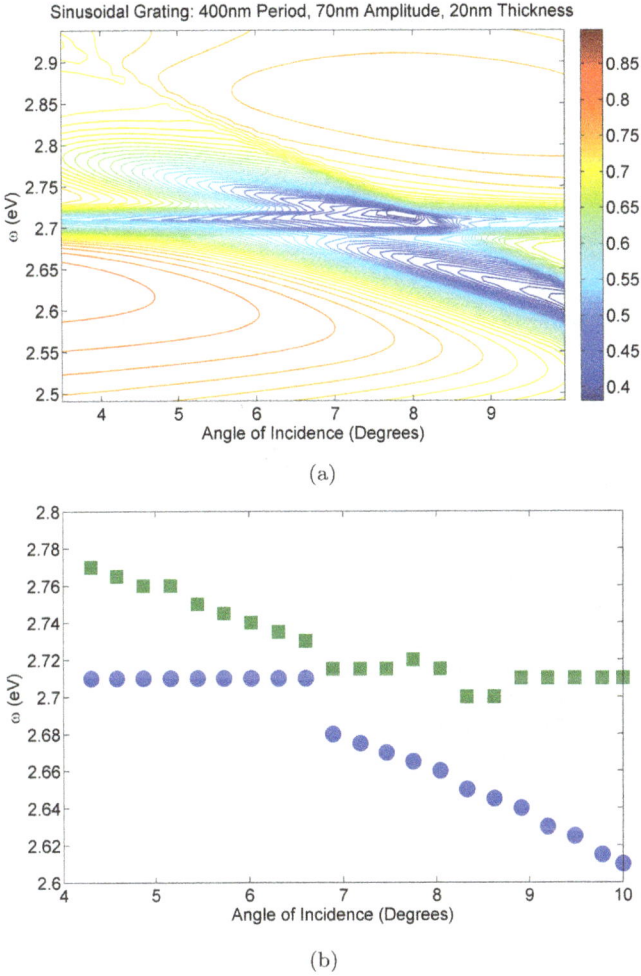

Fig. 8. Avoided crossing is a signature of the strong coupling regime and is demonstrated here. Panel (a): avoided crossing is observed when the incident wave is tilted, causing an SPP resonance to split and sweep past the emitter resonance. The quantity T + R is indicated by the contours. Panel (b): upper polariton (squares) and lower polariton (circles) energies as a function of incident angle. The minimum energy separation is on the order of 20 meV.

splits and as the incident wave is tilted further, this split resonance sweeps across the emitter resonance.

It is pointed out in (P. Torma, 2014) that "one of the attractions of the combination of having excitons as one of the oscillators and plasmon modes as the other is the very extensive control we have over the plasmon modes supported by metallic nanostructures". The geometry of the metallic structure can be adjusted to tune the SPP features as desired. In order to investigate strong coupling, one can prepare nanoscale slit arrays of varying periods with emitters deposited on top.

This effectively "sweeps" the SPP resonance through the emitter resonance. The simulations performed in this work demonstrate the possibility of sweeping the SPP resonance much more efficiently: one can tilt the incident wave, causing an SPP resonance to split. As the incident wave is tilted, the split SPP feature sweeps past the emitter resonance, eliminating the need to use multiple nanostructures with different parameters.

References

1. H. Raether, *Surface Plasmons on Smooth and Rough Surfaces and on Gratings* (Springer, 1988) Springer Tracts in Modern Physics.
2. F. J. Garcia de Abajo, Rev. Mod. Phys. **79**, 1267 (2007).
3. A. Salomon, R. Gordon, Y. Prior, T. Seideman and M. Sukharev, Phys. Rev. Lett. **109**, 073002 (2012).
4. M. Sukharev, P.R. Sievert, T. Seideman and J.B. Ketterson, J. Chem. Phys. **131**, 034708 (2009).
5. L. Allen and J. H. Eberly, *Optical Resonance and Two-Level Atoms* (John Wiley & Sons, Inc., 1975).
6. P. Torma and W. L. Barnes, arXiv:1405.1661 (2014).
7. P. Vasa, W. Wang, R. Pomraenke, M. Lammers, M. Maiuri, C. Manzoni, G. Cerullo, C. Lienau, Nat. Photonics **7**, 128 (2013).
8. J. Bellessa, C. Bonnand and J.C. Plenet, Phys. Rev. Lett. **93**, 036404 (2004).
9. I. Pockrand, A. Brillante, D. Mobius, J. Chem. Phys. **77**, 6289 (1982).
10. E. Le Ru and P. Etchegoin, *Principles of Surface Enhanced Raman Spectroscopy and Related Plasmonic Effects* (Elsevier 2009).
11. M. Sukharev, *Plasmonics: Computational Approach* in *Mathematical Optics: Classical, Quantum and Computational Methods* (2012).
12. S. K. Gray and T. Kupka, Phys. Rev. B **68**, 045415 (2003).
13. A. Taflove and S. C. Hagness, *Computational Electrodynamics: The Finite-Difference Time-Domain Method*, 3rd ed. (Artech House, Boston, 2005).
14. K. Yee, IEEE T. Antenn. Propag. **14**, 302 (1966).
15. A. Aminian and Y. Rahmat-Samji, IEEE T. Antenn. Propag. **54**, 1818 (2006).
16. P. Harms, R. Mittra, W. Ko, IEEE T. Antenn. Propag. **9**, 1317 (1994).
17. R. T. Lee and G. S. Smith, Microw. Opt. Techn. Let. **45**, 472 (2005).
18. D. J. Griffiths, *Introduction to Electrodynamics* (Prentice Hall, 1999).
19. I. Avrutsky, Y. Zhao and V. Kochergin, Opt. Lett. **25**, 595 (2000).
20. A. Hessel, A. A. Oliner, Appl. Optics **4**, 1275 (1965).

IV
Fundamental Physics

Chapter 8

An Introduction to Boson-Sampling

Bryan T. Gard

Hearne Institute for Theoretical Physics and Department of Physics & Astronomy,
Louisiana State University, Baton Rouge, LA 70803, USA

Keith R. Motes

Centre for Engineered Quantum Systems, Department of Physics and Astronomy,
Macquarie University, Sydney NSW 2113, Australia

Jonathan P. Olson

Hearne Institute for Theoretical Physics and Department of Physics & Astronomy,
Louisiana State University, Baton Rouge, LA 70803, USA

Peter P. Rohde

Centre for Engineered Quantum Systems, Department of Physics and Astronomy,
Macquarie University, Sydney NSW 2113, Australia
Centre for Quantum Computation and Intelligent Systems (QCIS),
Faculty of Engineering and Information Technology, University of Technology,
Sydney, NSW 2007, Australia

Jonathan P. Dowling

Hearne Institute for Theoretical Physics and Department of Physics & Astronomy,
Louisiana State University, Baton Rouge, LA 70803, USA

Boson-sampling is a simplified model for quantum computing that may hold the key to implementing the first ever post-classical quantum computer. Boson-sampling is a non-universal quantum computer that is significantly more straightforward to build than any universal quantum computer proposed so far. We begin this chapter by motivating boson-sampling and discussing the history of linear optics quantum computing. We then summarize the boson-sampling formalism, discuss what a sampling problem is, explain why boson-sampling is easier than linear optics quantum computing, and discuss the Extended Church-Turing thesis. Next, sampling with other classes of quantum optical states is analyzed. Finally, we discuss the feasibility of building a boson-sampling device using existing technology.

1. Introduction

1.1. *Motivation for Linear Optics Quantum Computing and Boson-sampling*

To-date, many different physical implementations and models for quantum computing have been proposed. These implementations include atom and ion trap quantum computing, superconducting qubits, nuclear magnetic resonance, quantum dots, nuclear spin, and optical quantum computing. When describing an implementation, one can use various models of computation. These include the gate model [1], cluster (or graph) states [2, 3], topological, adiabatic [4], quantum random walks [5], quantum Turing machines [6], permutational [7], and the one-clean qubit models [8]. The most familiar and intuitive model is the gate model as it is most analogous to the classical circuit model of computation. We use this gate model in order to describe linear optics quantum computing (LOQC) and eventually a special purpose implementation, boson-sampling.

As stated, there exist many choices of experimental implementations and computational models. But which model is likely to yield the first, scalable, demonstrated quantum computer? The answer is likely not just one, but a composite of different choices for the various required components of a quantum computer. For this discussion, we will focus on LOQC for the allure of simple implementation of boson-sampling.

There is a long history in the physics community of investigations into the use of linear interferometers, particularly linear optics interferometers, as a type of quantum information processor. In most of the early research, the consensus was that a linear optics interferometer (alone) could not be used to make a universal quantum computer regardless of the input states. For example, in 1993 (a year before Shor's discovery of the now-famous quantum factoring algorithm) there appeared a paper by Černý that proposed using a linear interferometer to solve **NP**-complete problems in polynomial time, but the scheme suffered from an exponential overhead in energy [9]. Similarly, in 1996, Clauser & Dowling showed that a linear optics Talbot interferometer could be used to factor integers in polynomial time but with either an exponential overhead in energy or physical size [10]. Also in 1996, Cerf, Adami & Kwiat showed how to construct a programmable linear optics interferometer that could perform any universal logic gate with single photon inputs. This scheme too suffered an exponential overhead in spatial dimension. In 2002, Bartlett *et al.* showed that even with quadratic nonlinearities any interferometer that processes only Gaussian state inputs can be efficiently simulated classically. This comprised a continuous variable analog of the Gottesman-Knill theorem for discrete variables in the ordinary circuit quantum computation model [11].

This litany of no-go theorems led to the widespread belief that linear interferometry alone could not provide a path to universal quantum computation and that, as a corollary, all passive linear optics interferometers were thought to be efficiently simulatable on a classical computer. For completeness we will introduce the LOQC approach of Knill, Laflamme & Milburn (KLM) [12, 13] in the following section, but the remaining focus of this chapter is instead on boson-sampling. This is because, for the KLM scheme's set of universal gates, one requires intermediate measurements on ancilla photons with a feed-forward mechanism that imparts a type of effective Kerr nonlinearity on the system [14]. We explicitly only discuss linear optics implementations due to the fact that present-day nonlinear Kerr media exhibit very poor efficiency [1] and very weak non-linearities.

It came as a surprise to many in the quantum optics community when Aaronson & Arkhipov (AA) argued that, in general, the operation of a passive 'linear' optics interferometer with Fock state inputs cannot likely be simulated by a classical computer [15]. In particular, if one samples the output distribution utilizing photon-number discriminating detectors, one cannot predict the outcome with a classical computer without an exponential overhead in time or resources. This has become known as the boson-sampling problem [16].

Gard, *et al.* independently reached the same conclusion in the context of trying to simulate multi-photon coincidence counts in the output of a linear optics implementation of a quantum random walk with multi-photon walkers [17]. In follow up papers, Gard *et al.* [18], as well as Motes *et al.* [19], argued from a physical (as opposed to a computational complexity) point of view that this difficulty to simulate such interferometers arose from two necessary requirements: (1) The photons 'interact' at the beamsplitters via a Hong-Ou-Mandel effect that gives rise to an exponentially large Hilbert space in the number-path degrees of freedom (ruling out a brute force simulation of the interferometer); and (2) That the simulation of the interferometer is tied to computing the permanent of a large matrix with complex entries, a problem known to be in the complexity class #**P**-hard. This complexity class is not only thought to be intractable for classical computers, but even for universal quantum computers [20]. While the first requirement is a necessary condition, it is not by itself sufficient to imply an intractable simulation. As a counterexample, the Gottesman-Knill theorem gives examples of quantum circuits where gates in the Clifford algebra class generate exponentially large amounts of qubit entanglement but are nevertheless classically simulatable. Since there are sometimes shortcuts through the exponential Hilbert space, by tying the simulation to the problem of permanent computation we expect it is very unlikely that any such shortcuts exist. In contrast, the equivalent sampling problem with fermions rather than bosons is known to be classically easy to simulate, as the problem relates to matrix determinants rather than permanents, which are known to be in the complexity class **P**, which is known to be efficiently classically simulatable. [18].

Since the first appearance of the AA paper in 2010 there has been an explosion of research into the field of boson-sampling. As we will discuss below, there have been a number of experiments utilizing three photons from spontaneous parametric down conversion (SPDC) sources [21–26] (although the validity of these experiments are under debate as not all three photons were heralded single photons [27]). The experimental work has continued in parallel to a number of theoretical developments considering the effects of loss, noise, decoherence, non-Fock inputs, scalability of SPDC sources, ion-trap implementations, and so forth [19, 28–32]. We will discuss and summarize these results and more in the sections below.

Why is boson-sampling getting so much attention? What is it good for? Boson-sampling is an example of a computationally complex mathematical problem that cannot be efficiently simulated on a classical computer, but with significantly reduced experimental requirements compared to universal quantum computing schemes. It is the first interesting example of a realistic post-classical computing paradigm, though the true scope and power of such machines is not yet fully understood.

Is a passive linear interferometer good for anything other than simply implementing boson-sampling? The problem itself, other than being a computational curiosity, has no known algorithmic applications or killer apps such as integer factorization. However, it was recently shown that the essence of boson sampling can be applied to quantum metrology [33]. Prior to Shor's algorithm, the same question was asked of a universal quantum computer. Feynman's work in the 1980s had hypothesized that an ordinary quantum computer could be used to carry out certain physics simulations without the exponential overhead required on a classical computer. This hypothesis was not proved until Lloyd's work in 1996 [34, 35].

Whilst the first exponential speedup advantage for a quantum computer was the Deutsch-Jozsa algorithm, discovered in 1992, this problem also had no practical applications [36]. In many ways, the boson-sampling quantum computer is akin to the ordinary circuit-based quantum computer pre-Shor. Perhaps passive linear optics interferometers, now that this hidden computational power has been uncovered, can be shown to be good for something else besides the boson-sampling problem? This potential is what has captured the imagination of many researchers in the field. While in the following section we arrive at our boson-sampling scheme by way of LOQC, we still maintain several of its benefits. Namely, there are no requirements for excessive cooling of the optical elements, we have long coherence times compared to the basic gate operations, and relatively simple to understand noise sources.

One final caveat — in almost all papers on the topic of boson-sampling, the interferometer is described as a passive *linear* device with non-interacting bosons (photons in this case). However, the Hong-Ou-Mandel effect (followed by a projective measurement) imparts an effective nonlinearity and hence an effective interaction at each beamsplitter. The presence or absence of a photon in one input mode radically changes the output state of a second photon in another input mode. This 'interaction' between indistinguishable particles, known as the exchange interaction, arises simply from the demand that the multi-particle wavefunction be properly symmetrized. While not a 'force' in the usual sense, it can give rise to quite noticeable effects. For example, the bound state of the neutral hydrogen molecule (the most common molecule in our Universe) arises from just such an exchange interaction. It is therefore a misnomer to describe these interferometers as linear devices with non-interacting bosons. The exchange interaction is just as real as tagging on an additional term in a Hamiltonian. If one adds post-selection in the number basis to the mix, this imparts an effective Kerr-like nonlinearity between the bosons to boot [14].

1.2. *Introduction to Linear Optics Quantum Computing*

We begin by defining some terminology and notation. The smallest amount of data that we can deal with in quantum computing, analogous to the classical bit, is the quantum qubit. A qubit is defined as a unit vector in the complex two-dimensional vector space $\mathbb{C}^2$. More simply, it can be represented in terms of the basis

$$\begin{bmatrix} 1 \\ 0 \end{bmatrix} = |0\rangle, \begin{bmatrix} 0 \\ 1 \end{bmatrix} = |1\rangle \tag{1}$$

for a zero and one qubit, respectively. This constitutes a quantum superposition to which classical bits have no analog. A general representation of a qubit is thus

$$|\psi\rangle = \cos\left(\frac{\theta}{2}\right)|0\rangle + e^{i\phi}\sin\left(\frac{\theta}{2}\right)|1\rangle. \tag{2}$$

If we let $\theta = \frac{\pi}{2}$ and $\phi = 0$, we obtain a particular qubit of the form

$$|\psi\rangle = \frac{1}{\sqrt{2}}\left(|0\rangle + |1\rangle\right). \tag{3}$$

In this superposition, our state (once measured) has a 1/2 probability of being in state zero and a 1/2 probability of being in state one. These superpositions can also be depicted on the Bloch sphere as shown in Fig. 1. One may consider this superposition as being in both states at the same time. However, once we measure the state in the logical basis, it collapses the superposition and takes either the value of zero or one. It is the act of measurement that forces the state to 'choose' a zero or one. Thus from a classical perspective, there is an attribute of these superposition states to contain some 'hidden' quantum information.

Also analogous to classical computing, we need a set of logic gates to perform operations on our quantum states [1]. Some of the most common gates are defined

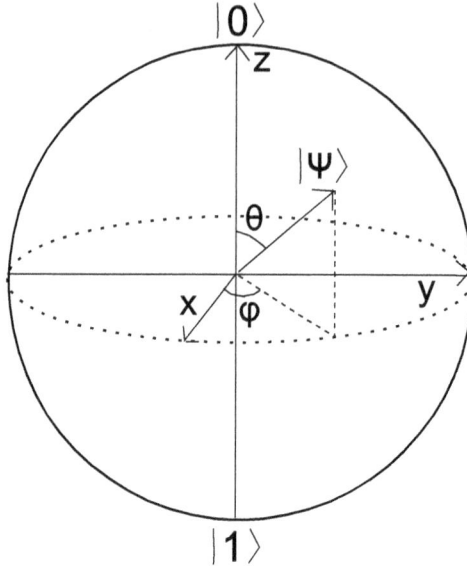

Fig. 1.   Bloch sphere showing a way to visualize the rotations that the Pauli matrices apply to a state. Pure states lie on the sphere while mixed states are contained within the sphere.

as

$$
\text{Controlled-NOT (CNOT):} \quad \begin{bmatrix} 1 & 0 & 0 & 0 \\ 0 & 1 & 0 & 0 \\ 0 & 0 & 0 & 1 \\ 0 & 0 & 1 & 0 \end{bmatrix} \tag{4}
$$

$$
\text{Hadamard (H):} \quad \frac{1}{\sqrt{2}} \begin{bmatrix} 1 & 1 \\ 1 & -1 \end{bmatrix}
$$

$$
\text{Pauli-X } (\sigma_x): \quad \begin{bmatrix} 0 & 1 \\ 1 & 0 \end{bmatrix}
$$

$$
\text{Pauli-Y } (\sigma_y): \quad \begin{bmatrix} 0 & -i \\ i & 0 \end{bmatrix}
$$

$$
\text{Pauli-Z } (\sigma_z): \quad \begin{bmatrix} 1 & 0 \\ 0 & -1 \end{bmatrix}
$$

$$
\text{Phase:} \quad \begin{bmatrix} 1 & 0 \\ 0 & i \end{bmatrix}
$$

$$
\frac{\pi}{8}: \quad \begin{bmatrix} 1 & 0 \\ 0 & e^{i\pi/4} \end{bmatrix}.
$$

The first of these, the CNOT gate, is a maximally entangling two-qubit gate, which is the quantum equivalent of the classical XOR gate. The latter gates are single qubit gates, which implement rotations on the Bloch sphere. The single qubit gates may be trivially implemented using waveplates in quantum optics, whilst the CNOT gate is far more challenging, requiring an effective Kerr non-linearity.

These gates form one possible choice of a universal gate set. A particular sequence of gates from a universal gate set can approximate the action of any other gate to arbitrary precision. So far, there are three classes of problems to which quantum computation outperforms classical computing. The first such class contains algorithms that make use of the quantum Fourier transform (such as Shor's algorithm for factoring and discrete logarithms). For $N = 2^n$ numbers, a classical fast Fourier transform would require $N \log N \approx 2^n n$ steps while a quantum computer could do this same transform in only $\log^2 N \approx n^2$ steps [1].

Another class consists of quantum search algorithms which make use of superposition to speed search times. The most well known example of such an algorithm was discovered by Grover [37], where in a search of an unstructured database of N elements, one wants to find an element of that search space satisfying a specific property. On a classical computer this search would require $O(N)$ operations, whilst a quantum search could accomplish this in $O(\sqrt{N})$ operations.

The third class is quantum simulation, where one simply attempts to simulate the evolution of a specific quantum system. It is not a surprise that this class would generally require a quantum computer to simulate efficiently. For a classical computer to simulate a typical quantum system with n distinct components, it would require $O(\exp(n))$ resources. A quantum computer would only require $O(n)$ qubits of memory however, where the proportionality constant depends on the choice of the physical system being simulated. We thus $O(\text{poly}(n))$, for the most general case, reduce an exponential resource use to only a polynomical one! For further discussion on quantum optics and quantum information processes see Refs. [38–40].

In computational terms, a computation is considered efficient if the required resources and time scale at most polynomially with the size of the input. With only linear optical elements such as beamsplitters, phase-shifters, photodetectors, and feedback from photodetector outputs, it can be shown that one can achieve this efficiency. Using only linear optical elements, it can be shown that we can implement [12],

(1) Non-deterministic quantum computation
(2) Efficiency of implementing of success of quantum gates approach unity
(3) Coding methods that achieve fault tolerance

Discussion of linear optics quantum gate efficiency, such as beamsplitters and the controlled phase gate are discussed in Ref. [41] with the description of entanglement power and entanglement efficiency.

1.3. *Linear Optics Quantum Computing*

In general, to fully achieve a true quantum computer we require a way to prepare quantum states, perform a universal gate set on the qubits, and measure the output state.

In order to generate a quantum state we use a single photon source which in the Fock basis, adds a photon to the vacuum state $|0\rangle$ and thus sets any vacuum mode to the $|1\rangle$ state. This process is non-deterministic but is sufficient for quantum computing.

The simplest optical elements are phase-shifters and beamsplitters. These elements are used to act as gate operations on our prepared states. Since both of these transformations are unitary we can write each of these elements in terms of their unitary matrix. A phase-shifter's unitary, acting on a single mode, is simply, $\hat{P}_\phi = e^{i\hat{n}\phi}$, where $\hat{n}$ is the number operator. The unitary matrix for a beamsplitter is given by

$$B_{\theta,\phi} = \begin{pmatrix} \cos\theta & -e^{i\phi}\sin\theta \\ e^{-i\phi}\sin\theta & \cos\theta \end{pmatrix}, \tag{5}$$

in the basis of optical modes, where ϕ gives the phase relationship and θ stipulates the bias of the beamsplitter.

In order to measure the state, we use photodetectors which destructively determine if a mode contains a photon or not. For states with more than one photon then, we need a photon counting detector, which can be implemented by using a series of beamsplitters and photodetectors. The beamsplitters act so that the photons are spread evenly over N modes, with each mode containing a photodetector. The probability of under-counting given that the photon number is k is at most $k(k-1)/(2N)$. This is referred to as multiplexed photodetection [42–46]. Another alternative is to use photon-number-resolving detectors.

In addition to these single qubit rotations we also require a nonlinear sign-flip (NS) gate [12]. This gate implements the transformation

$$\text{NS} : \alpha_0|0\rangle + \alpha_1|1\rangle + \alpha_2|2\rangle \rightarrow \alpha_0|0\rangle + \alpha_1|1\rangle - \alpha_2|2\rangle, \tag{6}$$

and is the basis (along with ancilla notes) of implementing the CNOT gate. The two qubit CNOT gate along with the previously discussed single qubit gates form the required universal gate set to perform quantum computing. One only needs a set of one- and two-qubit universal gates in order to construct general multi-qubit gates. Specifically we only require the Hadamard, phase, $\pi/8$ and CNOT gates [1].

Using just linear optics and photodetection, implementing the NS gate is non-deterministic, which implies that with multiple gates in our circuit, the success probability of the computation drops exponentially with the number of gates. To overcome this, another useful tool in LOQC is the use of quantum gate teleportation to increase the probability of success of non-deterministic gates [12, 47]. Here we use two Bell pairs (maximally entangled two-qubit states) as a resource to teleport the action of a gate onto two qubits. This teleportation trick increases the success probability of the non-deterministic gate, but is itself non-deterministic. However, by concatenating the teleportation protocol we are able to increase the success probability of a non-deterministic gate asymptotically close to unity [12], enabling efficient large-scale computation.

1.4. *Why is Linear Optics Quantum Computing Hard?*

All of this may lead one to ask, "if this scheme, using only linear elements is so simple, what's the hold up in implementing it?" To implement this scheme we require a myriad of technicalities. These include synchronization of pulses, mode-matching, quickly controllable delay lines, tunable beamsplitters and phase-shifters, single-photon sources, and accurate, fast, single-photon detectors. Most of this list is not terribly unrealistic to adhere to but current efficiencies of photodetectors are not at the point at which they may realistically implement the teleportation and the more complex gate operations (two qubit gates). The feedback control of these detectors must also be extremely fast in order to select proper state preparation before photon loss becomes an issue.

As an example, if we investigate actual implementation of a teleported (i.e high success probability) CNOT gate, which requires many individual non-deterministic CNOT gates, we can attain a probability of success in implementing this entangling operation of 95% with approximately 300 successful CNOT gates which translates to an excessively large number ($> 10^4$) of optical elements. Whilst this may seem daunting, recent approaches using cluster states have reduced experimental requirements by orders of magnitude [48, 49], but nonetheless the experimental requirements are substantial and still require challenging technologies such as fast-feedforward and dynamic control.

Without accurate implementation of these protocols we likely lose our claim to universality, but we still retain our ability to investigate some interesting problems. This realm of LOQC without fast feedback control or unrealistically accurate photodetectors leads us into boson-sampling.

2. The Boson-sampling Formalism

Unlike full LOQC, which requires active elements, the boson-sampling model is strictly passive, requiring only single-photon sources, passive linear optics (i.e beamsplitters and phase-shifters), and photodetection. No quantum memory or feedforward is required.

We begin by preparing an input state comprising n single photons in m modes,

$$|\psi_{\text{in}}\rangle = |1_1, \ldots, 1_n, 0_{n+1}, \ldots, 0_m\rangle$$
$$= \hat{a}_1^\dagger \ldots \hat{a}_n^\dagger |0_1, \ldots, 0_m\rangle, \tag{7}$$

where $\hat{a}_i^\dagger$ is the photon creation operator in the ith mode. It is assumed that the number of modes scales quadratically with the number of photons, $m = O(n^2)$. The input state is evolved via a passive linear optics network, which implements a unitary map on the creation operators,

$$\hat{U}\hat{a}_i^\dagger\hat{U}^\dagger = \sum_{j=1}^m U_{i,j}\hat{a}_j^\dagger, \tag{8}$$

where $\hat{U}$ is a unitary matrix characterizing the linear optics network. It was shown by Reck *et al.* [50] that any $\hat{U}$ may be efficiently decomposed into $O(m^2)$ optical elements. The output state is a superposition of the different configurations of how the n photons could have arrived in the output modes,

$$|\psi_{\text{out}}\rangle = \sum_S \gamma_S |n_1^{(S)}, \ldots, n_m^{(S)}\rangle, \tag{9}$$

where S is a configuration, $n_i^{(S)}$ is the number of photons in the ith mode associated with configuration S, and γ_S is the amplitude associated with configuration S. The probability of measuring configuration S is given by $P_S = |\gamma_S|^2$. The full model is illustrated in Fig. 2.

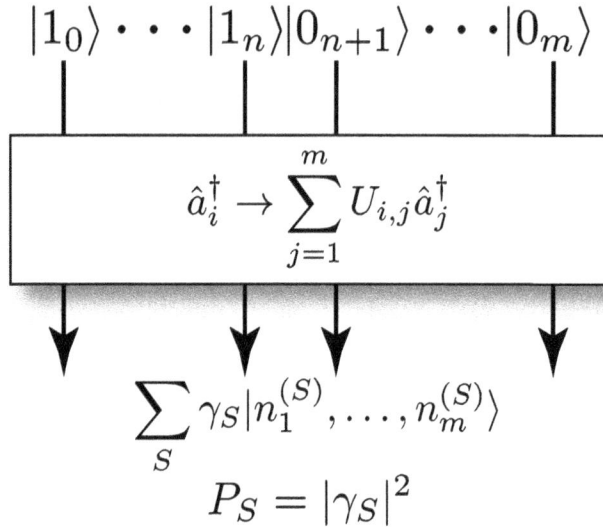

Fig. 2.   The boson-sampling model. n single photons are prepared in m optical modes. These are evolved via a passive linear optics network $\hat{U}$. Finally the output statistics are sampled via coincidence photodetection. The experiment is repeated many times, partially reconstructing the output distribution.

It was shown by Scheel [51] that the amplitudes γ_S are related to matrix permanents,

$$\gamma_S = \frac{\text{Per}(U_S)}{\sqrt{n_1^{(S)}! \ldots n_m^{(S)}!}}, \tag{10}$$

where U_S is an $n \times n$ sub-matrix of U that depends on the specific configuration S, the, and $\text{Per}(U_S)$ is the permanent of U_S.

Let us examine this relationship with the permanent more closely. Consider Fig. 3. Here the first two modes have single photons, with the remaining modes in the vacuum state. Let us consider the amplitude of measuring one photon at output mode 2 and another at output mode 3. Then there are two ways in which this could occur. Either the first photon reaches mode 2 and the second mode 3, or vice versa, i.e the photons pass straight through, or swap. Therefore there are $2! = 2$ ways in which the photons could reach the outputs. Thus, this amplitude may be written as,

$$\gamma_{\{2,3\}} = \underbrace{U_{1,2}U_{2,3}}_{\text{photons don't swap}} + \underbrace{U_{1,3}U_{2,2}}_{\text{photons swap}}$$

$$= \text{Per} \begin{bmatrix} U_{1,2} & U_{2,2} \\ U_{1,3} & U_{2,3} \end{bmatrix}, \tag{11}$$

which is a 2×2 matrix permanent.

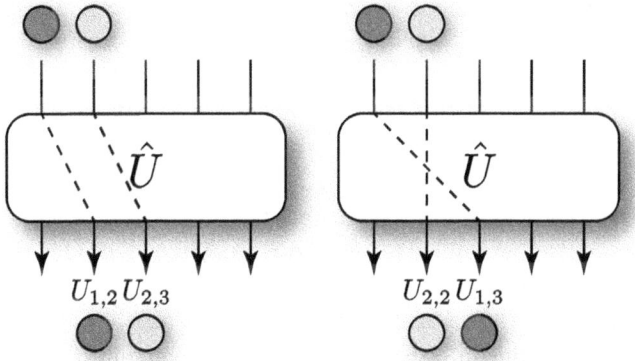

Fig. 3. Two-photon boson-sampling, where we wish to calculate the amplitude of measuring a photon at each of the output modes 2 and 3. There are two ways in which this may occur – either the photons pass straight through, or swap, yielding a sum of two paths.

As a slightly more complex example, consider the three photon case shown in Fig. 4. Now we see that there are $3! = 6$ ways in which the three photons could reach the outputs, and the associated amplitude is given by a 3×3 matrix permanent,

$$\gamma_{\{1,2,3\}} = U_{1,1}U_{2,2}U_{3,3} + U_{1,1}U_{3,2}U_{2,3}$$
$$+ U_{2,1}U_{1,2}U_{3,3} + U_{2,1}U_{3,2}U_{1,3}$$
$$+ U_{3,1}U_{1,2}U_{2,3} + U_{3,1}U_{2,2}U_{1,3}$$
$$= \text{Per} \begin{bmatrix} U_{1,1} & U_{2,1} & U_{3,1} \\ U_{1,2} & U_{2,2} & U_{3,2} \\ U_{1,3} & U_{2,3} & U_{3,3} \end{bmatrix}. \tag{12}$$

In general, with n photons, there will be $n!$ ways in which the photons could reach the outputs (assuming they all arrive at distinct outputs), and the associated amplitude will relate to an $n \times n$ matrix permanent. Calculating matrix permanents, in general is known to be #**P**-hard, even harder than **NP**-complete, and the best known algorithm is by Ryser [20], requiring $O(2^n n^2)$ runtime. Thus, we can

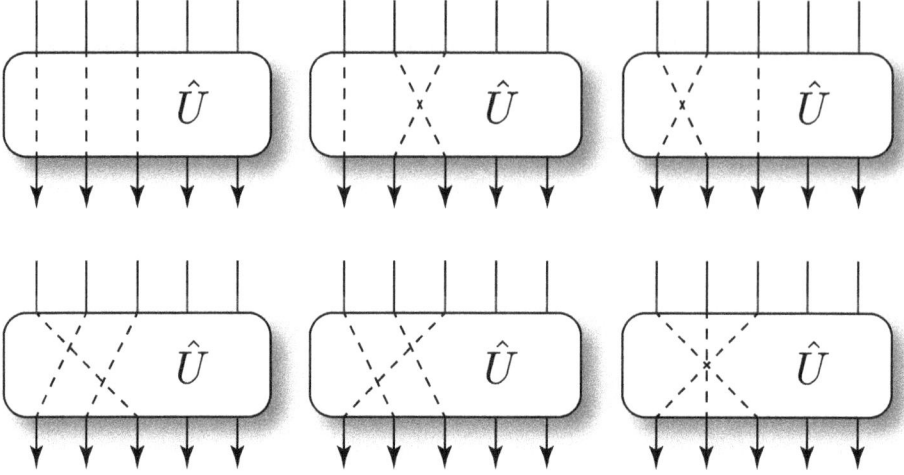

Fig. 4.   Three photon boson-sampling, where we wish to calculate the amplitude of measuring a photon at each of the output modes 1, 2 and 3. There are now 3! = 6 possible routes for this to occur.

immediately see that if boson-sampling were to be classically simulated by exactly calculating the matrix permanents, it would require exponential classical resources.

Because the number of modes scales quadratically with the number of photons, for large systems we are statistically guaranteed that all photons will arrive at different output modes. This implies that in this regime on/off (or 'bucket') detectors will suffice, and photon-number resolution is not necessary, a further experimental simplification compared to full-fledged LOQC.

The number of configurations in the output modes scales as,

$$|S| = \binom{n + m - 1}{n},\tag{13}$$

which is super-exponential in n. Thus, with an 'efficient' (i.e polynomial) number of trials, we are unlikely to sample from a given configuration more than once. This implies that we are unable to determine any given P_S with more than binary accuracy. Thus, boson-sampling does *not* let us *calculate* matrix permanents, as doing so would require determining amplitudes with a high level of precision, which would require an exponential number of measurements.

The experiment is repeated many times, each time performing a coincidence photodetection at the output modes. Thus, after each run we sample from the distribution P_S. This yields a so-called *sampling problem*, whereby the goal is to sample a statistical distribution using a finite number of measurements. This is in contrast to well-known *decision problems*, such as Shor's algorithm [52], which provide a well-defined answer to a well-posed question. Because boson-sampling is a sampling problem, finding a computational application is further complicated — if every time we run the device we obtain a different outcome, how does the outcome answer a well-defined question, and how do we map it to a problem of interest? This is one of the central challenges of boson-sampling —what can we do with it?

This sampling problem was shown by AA to likely be a computationally hard problem. That is, sampling the statistical distribution at the output to the boson-sampling device is computationally hard. However, whilst shown to be computationally hard, no known algorithmic applications for boson-sampling have been described. Thus, boson-sampling acts as an interesting proof-of-principle demonstration that passive linear optics can outperform classical computers, but, based on present understanding, does not solve a problem of practical interest.

2.1. *Errors in Boson-sampling*

There are a number of subtleties inherent in boson-sampling, often arising because one considers *estimation* of a distribution. In particular, we have shown that the amplitude of a given configuration of photons is tied to computing the permanent of a matrix. While finding the exact permanent of a binary matrix is known to be #**P**-hard, one can efficiently *estimate* the permanent of a matrix with real, non-negative entries [53, 54]. If efficient estimation were also possible for complex-valued unitaries, one could pass off the boson-sampling problem as simply a type of singularity —a mathematical anomaly occurring only when one tries to know exact values. Since implementing a physical system such as boson-sampling is bound to have some kind of error, one could not hope to experimentally achieve the true distribution anyway.

In their paper, AA went to great lengths to prove the robustness of their result in the presence of error. Thus, even attempting to estimate the output distribution of a boson-sampling machine is likely computationally hard. (As one might expect, this correlates with the result that complex-valued matrix permanents cannot in general even be efficiently estimated unless exact permanents can as well [54].) In a sampling regime, one cannot tolerate so much error without eventually deviating too far from the desired distribution or requiring so many samples as to make the algorithm inefficient. For example, one could try postselecting to remove errant samples where photon-loss occurred. However, if the probability were to scale on the order of $O(1/\exp(n))$, we would lose any hope of scaling for large n.

What then is an acceptable level of error? Suppose we fix an error threshold ϵ (i.e. we set ϵ to be the maximum allowable variation distance from the true distribution). Then so long as we have a success probably P where $P > 1/\text{poly}(n)$, we could correctly (within ϵ) and efficiently sample a boson-sampling device with arbitrarily large photon number. More generally, if we wish also to scale ϵ smaller, $P > 1/\text{poly}(n, 1/\epsilon)$ [15]. As we discuss in the next section, however, we cannot experimentally determine whether this is achievable for asymptotic n.

3. Boson-sampling and the Extended Church-Turing Thesis

Any model for quantum computation is subject to errors of some form. In the conventional circuit model, this includes errors such as dephasing. In linear optics, this includes photon loss and mode-mismatch. Let us consider a very generic error model for boson-sampling, where the single-photon states are the desired single photon with probability p, otherwise are in some erroneous state [55]. This erroneous state could, for example, comprise terms with the wrong photon number (such as loss or second order excitations), or mode-mismatch. Then our input state is of the form,

$$\hat{\rho}_{\text{in}} = \left(\bigotimes_{i=1}^{n} [p|1\rangle\langle 1| + (1-p)\hat{\rho}_{\text{error}}^{(i)}] \right) \otimes [|0\rangle\langle 0|]^{\otimes^{m-n}}, \tag{14}$$

where $\hat{\rho}_{\text{error}}^{(i)}$ may be different for each input mode i. This is an independent error model, whereby each state is independently subject to an error channel. p stipulates the fidelity of the single photon states. When $p = 1$, the states are perfect single photons, and when $p < 1$, the state contains erroneous terms. We desire to sample from the distribution of Eq. (7), whereby none of the input states are erroneous. This occurs with probability p^n.

Let P be the probability that upon performing boson-sampling we have sampled from the correct distribution, otherwise we sample from noise. The complexity proof provided by AA only considered the regime where $P > 1/\text{poly}(n)$. Thus, for computational hardness, we require $p^n > 1/\text{poly}(n)$. Clearly in the asymptotic limit of large n, this bound can never be satisfied for any $p < 1$. Thus, with this independent error model, boson-sampling will always fail in the asymptotic limit.

Numerous authors [56–60] have claimed that large-scale demonstrations of boson-sampling could provide elucidation on the validity of the Extended Church-Turing (ECT) thesis —the statement that any physical system may be efficiently simulated on a Turing machine. However, it must be noted that the ECT thesis is by definition an asymptotic statement about arbitrarily large systems. Because the required error bound for boson-sampling is never satisfied in this limit, it is clear that boson-sampling cannot elucidate the validity of the ECT thesis as asymptotically large boson-sampling devices must fail under an independent error model.

This concern might be overcome in the future with either (1) a loosening of the error bound to $1/\exp(n)$, or (2) the development of fault-tolerance techniques for boson-sampling. However, to-date no such developments have been made. Thus, based on *present* understanding, boson-sampling will not answer the question as to whether the ECT thesis is correct or not. However, this is distinct from the question 'will boson-sampling yield *post-classical* computation?'. The answer to this question may very well be affirmative, as this only requires a finite sized device, just big enough to beat the best classical computers.

4. Boson-sampling with Other Classes of Quantum Optical States

Recall that the original proposal for boson-sampling by AA proceeds in three steps:

(1) Input — Vacuum and single photon Fock states
(2) Evolve — Via a passive linear interferometer
(3) Measure — Using an on-off photodetector

Together, this process proves to be classically hard to simulate. While this process is lauded for its relative simplicity compared to universal quantum computation, it is natural to ask if there are other systems that share these same properties. That is, one can consider a generalization of boson-sampling by replacing one or more of the above three procedures with an analogous one. The difficulty, as was the case with the original proposal, is proving that the system is classically intractable.

One could attempt to construct a complexity proof directly, but a rigorous proof may be hard to achieve; even AA's result for the approximate boson-sampling case relies on several likely yet unproven lemmas. A simpler approach is to show that a system is equivalent to boson-sampling, implementing the same logical problem. In other words, with only small overhead, the resulting statistics of one system could be used to compute the statistics of the other. Informally, we might say a problem is 'boson-sampling hard' if it is at least as hard to compute as a boson-sampling distribution. By showing this property of a system, one can extend AA's result to a more general class of problems.

One straightforward instance of such a proof was given by Seshadreesan *et. al* [61] by generalizing the input to coherent states (instead of vacuum) and photon-added coherent states (instead of single photon Fock states). More precisely, the input takes the form

$$|\psi_{in}\rangle = \hat{a}_1^\dagger \ldots \hat{a}_n^\dagger |\alpha_1 \ldots \alpha_m\rangle. \tag{15}$$

Without changing the measurement scheme, intuitively one might expect this would work only when the coherent state has a small amplitude $|\alpha|$ since large amplitude coherent states tend to act 'more classically' than other quantum states. Additionally, it would likely become difficult to distinguish between photons contributed from a large amplitude coherent state and a single photon contributed from the creation operators. Indeed, for coherent states the amplitude must scale according to $|\alpha| < 1/\text{poly}(n)$ for AA's result to hold. A similar result holds for any general separable state close to vacuum. It remains an open question whether other measurement schemes (such as homodyne detection) could be used to perform analogous boson-sampling on coherent or other quantum states.

Another class of states considered by Rohde *et al.* [62] was cat states — superpositions of coherent states ($|\alpha\rangle \pm |-\alpha\rangle$). It was shown that for $\alpha \to 0$ this yields ideal boson-sampling, for small but non-zero amplitude it is provably computationally hard by treating the cat state as an error model, and for general α strong evidence was provided that the problem is hard by relating the amplitudes to a permanent-like function. Intuitively, one might expect that there is a plethora of non-Fock quantum states of light that yield computationally hard sampling problems, a question that warrants future research.

5. How to Build a Boson-sampling Device

In this section we explain the basic components required to build a boson-sampling device. This device consists of three basic components: (1) single-photon sources; (2) linear optics networks; and, (3) photodetectors. Each of these present their own engineering challenges and there are a range of technologies that could be employed for each of these components. However, although boson-sampling is much easier to implement than full-scale LOQC, it remains challenging to build a post-classical boson-sampling device. While challenging, a realizable post-classical boson-sampling device is foreseeable in the near future.

5.1. Photon Sources

The first engineering challenge is to prepare an input state of the form of Eq. 7. This state may be generated using various photon source technologies. For a review of many of the photon sources see Ref. [63]. Presently, the most commonly employed photon source technology is spontaneous parametric down-conversion (SPDC).

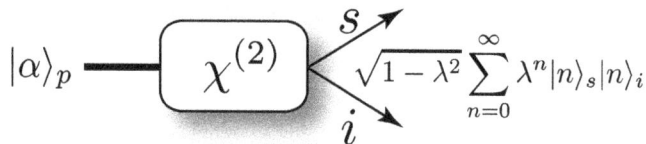

Fig. 5.   Spontaneous parametric down-conversion (SPDC) source. A crystal with a second order non-linearity, $\chi^{(2)}$, is pumped with a classical coherent light source $|\alpha\rangle_p$. The source then probabilistically emits photon pairs into the signal and idler modes, including vacuum $|0\rangle|0\rangle$ and higher order terms where multiple pairs are emitted.

The SPDC source works by first pumping a non-linear crystal with a coherent state $|\alpha\rangle$ as shown in Fig. 5. A coherent state is well approximated by a laser source. With some probability one of the laser photons interacts with the crystal and emits an entangled superposition of photons across two output modes, the *signal* and *idler*. The output of an SPDC source is of the form [64],

$$|\Psi_{\text{SPDC}}\rangle = \sqrt{1-\lambda^2} \sum_{n=0}^{\infty} \lambda^n |n\rangle_s |n\rangle_i, \tag{16}$$

where $0 \leq \lambda \leq 1$ is the squeezing parameter, n is the number of photons, s represents the signal mode, and i represents the idler mode. For boson-sampling, we are interested in the $|1\rangle_s|1\rangle_i$ term of this superposition since we require single photons at the input of the first n modes. The signal photons are measured by a photodetector and because of the correlation in photon-number, we know that a photon is also present in the idler mode. The idler photons are then routed into one of the input ports of the boson-sampling device using a multiplexer [45, 46, 65].

There are several problems associated with SPDC sources, which limit the scalability of boson-sampling. The major problem is higher order photon-number terms. In the boson-sampling model we only want the $|1\rangle_s|1\rangle_i$ term, which is far from deterministic. The SPDC source is going to emit the zero-photon term with highest probability and emit higher order terms with exponentially decreasing probability. If the heralding photodetector does not have unit efficiency, then the heralded mode may contain higher order photon-number terms.

It was recently shown by Motes *et al.* [66] that SPDC sources are scalable in the asymptotic limit for boson-sampling using a multiplexing device. Specifically, if the photodetection efficiency is sufficient to guarantee post-selection at the output of the boson-sampling device with high probability, then the heralded SPDC photons also have asymptotically high fidelity. The boson-sampling architecture with multiplexing is shown in Fig. 6. Furthermore, it was shown by Lund *et al.* [67] that one can do away with the multiplexer altogether, simply routing the SPDC outputs directly to the interferometer (which has become known as 'scattershot' boson-sampling), which still yields an equivalently hard sampling problem, a massive experimental simplification.

Another problem is that photons from SPDC sources have uncertainty in their temporal distribution. If a boson-sampling device is built using multiple SPDC sources it is difficult to temporally align each of the n photons entering the device. This is called temporal mismatch. The error term associated with this scales exponentially with n, yielding an error model consistent with Eq. (14), which undermines operation in the asymptotic limit.

5.2. *Linear Optics Networks*

After the input state has been prepared it is evolved via a linear optics network, $\hat{U}$. $\hat{U}$ transforms the input state as per Eq. (8) and may be completely characterized before the experiment using coherent state inputs [68]. $\hat{U}$ is composed of an array of discrete elements, namely, beamsplitters and phase-shifters. A beamsplitter with phase-shifters may be represented as a two-mode unitary of the form [64],

$$U_{\mathrm{BS}}(t) = \begin{pmatrix} e^{i(\alpha-\frac{\beta}{2}-\frac{\gamma}{2})}\cos\left(\frac{\delta}{2}\right) & -e^{i(\alpha-\frac{\beta}{2}+\frac{\gamma}{2})}\sin\left(\frac{\delta}{2}\right) \\ e^{i(\alpha+\frac{\beta}{2}-\frac{\gamma}{2})}\sin\left(\frac{\delta}{2}\right) & e^{i(\alpha+\frac{\beta}{2}+\frac{\gamma}{2})}\cos\left(\frac{\delta}{2}\right) \end{pmatrix}, \tag{17}$$

where $0 \le \alpha \le 2\pi$ and $0 \le \{\beta, \gamma, \delta\} \le \pi$ are arbitrary phases.

For a $\hat{U}$ that implements a classically hard problem one would need hundreds of discrete optical elements. Constructing an arbitrary $\hat{U}$ using the traditional linear optics approach of setting and aligning each optical element would be extremely

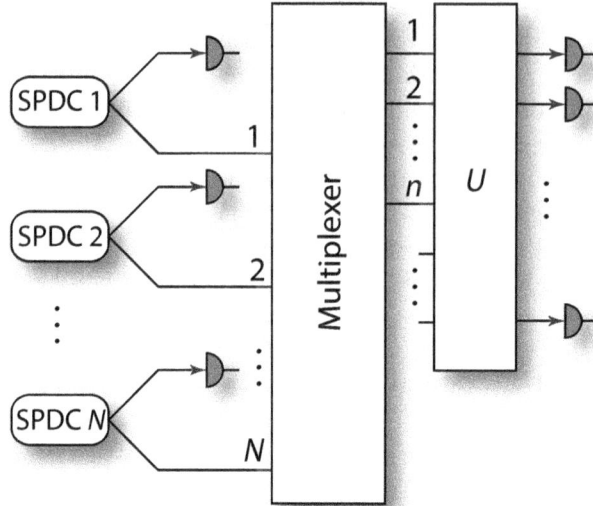

Fig. 6.   Boson-sampling architecture using SPDC sources with an active multiplexer. N sources operate in parallel, each heralded by an inefficient single-photon number-resolving detector. It is assumed that $N \gg n$, which guarantees that at least n photons will be heralded. The multiplexer dynamically routes the successfully heralded modes to the first n modes of the unitary network $\hat{U}$. Finally, photodetection is performed and the output is post-selected on the detection on all n photons.

cumbersome. Thus, using discrete optical elements is not a very promising route towards scalable boson-sampling.

   One method to simplify the construction of the linear optics network is to use integrated waveguides. Quantum interference was first demonstrated with this technology by Peruzzo *et al.* [69]. This technology requires more frugal space requirements, is more optically stable, and far easier to manufacture, allowing the entire linear optics network to be integrated onto a small chip [70–72]. The main issue with integrated waveguides is achieving sufficiently low loss rates inside of the waveguide and in the coupling of the waveguide to the photon sources and photodetectors. Presently, the loss rates in these devices are extremely high and thus post-selection upon n photons at the output occurs with very low probability. It is foreseeable that photon sources and photodetectors will eventually be integrated into the waveguide which would eliminate coupling loss rates, substantially improving scalability.

   Another potential route to simplifying the linear optics network is to use time-bin encoding in a loop based architecture [73]. The major advantage of this architecture is that it only requires two delay loops, two on/off switches, and one controllable beamsplitter as shown in Fig. 7. This possibility eliminates the problem of aligning hundreds of optical elements and has fixed experimental complexity, irrespective of the size of the boson-sampling device. A major problem with this architecture however is that it remains difficult to control a dynamic beamsplitter with high fidelity at a rate that is on the order of the time-bin width τ.

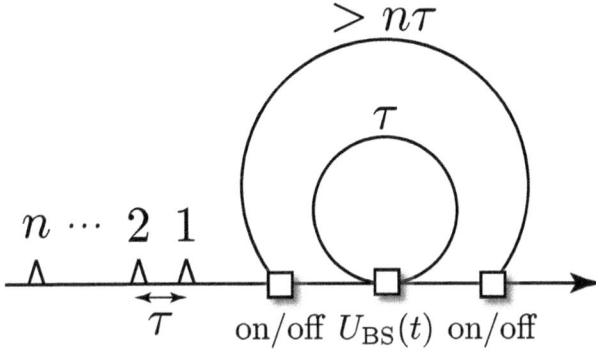

Fig. 7. Time-bin encoding architecture for implementing a boson-sampling device. Single photons arrive in a train of time-bins instead of in spatial modes. Each time-bin corresponds to spatial modes in the boson-sampling scheme and are separated by time τ. The photon train is coupled into the loop by the first switch. The photons then traverse the inner loop such that each time-bin may interact. The first (last) photon is coupled completely in (out). The outer loop allows an arbitrary number of the smaller loops to be applied consecutively which is determined by the third switch. Finally, the photon train is measured at the output using time-resolved detection.

5.3. *Photodetection*

The final requirement in the boson-sampling device is sampling the output distribution as shown in Eq. (9). With linear optics this is done using photodetectors. For a review on various types of photodetection see Ref. [63].

There are two general types of photodetectors — photon-number resolving detectors and bucket detectors. The former counts the number of incident photons. These are much more difficult to make and more expensive in general than bucket detectors. Bucket detectors, on the other hand, simply trigger if any non-zero number of photons are incident on the detector. As discussed earlier, in the limit of large boson-sampling devices, we are statistically guaranteed that we never measure more than one photon per mode, since the number of modes scales as $m = O(n^2)$. Thus, bucket detectors are sufficient for large boson-sampling devices, a significant experimental simplification compared to universal LOQC protocols.

Many photodetector designs use superconductivity to measure photons. Superconductivity is an extreme state where electrical current flows with zero resistance. It occurs in conductive materials when a certain critical temperature is reached. This critical temperature is far from occurring naturally on Earth and thus high-tech and expensive lab equipment is required. For many materials this temperature is close to absolute zero. Such extreme conditions are required for detecting single photons because single photons are themselves extreme. They are after all the smallest unit of light.

While there are several variations of superconductive photodetectors, they tend to work similarly to the one shown in Fig. 8. The idea is that a superconductor is cooled to a point just below its critical temperature. Current is then applied through the superconductor which experiences zero resistance. If there is no resistance, then there is no voltage drop across the superconductor and our conductance measurement reads infinity. Then the photon or photons that are to be measured will hit the superconductor and be absorbed. Each photon that is absorbed by the superconductor imparts energy $h\nu$ onto it, where h is Planck's constant and ν is the frequency of the photon. This heats the superconductor above its critical temperature. The conductance measurement will then change according to the absorbed photon, thus informing the measurer that a photon was detected. This scheme may be able to count several photons since the conductance will change proportionally to the number of absorbed photons. However, if too many photons are absorbed all properties of superconductivity are lost and thus number-resolution is lost.

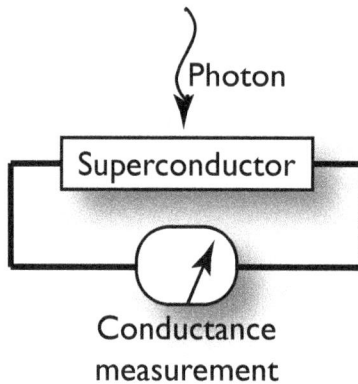

Fig. 8. Basic design for a superconducting photon detector. A current is passed through a superconductor and the conductance is monitored. Photons impart energy on the superconductor that is just cooled to its critical temperature. This added photon energy causes a measurable change in the conductance allowing for the detection of photons.

Photodetectors may be used to help overcome the problem of temporal mismatch. Such detectors must have the ability to record the time at which the photon arrived. If we post-select upon detecting all n output photons in the same time-window Δt then we can assume that their temporal distribution overlaps sufficiently to yield a classically hard sampling problem. This method however is not reliable for scalable boson-sampling. If the temporal distributions are not sufficiently overlapping, then the probability of post-selecting all n photons in the same time-window decreases exponentially with n. However, if the sources are producing nearly identical photons in the time domain then this method would be a practical cross check.

As the distinguishability of photons varies the complexity of sampling the output distribution also varies. A theoretical framework was developed by Tillmann *et al.* [74] that describes the transition probabilities of photons with arbitrary distinguishability through the linear optical network. The output distribution of boson-sampling with distinguishable photons is then given by matrix immanants, thus affecting the computational complexity of the output distribution. They also test this experimentally by tuning the temporal mismatch of their input photons. This boson-sampling experiment is unique in that it is the first to use distinguishable photons at the input. A similar situation, of boson-sampling with photons of arbitrary spectral structure, was considered by Rohde [75].

6. Conclusion

In this chapter we have given an introduction to the rapidly evolving field of passive linear optical quantum computation, a new model of quantum computation that, while not universal, nevertheless can carry out efficiently — at least! — the interesting mathematical problem of boson-sampling. Since a boson-sampling output is strongly believed to be inefficient to verify on a universal quantum computer, much less a classical computer, the passive linear optical approach to quantum computation really is something new and quite different than the usual quantum computer idea. While at the present the boson-sampling problem does not have any known practical uses — but has found an application in metrology — it nevertheless provides us a new window into the hidden computational power of quantum mechanical devices.

Often researchers in the field of quantum computation are asked by colleagues in other fields, or the popular press, "When will the quantum computer be built?" The answer depends greatly upon what exactly it is you mean by a 'quantum computer'. We turn this question around and ask our colleagues in the field of computer science, "Well first you tell us when the classical computer was built." One would think that at least this would have a universally agreed upon answer, but that is not the case. Many will reply that it was the ENIAC, a digital machine that came online in 1945 that was Turing complete, reprogrammable, electronic, digital, and had a memory. Originally designed to compute US Army artillery tables it was quickly turned to simulating H-bomb explosions. Others hold out for the Atanasoff–Berry Computer (ABC) that came online in 1941 and which was also electronic, digital, and programmable, but not universal and lacked the equivalent of today's RAM. To muddy the waters, some implore that we consider the British electronic Colossus computers that were first constructed in 1943 but only declassified in the 1970s. However they were not at time of operation Turing complete but could apparently have been made so. The list goes on and on [76]. Why no agreement on when the classical computer was built? It is because experts in the field disagree on exactly what is a classical computer. Some even hold out for the all-mechanical Babbage machine of the 1800s. All we can say is that in 1850 the classical computer certainly had not been built and by 1950 it certainly had but that nobody can agree on a precise date or a precise computer.

We should not expect any more or less of the future history of the quantum computer. We have ion trappers painstakingly assembling a universal quantum computer a qubit at a time. But then we have the company D-Wave the makes non-universal quantum machine that nevertheless implements a watered-down version of adiabatic quantum computing, called quantum simulated annealing, that appears to show some polynomial improvement on particular problems such as structured search or pattern matching algorithms. Then here comes the passive linear optical device that implements boson-sampling efficiently. What to think about that? The device is not universal, in that it cannot solve every problem a universal quantum computer can solve, but nevertheless it can solve the boson-sampling problem efficiently. In some sense the passive linear optics model of computation is as powerful as a universal quantum computer, but only when it comes to this one problem. It is a bit like the ABC machine, which was not universal but really good at solving linear sets of equations, or like the Colossus machines that were also not universal but really good at cracking a particular type of German cipher called Enigma. Those problems were certainly useful and it remains to be seen if our non-universal passive linear optical machine has the ability to solve some useful problems efficiently as well [77]. But in the meantime it is certainly interesting to be working at this interface of quantum optics and quantum computation where the work of AA and others have provided us with a totally new ball game. Who knows what other secrets lay hidden in simple interferometers? All we can say for sure here is that there is a great future in photons.

To recap this chapter, in Section I, we have reviewed universal quantum computation in general and then particularly with photons, the so-called linear optical approach. While universal it is unwieldy given the overhead associated with active optical elements. Far simpler is the passive linear optics interferometer with no moving parts, and to such a machine there fits quite nicely the boson-sampling problem, as outlined in Section II. Here we also discussed the error tolerance required for such a machine in order for it to show a computational speedup. In Section III we take a detour through quantum computer complexity theory and discuss just what experiments with boson-sampling do and do not prove in that realm. Specifically, we argue that, contrary to a popular claim made in the literature, boson-sampling cannot disprove or even provide evidence against the Extended Church-Turing thesis. The original boson-sampling paradigm of AA admitted only single photons as input. In Section IV we discuss what happens if that assumption is relaxed. We surmise that any passive linear optical network (that implements a random unitary from the Haar class) fed with negative Wigner function states yields a hard sampling problem. In Section V we review how the quantum mechanics out there actually build such devices and we delve into a number of hardware and implementation issues. Finally in this Section VI the section refers to itself in true Gödelian fashion by concluding: Now if we were to take the set of all book chapter conclusions that conclude themselves. . .

Acknowledgments

KRM and PPR would like to acknowledge support from the the the Australian Research Council Centre of Excellence for Engineered Quantum Systems (Project number CE110001013). JPD would like to acknowledge support from the Air Force Office of Scientific Research and the Army Research Office. BTG would like to acknowledge support from the National Physical Science Consortium & National Institute of Standards and Technology graduate fellowship program and the National Science Foundation. JPO would like to acknowledge support from a Louisiana State University System Board of Regents Fellowship.

References

[1] M. Nielsen and I. Chuang, *Quantum Computation and Quantum Information.* (Cambridge University Press, 2000).

[2] R. Raussendorf and H. J. Briegel, A one-way quantum computer, *Phys. Rev. Lett.* **86**, 5188, (2001).

[3] R. Raussendorf, D. E. Browne, and H. J. Briegel, Measurement-based quantum computation on cluster states, *Phys. Rev. A.* **68**, 022312, (2003).

[4] E. Farhi, J. Goldstone, S. Gutmann, J. Lapan, A. Lundgren, and D. Preda, A quantum adiabatic evolution algorithm applied to random instances of an NP-complete problem, *Science.* **292**, 472, (2001).

[5] Y. Aharonov, L. Davidovich, and N. Zagury, Quantum random walks, *Phys. Rev. A.* **48**, 1687, (1993).

[6] E. Bernstein and U. Vazirani, Quantum complexity theory, *Usiam Journal on Computing.* **26**, 1411, (1997).

[7] S. P. Jordan, Permutational quantum computing, *Quantum Information & Computation.* **10**, 470, (2010).

[8] O. Moussa, C. Ryan, D. Cory, and R. Laflamme, Testing contextuality on quantum ensembles with one clean qubit, *Phys. Rev. Lett.* **104**, 160501, (2010).

[9] V. Cerny, Quantum computers and intractable (NP-complete) computing problems, *Phys. Rev. A.* **48**, 116–119, (1993).

[10] J. Clauser and J. Dowling, Factoring integers with young's N-slit interferometer, *Phys. Rev. A.* **53**, 4587–4590, (1996).

[11] S. D. Bartlett, B. C. Sanders, S. L. Braunstein, and K. Nemoto, Efficient classical simulation of continuous variable quantum information processes, *Phys. Rev. Lett.* **88**, 097904, (2002).

[12] E. Knill, R. Laflamme, and G. Milburn, A scheme for efficient quantum computation with linear optics, *Nature.* **409**, 46–52, (2001).

[13] P. Kok, W. J. Munro, K. Nemoto, T. C. Ralph, J. P. Dowling, and G. J. Milburn, Linear optical quantum computing with photonic qubits, *Reviews of Modern Physics.* **79**, 135–174, (2007).

[14] G. Lapaire, P. Kok, J. P. Dowling, and J. E. Sipe, Conditional linear-optical measurement schemes generate effective photon nonlinearities, *Phys. Rev. A.* **68**, 042314, (2003).

[15] S. Aaronson and A. Arkhipov, The computational complexity of linear optics, *Theory of Computing.* **9 (4)**, 143–252, (2013).

[16] C. Goglin, M. Kliesch, L. Aalita, and J. Eiset, Boson-sampling in the light of sample complexity. arXiv:1306.3995, (2013).

[17] B. T. Gard, R. M. Cross, P. M. Anisimov, H. Lee, and J. P. Dowling, Quantum random walks with multiphoton interference and high-order correlation functions, *J. Optical Soc. Am. B.* **30**, 1538–1545, (2013).

[18] B. T. Gard, J. P. Olson, R. M. Cross, M. B. Kim, H. Lee, and J. P. Dowling, Inefficiency of classically simulating linear optical quantum computing with fock-state inputs *Phys. Rev. A.* **89**, 022328 (2014).

[19] K. R. Motes, J. P. Dowling, and P. P. Rohde, Spontaneous parametric down-conversion photon sources are scalable in the asymptotic limit for boson sampling, *Phys. Rev. A.* **88**, 063822 (2013).

[20] H. J. Ryser, *Combinatorial Mathematics, Carus Mathematical Monograph No. 14.* (1963).

[21] T. C. Ralph, Quantum computation: Boson sampling on a chip, *Nature Photonics.* **7**, 514–515, (2013).

[22] M. A. Broome, A. Fedrizzi, S. Rahimi-Keshari, J. Dove, S. Aaronson, T. C. Ralph, and A. G. White, Photonic boson sampling in a tunable circuit, *Science.* **339**, 794–798, (2013).

[23] J. B. Spring, B. J. Metcalf, P. C. Humphreys, W. S. Kolthammer, X.-M. Jin, M. Barbieri, A. Datta, N. Thomas-Peter, N. K. Langford, D. Kundys, J. C. Gates, B. J. Smith, P. G. R. Smith, and I. A. Walmsley, Boson sampling on a photonic chip, *Science.* **339**, 798–801, (2013).

[24] Anonymous, The rise of the boson-sampling computer, *Photonics Spectra.* **47**, 33–34, (2013).

[25] M. Tillmann, B. Dakic, R. Heilmann, S. Nolte, A. Szameit, and P. Walther, Experimental boson sampling, *Nature Photonics.* **7**, 540–544, (2013).

[26] A. Crespi, R. Osellame, R. Ramponi, D. J. Brod, E. F. Galvao, N. Spagnolo, C. Vitelli, E. Maiorino, P. Mataloni, and F. Sciarrino, Integrated multimode interferometers with arbitrary designs for photonic boson sampling, *Nature Photonics.* **7**, 545–549, (2013).

[27] J. P. Dowling. (2013). URL `http://quantumpundit.blogspot.com/2013/07/sampling-schmampling.html`.

[28] P. P. Rohde, Optical quantum computing with photons of arbitrarily low fidelity and purity, *Phys. Rev. A.* **86**, 052321 (2012).

[29] P. P. Rohde and T. C. Ralph, Error tolerance of the boson-sampling model for linear optics quantum computing, *Phys. Rev. A.* **85**, 022332 (2012).

[30] Z. Jiang, M. D. Lang, and C. M. Caves, Mixing nonclassical pure states in a linear-optical network almost always generates modal entanglement, *Phys. Rev. A.* **88**, 044301 (2013).

[31] V. S. Shchesnovich, Sufficient condition for the mode mismatch of single photons for scalability of the boson-sampling computer, *Phys. Rev. A.* **89**, 022333 (2014).

[32] H. Lon and D. F. V. James, Proposal for a scalable universal bosonic simulator using individually trapped ions, *Phys. Rev. A.* **85**, 062329, (2012).

[33] K. R. Motes, J. P. Dowling, S. J. Olson, and P. P. Rohde, Linear optical quantum metrology with single photons — exploiting spontaneously generated entanglement to beat the shotnoise limit, arXiv:1501.01067, (2015).

[34] R. P. Feynman, Simulating physics with computers, *Int. J. Theor. Phys.* **21**, 467–488, (1982).

[35] S. Lloyd, Universal quantum simulators, *Science.* **273**, 1073–1078, (1996).

[36] D. D and J. R, Rapid solution of problems by quantum computation, *Proceedings of the Royal Society of London Series A-Mathematical Physical and Engineering Sciences.* **439**, 553–558, (1992).

[37] L. K. Grover, *Proc. 28th Annual ACM Symposium on the Theory of Computing (STOC).* p. 212, (1996).

[38] D. F. Walls and G. J. Milburn, *Quantum Optics.* (Springer, Berlin, 1994).

[39] D. Aharonov, *Annual Reviews of Computational Physics VI (ed. Stauffer, D.).* (World Scientific, Singapore, 1999).

[40] D. DiVincenzo, The physical implementation of quantum computation, *Fort. Phys.* **48**, 771–783, (2000).

[41] K. Lemr, A. Černoch, J. Soubusta, and M. Dušek, Entangling efficiency of linear-optical quantum gates, *Phys. Rev. A.* **86**, 032321, (2012).

[42] P. P. Rohde, J. G. Webb, E. H. Huntington, and T. C. Ralph, Comparison of architectures for approximating number-resolving photo-detection using non-number-resolving detectors, *New J. Phys.* **9**, 233 (2007). arXiv:0705.4003.

[43] M. J. Fitch, B. C. Jacobs, T. B. Pittman, and J. D. Franson, Photon number resolution using time-multiplexed single-photon detectors, *Phys. Rev. A.* **68**, 043814, (2003).

[44] D. Achilles, C. Silberhorn, C. Sliwa, K. Banaszek, I. A. Walmsley, M. J. Fitch, B. C. Jacobs, T. B. Pittman, and J. D. Franson, Photon number resolving detection using time-multiplexing, *J. Mod. Opt.* **51**, 1499, (2004).

[45] T. Meany, L. A. Ngah, M. J. Collins, A. S. Clark, R. J. Williams, B. J. Eggleton, M. J. Steel, M. J. Withford, O. Alibart, and S. Tanzilli, Hybrid photonic circuit for multiplexed heralded single photons, *Laser & Photonics Reviews.* (2014). ISSN 1863-8899.

[46] X. S. Ma, S. Zotter, J. Kofler, T. Jennewein, and A. Zeilinger, Experimental generation of single photons via active multiplexing, *Phys. Rev. A.* **83**, 043814, (2011).

[47] D. Gottesman and I. L. Chuang, Demonstrating the viability of universal quantum computation using teleportation and single-qubit operations, *Nature (London).* **402**, 390, (1999).

[48] M. A. Nielsen, Optical quantum computation using cluster states, *Phys. Rev. Lett.* **93**, 040503, (2004).

[49] D. E. Browne and T. Rudolph, Resource-efficient linear optics quantum computation, *Phys. Rev. Lett.* **95**, 010501, (2005).

[50] M. Reck, A. Zeilinger, H. J. Bernstein, and P. Bertani, Experimental realization of any discrete unitary operator, *Phys. Rev. Lett.* **73**, 58, (1994).

[51] S. Scheel, Permanents in linear optical networks, *Acta Physica Slovaca* **58**, 675, (2008).

[52] P. W. Shor, Polynomial-time algorithms for prime factorization and discrete logarithms on a quantum computer, *SIAM J. Comput.* **26**, 1484, (1997).

[53] L. G. Valiant, The complexity of computing the permanent, *Theoretical Computer Science.* **8**, 189, (1979).

[54] M. Jerrum, A. Sinclair, and E. Vigoda, A polynomial-time approximation algorithm for the permanent of a matrix with nonnegative entries, *Journal of the ACM.* **51**, 673, (2004).

[55] P. P. Rohde, K. R. Motes, P. A. Knott, and W. J. Munro, Will boson-sampling ever disprove the extended church-turing thesis? (2014). arXiv:1401.2199.

[56] M. A. Broome, A. Fedrizzi, S. Rahimi-Keshari, J. Dove, S. Aaronson, T. C. Ralph, and A. G. White, Photonic boson sampling in a tunable circuit, *Science.* **339**, 6121, (2013).

[57] C. Shen, Z. Zhang, and L.-M. Duan, Scalable implementation of boson sampling with trapped ions, *Phys. Rev. Lett.* **112**, 050504, (2014).

[58] S. Aaronson and A. Arkhipov. Bosonsampling is far from uniform, in Electronic Celloquium on Computational Complexity, page Report No. 135, (2013).

[59] V. S. Shchesnovich, Sufficient condition for the mode mismath of single photons for scalability of the boson sampler computer, *Phys. Rev. A.* **89**, 022333, (2014).

[60] M. C. Tichy, K. Mayer, A. Buchleitner, and K. Mølmer, Stringent and efficient assessment of boson-sampling devices, *Phys. Rev. Lett.* **113**, 020502, (2014).

[61] K. P. Seshadreesan, J. P. Olson, K. R. Motes, P. P. Rohde, and J. P. Dowling, Boson sampling with displaced single-photon Fock states versus single-photon-added coherent states: The quantum-classical divide and computational-complexity transitions in linear optics, *Phys. Rev. A.* **91**, 022334, (2015).

[62] P. P. Rohde, K. R. Motes, P. Knott, J. Fitzsimons, W. Munro, and J. P. Dowling, Sampling generalized cat states with linear optics is probably hard, *Phys. Rev. A.* **91**, 012342, (2015).

[63] M. D. Eisaman, J. Fan, A. Migdall, and S. V. Polyakov, Invited review article: Single-photon sources and detectors, *Review of Scientific Instruments.* **82**, 071101, (2011).

[64] C. C. Gerry and P. L. Knight, *Introductory quantum optics.* (Cambridge University Press, 2005).

[65] A. Migdall, D. Branning, and S. Castelletto, Tailoring single-photon and multiphoton probabilities of a single-photon on-demand source, *Phys. Rev. A.* **66** (5), 053805, (2002).

[66] K. R. Motes, J. P. Dowling, and P. P. Rohde, Spontaneous parametric down-conversion photon sources are scalable in the asymptotic limit for boson sampling, *Phys. Rev. A.* **88**, 063822, (2013).

[67] A. P. Lund, A. Laing, S. Rahimi-Keshari, T. Rudolph, J. L. O'Brien, and T. C. Ralph, Boson sampling from gaussian states, *Phys. Rev. Lett.* **113**, 100502, (2014).

[68] M. Reck, A. Zeilinger, H. J. Bernstein, and P. Bertani, Experimental realization of any discrete unitary operator, *Phys. Rev. Lett.* **73**, 58–61 (1994).

[69] A. Peruzzo, A. Laing, A. Politi, T. Rudolph, and J. L. O'Brien, Multimode quantum interference of photons in multiport integrated devices, *Nature Commun.* **2**, 224, (2011).

[70] A. Politi, M. J. Cryan, J. G. Rarity, S. Yu, and J. L. O'Brien, Silica-on-silicon waveguide quantum circuits, *Science.* **320**, 646, (2008).

[71] J. C. Matthews, A. Politi, A. Stefanov, and J. L. O'Brien, Manipulation of multiphoton entanglement in waveguide quantum circuits, *Nature Photonics.* **3**, 346, (2009).

[72] A. Politi, J. C. F. Matthews, and J. L. O'Brien, Shors quantum factoring algorithm on a photonic chip, *Science.* **325**, 1221, (2009).

[73] K. R. Motes, A. Gilchrist, J. P. Dowling, and P. P. Rohde, Scalable boson-sampling with time-bin encoding using a loop-based architecture, *Phys. Rev. Lett.* **113**, 120501, (2014).

[74] M. Tillmann, S. H. Tan, S. E. Stoeckl, B. C. Sanders, H. de Guise, R. Heilmann, S. Nolte, A. Szameit, and P. Walther. Bosonsampling with controllable distinguishability, (2014). arXiv:1403.3433.

[75] P. P. Rhode, Boson sampling with photons of arbitrary spectra structure, *Phys. Rev. A* **91**, 012307, (2015).

[76] P. E. Ceruzzi. (Cambridge, Mass: MIT Press, 2003).

[77] J. P. Dowling, *Schrödinger's Killer App —Race to Build the World's First Quantum Computer.* (Taylor & Francis, 2013).

Chapter 9

New Approach to Quantum Amplification by Superradiant Emission of Radiation

Gavriil Shchedrin, Yuri Rostovtsev, Xiwen Zhang, and Marlan O. Scully*

Texas A&M University, College Station, TX 77843, USA
University of North Texas, Denton, TX 76203, USA
Princeton University, Princeton, NJ 08544, USA
Baylor University, Waco, TX 76798, USA

We report a new analytical solution of a two–level system driven by an arbitrary time–dependent electromagnetic pulse. The new solution goes beyond the rotating wave approximation (RWA), and is governed by the complex pulse area and the time–dependent Stark shift phase. The present work is motivated by a new concept of the quantum amplification by superradiant emission of radiation (QASER). The original QASER mechanism was suppressed by the Stark effect induced by the strong off–resonant driving field. Furthermore the QASER description was based on the Rabi solution for a driven two–level system derived for a square driving pulse in the RWA. The present approach, based on a new solution for a driven two–level system, is valid for a general time–dependent field beyond the RWA. With a two–mode driving field that is detuned above and below the atomic transition frequency by the same amount Δ we achieve cancellation of the induced Stark shift. In this way we secure parametric excitation of a driven two–level atomic media without a time–dependent Stark shift.

1. Introduction

Simple models are at the heart of fundamental physics: the harmonic oscillator in classical mechanics, the ideal gas in statistical physics, and the two–level system in quantum mechanics. A two–level system, (e.g spin up–spin down system), driven by an electromagnetic pulse is the quintessential problem in nuclear magnetic resonance, laser physics and quantum optics.[1–3]

In this Chapter we report a new solution for a two–level system driven by an arbitrary (i.e. not a square pulse), time–dependent field beyond the rotating wave approximation (RWA). The present solution is based on the time evolution

---

*scully@tamu.edu

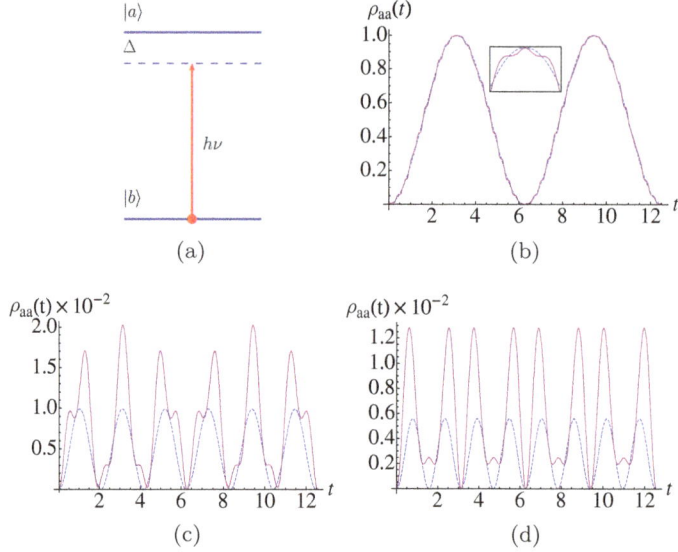

Fig. 1. (a) Two-level atom interacting with an electromagnetic field, $E(t) = E_0 \cos(\nu t)$ (b) Rabi solution of the density matrix element $\rho_{aa}(t)$ obtained from the Hamiltonian of a driven two–level system in the rotating wave approximation (RWA) (dashed) versus Rabi solution beyond the RWA (solid) in the specific case of driving field frequency ν resonant with the atomic transition frequency ω, $\nu = \omega = 10$, and Rabi frequency $\Omega_R = 1$. (c) Rabi solution of the density matrix element $\rho_{aa}(t)$ in the specific case of $\nu/\omega = 2/5$, and $\Omega_R = 3/10$, (d) Rabi solution of the density matrix element $\rho_{aa}(t)$ in the specific case of $\nu/\omega = 1/5$, and $\Omega_R = 3/10$. All the solid curves are based on the analytical solutions and are in excellent agreement with the numerical simulations.

operator technique enhanced by a group–theoretical analysis, which permits simple yet surprisingly accurate asymptotic solutions as shown in Fig. 1.

The present work is motivated by the new concept of quantum amplification by superradiant emission of radiation (QASER).[4,5] The QASER mechanism involves a coherent driving field interacting with an atomic media that results in a parametric resonance and leads to exponential gain for a resonant pulse seed. The original QASER effect was suppressed by the Stark shift induced by the strong off–resonant driving field. Furthermore, the QASER mechanism was analyzed within the RWA. In the RWA the key terms that depend on the difference frequencies between the atomic frequency ω and the field carrier frequency ν, are kept while the counter–rotating terms expressed in terms of the sum of the frequencies are neglected. In this case we have the well–known Rabi solution for a square driving pulse, used in QASER description.

A number of powerful methods have been developed which improve the Rabi solution for a driven two–level system.[6–14] The present approach, based on a new solution for a driven two–level system, is valid for a general time–dependent field beyond the RWA.

2. New Solution of a Driven Two–level System

We start from the exact Hamiltonian of a driven two–level system in the interaction picture,

$$H_{int}(t) = -\hbar \begin{pmatrix} 0 & \Omega(t)\exp[i\omega t] \\ \Omega^*(t)\exp[-i\omega t] & 0 \end{pmatrix}. \tag{1}$$

The light-matter interaction is given in terms of the time–dependent Rabi amplitude $\Omega(t) = \wp_{ab}E(t)/\hbar$, the matrix element of the dipole moment $\wp_{ab} = e\langle a|r|b\rangle$ between the ground state $|b\rangle$ and the excited state $|a\rangle$. The atomic transition frequency is $\omega = \omega_a - \omega_b$ and the detuning is $\Delta = \omega - \nu$. The initial state is prepared in the ground state,

$$|\psi(t_0)\rangle = |b\rangle. \tag{2}$$

Then the well–known Rabi solution[15] of a two–level system obtained in the rotating wave approximation for a square driving pulse is,

$$|\psi(t)\rangle = i\exp\left[\frac{i\Delta t}{2}\right]\frac{\Omega_R}{\mu}\sin\left[\frac{\mu t}{2}\right]|a\rangle \tag{3}$$

$$+ \exp\left[-\frac{i\Delta t}{2}\right]\left(\cos\left[\frac{\mu t}{2}\right] + \frac{i\Delta}{\mu}\sin\left[\frac{\mu t}{2}\right]\right)|b\rangle.$$

Here Ω_R is the Rabi frequency and

$$\mu = \sqrt{\Omega_R^2 + \Delta^2} \tag{4}$$

is the "detuned" Rabi frequency.

The simplest form of the new solution for a driven two–level system obtained from the time–evolution operator (42, 43) and depicted in Fig. 1 is

$$|\psi(t)\rangle = i\exp\left[-i\phi(t)\right]\sqrt{\frac{\theta}{\theta^*}}\sin\left[\sqrt{\theta\theta^*}\right]|a\rangle + \tag{5}$$

$$\exp\left[i\phi(t)\right]\cos\left[\sqrt{\theta\theta^*}\right]|b\rangle.$$

Here the complex pulse area $\theta(t)$ is

$$\theta(t) = \int_0^t dt\,\Omega(t)\exp[i\omega t] \tag{6}$$

and the Stark shift phase $\phi(t)$ is given by the field commutator,

$$i\phi(t) = \frac{1}{2}\int_0^t dt\left[\theta^*(t)\frac{\partial}{\partial t}\theta(t) - \theta(t)\frac{\partial}{\partial t}\theta^*(t)\right] = \tag{7}$$

$$\frac{1}{2}\int_0^t dt\,\theta(t)\theta^*(t)\frac{\partial}{\partial t}\left(\ln\left[\frac{\theta(t)}{\theta^*(t)}\right]\right)$$

3. Quantum Amplification by Superradiant Emission of Radiation

We shall emphasize the difference between QASER and laser mechanisms. The Maxwell-Schrödinger equation for the signal field $\Omega_s(t)$ is

$$\left(\frac{\partial}{\partial t} + c\frac{\partial}{\partial z}\right)\Omega_s(t) = i\Omega_a^2\rho_{ab}. \tag{8}$$

Here the collective atomic frequency Ω_a is given in terms of the atomic density n, atomic wavelength λ_{ab}, and spontaneous decay rate γ,

$$\Omega_a^2 = \frac{3}{8\pi}n\lambda_{ab}^2\gamma c \tag{9}$$

and the density matrix equation in the interaction picture is,

$$\frac{\partial\rho_{ab}}{\partial t} = -\gamma\rho_{ab} + i\Omega_s(t)(\rho_{bb} - \rho_{aa}). \tag{10}$$

In the steady state regime we obtain,

$$\rho_{ab} = \frac{i}{\gamma}\Omega_s(t)(\rho_{bb} - \rho_{aa}), \tag{11}$$

$$\frac{\partial}{\partial t}\Omega_s(t) = -\frac{\Omega_a^2}{\gamma}(\rho_{bb} - \rho_{aa})\Omega_s(t) \tag{12}$$

and with no population in the excited state $\rho_{aa} = 0$ we have exponential decay of the pulse (see Fig. 2a)

$$\Omega_s(t) = \exp\left[-\frac{\Omega_a^2}{\gamma}\rho_{bb}\right]\Omega_s(t_0) \tag{13}$$

However for the ultrashort pulses decay rate is negligible (i.e. $\gamma \ll \tau_p$) and pulse evolves as

$$\frac{\partial^2}{\partial t^2}\Omega_s(t) = -\Omega_a^2(\rho_{bb} - \rho_{aa})\Omega_s(t) \tag{14}$$

with gain G given by

$$G = \Omega_a\sqrt{\rho_{aa} - \rho_{bb}}. \tag{15}$$

The laser operates with positive gain that leads to the exponential growth of the electromagnetic pulse (see Fig. 2b). Therefore the necessary condition for the laser operation is the population inversion,

$$\rho_{aa} - \rho_{bb} > 0. \tag{16}$$

If we do not have any population in the excited state $\rho_{aa} = 0$, gain G is purely imaginary and the laser pulse oscillates with collective atomic frequency Ω_a (see Fig. 2a).

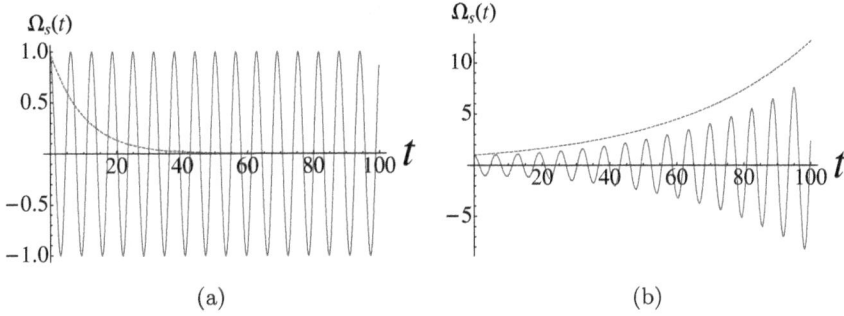

Fig. 2. (a) Exponential decay of the electromagnetic pulse propagating through an atomic media (dotted curve) vs oscillatory time evolution of ultrashort pulse (solid curve) (b) The exponential growth of the electromagnetic pulse with population inversion for the Laser operation (dotted curve) versus parametric resonance of the QASER mechanism (solid curve).

From Eq. (3) we obtain the population in the excited state $|a\rangle$,

$$\rho_{aa} = \left(\frac{\Omega_R}{\mu}\right)^2 \sin^2\left[\frac{\mu t}{2}\right]. \tag{17}$$

Together with conservation of the population

$$\rho_{aa} + \rho_{bb} = 1 \tag{18}$$

in the large detuned case we have the well-known Mathieu equation,

$$\frac{\partial^2}{\partial t^2}\Omega_s(t) + \Omega_a^2\left(1 + \left(\frac{\Omega_R}{\Delta}\right)^2 \cos(\Delta t)\right)\Omega_s(t) = 0. \tag{19}$$

As a result the parametric resonance[16] occurs if we match detuning Δ with the collective atomic frequency Ω_a (see Fig. 2b),

$$\Delta = 2\Omega_a \tag{20}$$

There are, however, a few problems associated with the present description of the QASER mechanism. First of all, we assumed a square driving pulse while the actual pulse has a Gaussian profile. Furthermore the strong off–resonant driving field induces a strong, time–dependent Stark shift which has not been taken into account. In order to address these issues we have developed a new solution for a driven two–level system. The new solution is valid for a general time–dependent field beyond the RWA and is governed by the complex pulse area and the time–dependent Stark shift phase. As a result it provides a new approach to quantum amplification by superradiant emission of radiation.

4. Collective Excitation with a Single Mode Driving Field

First we consider the collective excitation of an atomic media with a single drive field. The complex pulse area $\theta(t)$ is given by the sum of the two terms,

$$\theta(t) = \theta_s(t) + \theta_d(t),$$ (21)

$$\theta_s(t) = \int_0^t dt\, \Omega_s(t) \exp\left[i\omega t\right],$$

$$\theta_d(t) = \int_0^t dt\, \Omega_d(t) \exp\left[i\omega t\right].$$

We consider the driving field

$$\Omega_d(t) = \Omega_d \cos[\nu t]$$ (22)

for which the driving pulse area becomes,

$$\theta_d(t) = \int_0^t dt\, \Omega_d(t) \exp\left[i\omega t\right] =$$

$$\Omega_d \left(\frac{e^{\frac{1}{2}it(\omega-\nu)} \sin\left[\frac{1}{2}t(\omega-\nu)\right]}{\omega - \nu} + \frac{e^{\frac{1}{2}it(\omega+\nu)} \sin\left[\frac{1}{2}t(\omega+\nu)\right]}{\omega + \nu} \right)$$

$$\simeq e^{\frac{1}{2}it\Delta} \frac{\Omega_d}{\Delta} \sin\left[\frac{\Delta t}{2}\right].$$ (23)

Then the Stark shift phase can be calculated in a straightforward manner,

$$i\phi(t) = \frac{1}{2} \int_0^t dt \left[\theta^*(t) \frac{\partial}{\partial t}\theta(t) - \theta(t) \frac{\partial}{\partial t}\theta^*(t) \right] =$$ (24)

$$= i\frac{\Omega_d^2}{4}\frac{\omega}{\nu} \left(\frac{2\nu t + \sin[2\nu t]}{\omega^2 - \nu^2} \right) +$$

$$i\omega\Omega_d^2 \left(\frac{\nu \cos[\omega t]\sin[\nu t] - \omega \cos[\nu t]\sin[\omega t]}{(\omega^2 - \nu^2)^2} \right) \simeq \frac{i}{4}\frac{\Omega_d^2}{\Delta}t.$$

The Maxwell-Schrödinger equation for the signal field is

$$\left(\frac{\partial}{\partial t} + c\frac{\partial}{\partial z} \right) \frac{\partial}{\partial t}\theta_s(t) = i\Omega_a^2 \rho_{ab},$$ (25)

where the off–diagonal density matrix element is given by

$$\rho_{ab} = \frac{i}{2} \exp\left[-2i\phi(t)\right] \sqrt{\frac{\theta}{\theta^*}} \sin\left[2\sqrt{\theta\theta^*}\right],$$ (26)

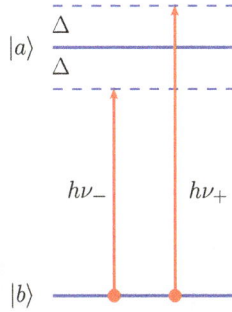

Fig. 3.   The two–mode driving field that is detuned above and below the resonance by the same amount Δ which provides cancellation of the induced Stark shift.

Finally the time evolution of the pulse is governed by,

$$\frac{\partial^2}{\partial t^2}\theta_s(t) = -\frac{\Omega_a^2}{2}\exp\left[-2i\phi(t)\right]\sqrt{\frac{\theta}{\theta^*}}\sin\left[2\sqrt{\theta\theta^*}\right].\tag{27}$$

The presence of the time-dependent Stark shift $\exp\left[-2i\phi(t)\right]$ prevents the parametric resonance.

5. Collective Excitation with Drive Fields Above-and-below Atomic Transition Frequency

The QASER mechanism with a single mode driving field is limited by the Stark effect induced by the strong off-resonant driving field. Now we present a new QASER approach that provides Stark–free parametric excitation of the atomic media driven by the field detuned above and below the atomic transition frequency by the same amount Δ.

We consider the driving field (see Fig. 3) in the form,

$$\Omega_d(t) = \Omega_d\left(\cos[(\omega+\Delta)t] + \cos[(\omega-\Delta)t]\right).\tag{28}$$

If we disregard the highly oscillating terms of the driving field in the interaction picture we obtain,

$$\Omega_d(t)\exp\left[i\omega t\right] = \tag{29}$$
$$\Omega_d\exp\left[i\omega t\right]\left(\cos[(\omega+\Delta)t] + \cos[(\omega-\Delta)t]\right) \simeq \Omega_d\cos(\Delta t)$$

and the driving pulse area becomes

$$\theta_d(t) = \int_0^t dt\,\Omega_d(t)\exp\left[i\omega t\right] = \frac{\Omega_d}{\Delta}\sin(\Delta t).\tag{30}$$

As a result we achieve the cancellation of the time–dependent Stark shift with the "above–and–below" driving scheme,

$$i\phi(t) = \int_0^t dt \left\{ \theta(t)\theta^*(t) \frac{\partial}{\partial t} \left(\ln \left[\frac{\theta(t)}{\theta^*(t)} \right] \right) \right\} = 0. \tag{31}$$

If we disregard the space propagation of the signal field the Maxwell–Schrödinger equation for the signal field becomes the generalized Mathieu equation for the pulse area,

$$\frac{\partial^2}{\partial t^2} \theta_s(t) = -\frac{\Omega_a^2}{2} \sqrt{\frac{\theta}{\theta^*}} \sin\left[2\sqrt{\theta\theta^*} \right]. \tag{32}$$

The oscillating part of the off-diagonal density matrix element can be simplified into

$$\sin\left[2\theta(t) \right] \simeq 2\theta_s(t) \cos\left[2\theta_d(t) \right] + \sin\left[2\theta_d(t) \right] \tag{33}$$

and in the leading order on the parameter,

$$\frac{\Omega_d}{\Delta} \ll 1, \tag{34}$$

the time-evolution of the signal field becomes

$$\frac{\partial^2}{\partial t^2} \theta_s(t) + \Omega_a^2 \left(1 + \left(\frac{\Omega_d}{\Delta} \right)^2 \cos\left(2\Delta t \right) \right) \theta_s(t) + \tag{35}$$

$$\Omega_a^2 \left(\frac{\Omega_d}{\Delta} \sin(\Delta t) \right) = 0.$$

With the ansatz,

$$\theta_s(t) = a(t)e^{it\Delta} + b(t)e^{-it\Delta} \tag{36}$$

and neglecting higher order derivatives we arrive at the system of equations,

$$2i\Delta \frac{\partial}{\partial t} a(t) + a(t) \left(\Omega_a^2 - \Delta^2 \right) + \frac{\Omega_a^2 \Omega_d^2}{2\Delta^2} b(t) - \frac{i\Omega_a^2 \Omega_d}{2\Delta} = 0, \tag{37}$$

$$-2i\Delta \frac{\partial}{\partial t} b(t) + b(t) \left(\Omega_a^2 - \Delta^2 \right) + \frac{\Omega_a^2 \Omega_d^2}{2\Delta^2} a(t) + \frac{i\Omega_a^2 \Omega_d}{2\Delta} = 0. \tag{38}$$

The system of equation has exponentially growing solution (see Fig. 4) with a real positive gain G given by,

$$G = \frac{\sqrt{\Omega_a^4 \Omega_d^4 - 4\Delta^4 \left(\Delta^2 - \Omega_a^2 \right)^2}}{4\Delta^3}. \tag{39}$$

The extremum condition for the gain is satisfied with detuning

$$\Delta = \sqrt[4]{\frac{\Omega_a^4}{2} + \frac{\Omega_a^4}{2} \sqrt{\Omega_a^4 - 3\Omega_d^4}} \simeq \Omega_a. \tag{40}$$

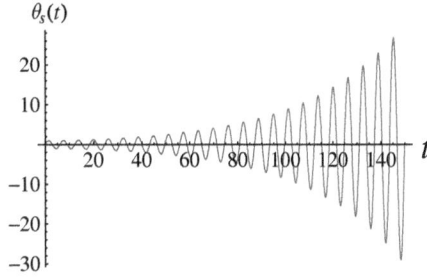

Fig. 4.   The effect of the parametric resonance for the signal pulse area $\theta_s(t)$. The parametric resonance occurs if the detuning of the driving field Δ matches with the collective resonance $\Delta = \Omega_a$.

Then the gain is given by

$$G_0 = \frac{\Omega_a^2}{4\Delta}\left(\frac{\Omega_d}{\Delta}\right)^2.$$

(41)

6. Time Evolution Operator of a Driven Two–level System

The new solution of a driven two–level system was obtained from the time evolution operator,

$$U(t) = \hat{T}\exp\left[-\frac{i}{\hbar}\int_0^t dt'\,H_{int}(t')\right]$$

(42)

$$\simeq \underbrace{\exp\left[-\frac{1}{2\hbar^2}\int_0^t dt'\int_0^{t'} dt''\,[H_{int}(t'),H_{int}(t'')]\right]}_{\text{Stark shift}}$$

$$\times \underbrace{\exp\left[-\frac{i}{\hbar}\int_0^t dt'\,H_{int}(t')\right]}_{\text{Field dynamics}}.$$

It can be reduced to the following matrix,

$$U(t) =$$

(43)

$$\begin{pmatrix} e^{[-i\phi(t)]}\cos\left[\sqrt{\theta\theta^*}\right] & ie^{[-i\phi(t)]}\sqrt{\frac{\theta}{\theta^*}}\sin\left[\sqrt{\theta\theta^*}\right] \\ ie^{[i\phi(t)]}\sqrt{\frac{\theta^*}{\theta}}\sin\left[\sqrt{\theta\theta^*}\right] & e^{[i\phi(t)]}\cos\left[\sqrt{\theta\theta^*}\right] \end{pmatrix}.$$

Projection of the time evolution operator on the ground state returns,

$$|\psi(t)\rangle = i\exp\left[-i\phi(t)\right]\sqrt{\frac{\theta}{\theta^*}}\sin\left[\sqrt{\theta\theta^*}\right]|a\rangle +$$

(44)

$$\exp\left[i\phi(t)\right]\cos\left[\sqrt{\theta\theta^*}\right]|b\rangle.$$

where the complex pulse area $\theta(t)$ is

$$\theta(t) = \int_0^t dt\, \Omega(t) \exp\left[i\omega t\right] \tag{45}$$

and the Stark shift phase $\phi(t)$ is given by the field commutator,

$$i\phi(t) = \frac{1}{2} \int_0^t dt\, \left[\theta^*(t)\frac{\partial}{\partial t}\theta(t) - \theta(t)\frac{\partial}{\partial t}\theta^*(t)\right] = \tag{46}$$

$$\int_0^t dt\, \theta(t)\theta^*(t)\frac{\partial}{\partial t}\left(\ln\left[\frac{\theta(t)}{\theta^*(t)}\right]\right).$$

In order to prove the new solution we consider the time derivative of the state vector,

$$i\hbar\frac{\partial}{\partial t}|\psi(t)\rangle = i\hbar\frac{\partial}{\partial t}U(t)|\psi(t_0)\rangle = i\hbar\frac{\partial}{\partial t}\exp\left[A(t)\right]|\psi(t_0)\rangle, \tag{47}$$

where the time–dependent operator $A(t)$ is given by

$$A(t) = -\frac{i}{\hbar}\int_0^t dt'\, H(t') - \frac{1}{2\hbar^2}\int_0^t dt'\int_0^{t'} dt''\,[H(t'), H(t'')]. \tag{48}$$

The derivative of a time–dependent matrix exponent is

$$\frac{\partial}{\partial t}\exp\left[A(t)\right] = \tag{49}$$

$$\left(\frac{\partial A(t)}{\partial t} + \frac{1}{2!}\left[A(t), \frac{\partial A(t)}{\partial t}\right] + \mathcal{O}\left(A^3\right)\right)\exp\left[A(t)\right].$$

First we note

$$\frac{\partial A(t)}{\partial t} = -\frac{i}{\hbar}H(t) - \frac{1}{2\hbar^2}[H(t), \int_0^t dt''\, H(t'')]. \tag{50}$$

Therefore the sum of the first two terms yields the Hamiltonian of a driven two–level system,

$$\frac{\partial A(t)}{\partial t} + \frac{1}{2!}\left[A(t), \frac{\partial A(t)}{\partial t}\right] = -\frac{i}{\hbar}H(t) + \mathcal{O}\left(A^3\right), \tag{51}$$

and we obtain the Schrödinger equation,

$$i\hbar\frac{\partial}{\partial t}|\psi(t)\rangle = i\hbar\frac{\partial}{\partial t}\exp\left[A(t)\right]|\psi(t_0)\rangle = \tag{52}$$

$$H(t)\exp\left[A(t)\right]|\psi(t_0)\rangle = H(t)|\psi(t)\rangle.$$

7. Conclusions

In conclusion we have considered the collective excitation of a driven two–level system. We have shown that the field–atom interaction described by system of the Maxwell–Schrödinger equations leads to a parametric resonance equation, known as the Mathieu equation, for the pulse area. We demonstrated that the parametric resonance occurs if we match the detuning of the driving field with the collective atomic frequency which leads to the exponential gain of the pulse area. As a result we achieved a quantum amplification by superradiant emission of radiation (QASER) for a driven two–level system. The original QASER mechanism was suppressed by the Stark effect induced by the strong off–resonant driving field. Moreover the original QASER description was based on the Rabi solution for a driven two–level system derived for a square driving pulse along with the rotating wave approximation (RWA). The present approach, based on a new solution for a driven two–level system, is valid for a general time–dependent field beyond the RWA. With a two–mode driving field that is detuned above and below the atomic transition frequency by the same amount Δ we achieved cancellation of the induced Stark shift. In this way we secured parametric excitation of a driven two–level atomic media without a time–dependent Stark shift.

The authors gratefully acknowledge stimulating discussions with Chris O'Brien, Wolfgang O. Schleich, and Anatoly Svidzinsky. The research was supported by the National Science Foundation Grants PHY-1241032 (INSPIRE CREATIV), PHY-1205868, PHY-1068554, EEC-0540832 (MIRTHE ERC) and the Robert A. Welch Foundation Award A-1261.

References

1. R. W. Boyd, *Nonlinear Optics*, Academic Press (2003).
2. P. Meystre and M. Sargent, *Elements of Quantum Optics*, Springer (2010).
3. E. T. Jaynes and F.W. Cummings, Proc. IEEE **51**, 89 (1963).
4. A. Svidzinsky, L. Yuan, and M. O. Scully, Phys. Rev. X **3**, 041001 (2013).
5. M. O. Scully, Laser Phys. **24**, 094014 (2014)
6. E. Barnes and S. Das Sarma, Phys. Rev. Lett. **109**, 060401, (2012).
7. N. Y. Yao, C. R. Laumann, A. V. Gorshkov, S. D. Bennett, E. Demler, P. Zoller, and M. D. Lukin, Phys. Rev. Lett. 109, 266804, (2012)
8. G. C. Hegerfeldt, Phys. Rev. Lett. 111, 260501, (2013).
9. N. Navon, S. Kotler, N. Akerman, Y. Glickman, I. Almog, and R. Ozeri, Phys. Rev. Lett. 111, 073001, (2013)
10. G. Oelsner, P. Macha, O. V. Astafiev, E. Ilichev, M. Grajcar, U. Hbner, B. I. Ivanov, P. Neilinger, and H.-G. Meyer, Phys. Rev. Lett. 110, 053602, (2013)
11. S. Gustavsson, J. Bylander, and W. D. Oliver, Phys. Rev. Lett. 110, 016603, (2013)
12. J. Restrepo, C. Ciuti, and I. Favero, Phys. Rev. Lett. 112, 013601, (2014)
13. Y. V. Rostovtsev, H. Eleuch, A. Svidzinsky, H. Li, V. Sautenkov, and M. O. Scully, Phys. Rev. A **79**, 063833 (2009).
14. P. K. Jha and Y.V. Rostovtsev, Phys. Rev. A **81**, 033827 (2010).
15. M. O. Scully and M. S. Zubairy, *Quantum Optics*, Cambridge (1997).
16. L. D. Landau and E.M. Lifshitz, *Mechanics*, Pergamon Press, (1969).

V

Ultrafast Dynamics
in Strong Laser Fields

Chapter 10

Circularly Polarized Attosecond Pulses and Molecular Atto-Magnetism

André D. Bandrauk* and Kai-Jun Yuan

Laboratoire de Chimie Théorique, Faculté des Sciences, Université de Sherbrooke, Sherbrooke, Québec, Canada, J1K 2R1

Various schemes are presented for the generation of circularly polarized molecular high-order harmonic generation (MHOHG) from molecules. In particular it is shown that combinations of counter-rotating circularly polarized pulses produce the lowest frequency Coriolis forces with the highest frequency recollisions, thus generating new harmonics which are the source of circular polarized attosecond pulses (CPAPs). These can be used to generate circularly polarized electronic currents in molecular media on attosecond time scale. Molecular attosecond currents allow then for the generation of ultrashort magnetic field pulses on the attosecond time scale, new tools for molecular atto-magnetism (MOLAM).

1. Introduction

Attosecond science opens the door to real time observations and control of electron dynamics on the electron's natural time scale, the attosecond (1 asec = 10^{-18} s).[1-3] The study of ultrafast electron motion in matter from molecules to materials is a new frontier of modern science due to the rapid evolution of laser technology, allowing for the synthesis and even shaping of ever shorter (few cycles) and more intense laser pulses. Since 152 asec is the classical period of revolution of the 1s electron in the ground state of the H atom, increasing laser intensity and/or nuclear charge as in the U atom ($Z = 92$), the energy band width $\Delta E = mc^2$ necessary to cover the electron (e^-)-positron (e^+) threshold for the creation of antimatter requires pulse durations of $\hbar/mc^2 \simeq 1$ zeptosecond (1 zps=10^{-21} s).[4] It is through the interaction of current ultrashort intense pulses of intensity $I_0 \geq 3.5 \times 10^{16}$ W/cm^2, corresponding to an electric field strength $E_0 \geq 5.14 \times 10^9$ V/cm, the atomic units (a.u.) of field intensity I_0 and field strength E_0, that one enters a new regime of laser-matter interaction, the highly nonlinear, nonperturbative regime where the electric **E**-magnetic **B** fields of intense pulses control electron-nuclear motion.[5] This regime allows for the generation of new coherent attosecond pulses

---

*Canada Research Chair, Computational Chemistry and Molecular Photonics; andre.bandrauk@usherbrooke.ca

for monitoring and controlling electrons in molecules[3,6] and zeptosecond pulses for nuclear fusion.[7,8]

The field of attosecond science is a new emerging science requiring improvement and development via theory and high level simulations the quality and stability of attosecond and even zeptosecond pulses. Quantum numerical simulations have been at the forefront of this research developing these new tools for tackling also photochemistry, photobiological and molecular photonics problems.[9] Molecular high-order harmonic generation (MHOHG) is now used to monitor electron dynamics in molecules,[3,5,10] to probe electrons at surface,[11] and to transport electron coherently to large distance[12] on attosecond-femtosecond time scales. The synthesis of attosecond pulses relies on the time-energy uncertainty principle, $\Delta E \Delta t \geq \hbar$, i.e., collecting together light coherent sources with an energy bandwidth ΔE gives a pulse with duration $\tau \simeq \hbar/\Delta E$. The shortest pulse, a delta function time pulse $\delta(t)$ used to imaging molecular orbitals[13] illustrates this simple principle:

$$\delta(t) = \frac{e^{i\phi}}{2\pi} \int_{-\infty}^{\infty} e^{i\omega t} dt. \tag{1}$$

Of note is that the strength of all electric field amplitudes $e^{i\omega t}$ is equal in addition to the phase $e^{i\phi}$. To date the most convenient source of such electric field amplitudes is high-order harmonic generation (HHG) for atoms and MHOHG in molecules. Furthermore, to date HHG and/or MHOHG are obtained solely using linearly polarized excitation electric pulses $E(t) = E_0 f(t) \cos(\omega t + \phi)$, where $f(t)$ is the pulse envelope, ϕ the carrier envelope phase (CEP), and E_0 the electric field amplitude which allows to define the maximum intensity $I_0 = cE_0^2/8\pi$.

Such linearly polarized pulses imply recollisions of ionized electron with the parent ion,[14,15] thus allowing to predict the maximum energy and order N in HHG as:

$$N\hbar\omega = I_p + 3.17U_p, \tag{2}$$

where $U_p = I_0/4m_e\omega^2$, is the pondremotive energy and I_p is the ionization potential, previously obtained in numerical simulations.[16] As pointed out by Corkum, for circularly polarized light, electron trajectories never return to the vicinity of the ion and electron-ion interactions are not important for such polarization.[14] Nevertheless circular polarization ionization in the long wavelength limit results in broad electron classical energy distributions, called above threshold ionization (ATI) spectra, with peaks at $2U_p$.[17] The general conclusion from such early work was that whereas the width of the energy distribution for linearly polarized pulses is determined by the phase angle ϕ at which the electron is ionized, the energy width for circularly polarized light is characteristic of the pulse envelope $f(t)$.[14]

Recent classical nonlinear dynamical theory and simulations show that in strong circularly polarized laser fields, key periodic classical orbits can drive recollisions due to Coulomb potentials.[18] Nonadiabatic theory of strong field atomic ionization has also shown that recollisions and correlated double and triple ionization are found to be possible with elliptical polarization.[19] Atomic ionization by strong elliptically polarized laser pulses has also been studied analytically and numerically[20-22] based on the original semi-classical model of Keldysh.[23] Contrary to atoms, molecules offer the opportunity of examining the effect of Coulomb multi-center effects on strong field ionization and laser induced recollisions. One important manifestation of such effects is laser induced electron diffraction (LIED)[13] by long pulses and recently by linearly and circularly polarized attosecond pulses.[24] MHOHG differs dramatically from atomic HHG by the emission of elliptically polarized MHOHG spectra even with linearly polarized laser pulses due to the nonspherical, but cylindrical symmetry in diatomic molecules.[25] Furthermore attosecond circular dichroism is inherent in MHOHG of aligned molecules due to multiple molecular electronic continua.[26] The effect of nuclear motion on MHOHG and on generation of attosecond pulses in linearly polarized intense laser pulses obtained from non-Born-Oppenheimer numerical solutions of the time-dependent Schrödinger equation (TDSE) for one dimension (1D) H_2 has shown that nuclear motion shortens attosecond pulse trains originating from the first electron ionization.[27] Another important difference between molecular and atomic HHG is the possibility in linear polarization for the ionized electron to recombine with neighboring ions at large internuclear distance thus extending MHOHG plateaus beyond the atomic $3.17U_p$ cut-off law [Eq. (2)] to $8U_p$.[15,28,29]

Strong field electron emission from aligned H_2^+ ions with circularly polarized laser pulses has shown complex laser driven electron dynamics due to the two-center Coulomb potentials.[30] Another source of the circularly polarized HHG spectra is ring current initial states[31] and non-zero initial velocity electrons.[32] In general, since circularly polarized laser pulses create electrons with "spinning" electron trajectories with very large radii, recollisions with parent ions is very unlikely in atoms and molecules at equilibrium, except at large internuclear distance,[33] thus making circularly polarized HHG very inefficient. Alternative, solutions to induce recollisions with circularly polarized pulses have been proposed using bicircular fields.[34-36] Milošević and Becker have analyzed the case of superposition of two coplanar counter-rotating circularly polarized fields. They have shown that superposing a few cycle circularly polarized pulse and a long circularly polarized counter-rotating pulse can generate a field that is linearly polarized for extremely short time and this can be used to produce single linearly polarized attosecond pulses[36] in atomic media. We have also studied MHOHG, i.e., in molecular media, and have shown that single circularly polarized pulses with time modulations of the envelope $f(t)$ gives rise to two different frequency circularly polarized pulses and always lead to recollisions with the molecular plane.[37,38] This is achieved by studying the laser induced electron dynamics in a rotating frame, which separates the dynamics under the influence of a *Coriolis* force and the *recollisions* potential. In both papers it is shown that the Coriolis force can be controlled by a static magnetic field of strength B (gauss). The effect of static magnetic and electric fields has also been examined theoretically for linear laser polarization with the condition that such a scenario will extend

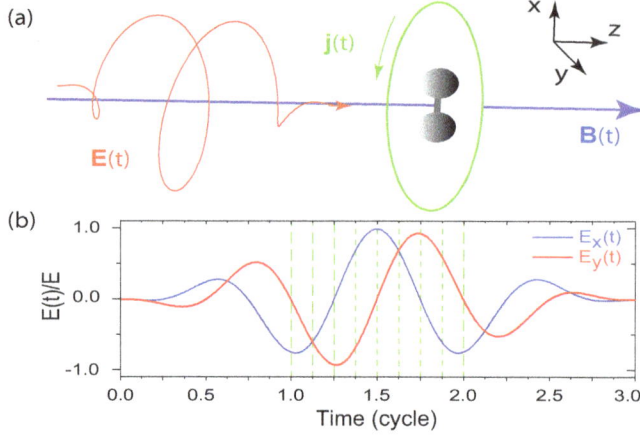

Fig. 1.   (a) Illustration of the attosecond magnetic fields $\mathbf{B}(\mathbf{r}, t)$ (blue line, along the z axis) for H_2^+ created by a few cycle circularly polarized attosecond UV pulse (red line). The green line represents the corresponding current $\mathbf{j}(\mathbf{r}, t)$ in the molecular (x, y) plane. The magnetic field $\mathbf{B}(\mathbf{r}, t)$ is perpendicular to the current $\mathbf{j}(\mathbf{r}, t)$. (b) Three cycle circularly polarized attosecond UV pulse $\mathbf{E}(t)$ at $\lambda = 50$ nm. Dashed lines correspond to the times in Table 1.

the maximum HHG energy cut-off beyond $3.17U_p$.[39] Using the ideas proposed previously by us,[37,38] we have recently shown new methods for generating circularly polarized MHOHG,[33,40] circularly polarized attosecond pulses,[41] and attosecond magnetic field pulses.[42] We note that other previous methods to produce sources of circularly polarized HHG relied on reflector phase-shifters of linearly polarized infrared laser light with an efficiency of a few percent, thus proposing such new pulses as new tools for the investigation of ultrafast magnetization dynamics.[43,44] In the next section we present and expand the theoretical ideas which propose a new science based on circularly polarized attosecond laser pulses: Atto-Magnetism.

2. Theoretical Molecular Model

We begin with the exact three-dimension (3D) Hamiltonian of H_2^+ with static (Born-Oppenheimer) nuclei adapted to the magnetic field B (gauss) problem described previously[37,45] (we use atomic units: $e = \hbar = m_e = 1$),

$$i\frac{\partial\psi(x, y, z, t)}{\partial t} = [H_0 + H_l + H_B]\,\psi(x, y, z, t), \tag{3}$$

where

$$H_0 = \frac{2m_p + 1}{4m_p}\left[\frac{\partial^2}{\partial x^2} + \frac{\partial^2}{\partial y^2} + \frac{\partial^2}{\partial z^2}\right] - \frac{1}{\sqrt{(x \pm R/2)^2 + y^2 + z^2}}, \tag{4}$$

$$H_l = \kappa E_0 \cos(\omega t)[x\cos(\bar{\omega}t) + y\sin(\bar{\omega}t)]$$
$$= \kappa \frac{E_0}{2}\{x\cos[(\omega + \bar{\omega})t] + y\sin[(\omega + \bar{\omega})t]\}$$
$$+\kappa \frac{E_0}{2}\{x\cos[(\omega - \bar{\omega})t] - y\sin[(\omega - \bar{\omega})t]\}, \tag{5}$$

$$H_B = \beta l_z + \frac{1}{2}\beta^2\rho^2 - \beta,$$
$$l_z = -i\left[x\frac{\partial}{\partial y} - y\frac{\partial}{\partial x}\right], \tag{6}$$
$$\rho = \sqrt{x^2 + y^2}.$$

$\kappa = (2m_p + 2)/(2m_p + 1)$, $m_p = 1837$ is the mass of the proton, $\beta = B/B_0$, and $B_0 = 2.35 \times 10^9$ G, the atomic unit of the magnetic field. We neglect here the effect of the magnetic field on the nuclear motion[46] since the resulting Lorentz force scales as $v/c = 1/137$ for one atomic unit (a.u.) of velocity and should therefore be negligible for the proton whose mass is 1837 a.u. Our model, Fig. 1(a) therefore consists of H_2^+ aligned with the x axis in the presence of a static magnetic field of strength B parallel to the internuclear axis R. The magnetic field induces angular momentum l_z perpendicular to the x molecular axis and a confining potential $\beta^2\rho^2/2$ around that axis (the diamagnetic energy). The parameter that characterizes the magnetic confinement is the Landau radius due to harmonic motion in that confining potential, i.e., $R_L = \sqrt{2/\beta}$. Thus a magnetic field of strength $\beta = 0.1$ (2.35×10^8 G) will induce a Landau radius $R_L \simeq 4.5$ a.u. in the (x, y) molecular plane (Fig. 1). The modulated field in Eq. (5) corresponds to two circularly polarized fields which differ by $2\bar{\omega}$ in frequency, co-rotating at frequency $\omega + \bar{\omega}$ and counter-rotating at frequency $\omega - \bar{\omega}$.

Applying the unitary transformation $\mathcal{T} = \exp(-i\bar{\omega}tl_z)$, which is a rotation around the $+z$ direction by the angle $\bar{\omega}t$, gives the new Hamiltonian

$$H' = H_0 + \frac{1}{2}\beta^2\rho^2 + (\beta - \bar{\omega})l_z + xE_0\cos(\omega t). \tag{7}$$

The new Hamiltonian exhibits a magnetic confinement potential $\frac{1}{2}\beta^2\rho^2$, a rotation term $(\beta - \bar{\omega})l_z$, and a linear driving field term $xE_0\cos(\omega t)$ in the frame rotating at frequency $\bar{\omega}$. This last term will give rise to a ponderomotive energy U_p. The model of MHOHG via recollisions of the ionized electron with the ion core predicts a maximum harmonic order N, Eq. (2). We note here that since x is not diagonal in the angular momentum quantum number m (around the z axis), the last laser-driving term in Eq. (7) couples different angular momentum terms, thus also pumping energy into circular motion.

Equations (3-7) demonstrate a universal behavior of modulated envelope circularly polarized pulses. Considering zero magnetic fields ($\beta = 0$), then at $\bar{\omega} = \omega$, Eq. (5) becomes a single circularly polarized field of frequency 2ω in the presence of a static electric field of strength $E_0/2$, which simplifies Eq. (7) to

$$H' = H_0 - \omega l_z + x E_0 \cos(\omega t), \qquad (8)$$

i.e., a Coriolis force of the same frequency ω as the recollisions electric field but half that of the incident circular pulse. We have used such a pulse combination to obtain circularly polarized MHOHG in H_2^+.[40] Setting $\omega \sim \bar{\omega}$ gives a combination of a fast $\omega + \bar{\omega}$ circularly polarized pulse with a nearly static Terahertz field of frequency $(\omega - \bar{\omega})$. Equation (7) then reduces at $\beta = 0$ to

$$H' = H_0 - \bar{\omega} l_z + x E_0 \cos(\omega t). \qquad (9)$$

This guarantees weaker Coriolis forces of lower frequency $\bar{\omega}$ in the presence of a strong linear ionizing field of higher frequency ω. The effect of the strong field E_0 is to force recollisions and produce a circularly polarized MHOHG spectrum.[41] It is to be noted that a single circularly polarized pulse results in the rotating frame with the Hamiltonian at $\omega = 0$, i.e., a Coriolis force with a static electric field E_0 only, Eq. (8). Such a Hamiltonian has been previously studied for Rydberg states in microwave fields, creating static classical periodic orbits.[47]

Static electric field induced polarizations have been studied in harmonic generation with the general condition that such fields result in elliptic dichroism in which the harmonic yield is different for right and left elliptically polarized laser fields.[48] In circularly polarized atomic HHG and molecular MHOHG, one requires the harmonic field components $E_x = E_y$ and maintain a phase difference of $\pi/2$[40–42] due to recollisions, involving therefore complex laser electron recollisions dynamics.[33] Equation (5) clearly predicts two different bicircular combinations. For $\omega > \bar{\omega}$, one obtains two *counter-rotating* circularly polarized laser pulses whereas for $\omega < \bar{\omega}$, one obtains two *co-rotating* pulses. Equation (7) predicts for all two cases a strong linearly polarized pulse which will induce recollisions at frequency ω and a corresponding ponderomotive energy U_p. We define next different frequency ratios $\bar{\omega}/\omega$ which will lead to various possible schemes for creating HHG or MHOHG, using Eqs. (5) and (7). Thus setting $\bar{\omega} = 2\omega$, one obtains neglecting magnetic fields $(\beta = 0)$,

$$\begin{aligned} E(t) = &\frac{E_0}{2} [x \cos(3\omega t) + y \sin(3\omega t)] \\ &+ \frac{E_0}{2} [x \cos(\omega t) + y \sin(\omega t)], \end{aligned} \qquad (10)$$

$$H'(t) = H_0 - 2\omega l_z + x E_0 \cos(\omega t). \qquad (11)$$

Such a 3ω and ω combination of co-rotating circularly polarized pulses results in recollisions at frequency ω but with a Coriolis force with 2ω frequency.

Setting $\omega = 2\bar{\omega}$ leads to $3\bar{\omega}$ and $\bar{\omega}$ counter-rotating circularly polarized pulses.

$$E(t) = \frac{E_0}{2} [x \cos(3\bar{\omega}t) + y \sin(3\bar{\omega}t)]$$

$$+ \frac{E_0}{2} [x \cos(\bar{\omega}t) - y \sin(\bar{\omega}t)], \tag{12}$$

$$H'(t) = H_0 - \bar{\omega}l_z + xE_0 \cos(2\bar{\omega}t), \tag{13}$$

with now the recollisions frequency $2\bar{\omega}$ is twice that of the Coriolis force. Comparing Eqs. (11) and (13), we note the large Coriolis frequency in the co-rotating system, Eq. (10).

Setting $\bar{\omega} = 3\omega$ leads to 4ω and 2ω co-rotating circularly polarized pulse scheme,

$$E(t) = \frac{E_0}{2} [x \cos(4\omega t) + y \sin(4\omega t)]$$

$$+ \frac{E_0}{2} [x \cos(2\omega t) + y \sin(2\omega t)], \tag{14}$$

$$H'(t) = H_0 - 3\omega l_z + xE_0 \cos(\omega t), \tag{15}$$

where the Coriolis force acts at three times the frequency of the recollisions force.

Setting $\omega = 3\bar{\omega}$ leads to a $4\bar{\omega}$ and $2\bar{\omega}$ counter-rotating scheme with the recollisions force at frequency $3\bar{\omega}$ whereas the Coriolis force has frequency $\bar{\omega}$ only. This scheme has been investigated in detail[34-36] but not in the rotating frame. Finally setting $2(\omega + \bar{\omega}) = 3(\omega - \bar{\omega})$, which corresponds to $\omega = 5\bar{\omega}$ leads to counter-rotating pulses with relative frequencies 3ω and 2ω, i.e., with $\omega = 2\bar{\omega}$,

$$E(t) = \frac{E_0}{2} [x \cos(6\bar{\omega}t) + y \sin(6\bar{\omega}t)]$$

$$+ \frac{E_0}{2} [x \cos(4\bar{\omega}t) - y \sin(4\bar{\omega}t)], \tag{16}$$

$$H'(t) = H_0 - \bar{\omega}l_z + xE_0 \cos(5\bar{\omega}t), \tag{17}$$

whereas the co-rotating combination for $\bar{\omega} = 5\omega$

$$E(t) = \frac{E_0}{2} [x \cos(6\omega t) + y \sin(6\omega t)]$$

$$+ \frac{E_0}{2} [x \cos(4\omega t) + y \sin(4\omega t)], \tag{18}$$

$$H'(t) = H_0 - 5\omega l_z + xE_0 \cos(\omega t). \tag{19}$$

In general setting $\omega = n\bar{\omega}$ leads to counter-rotating pulses with relative frequencies $(n + 1)\bar{\omega}$ and $(n - 1)\bar{\omega}$, Eq. (5) which in the rotating frame with frequency

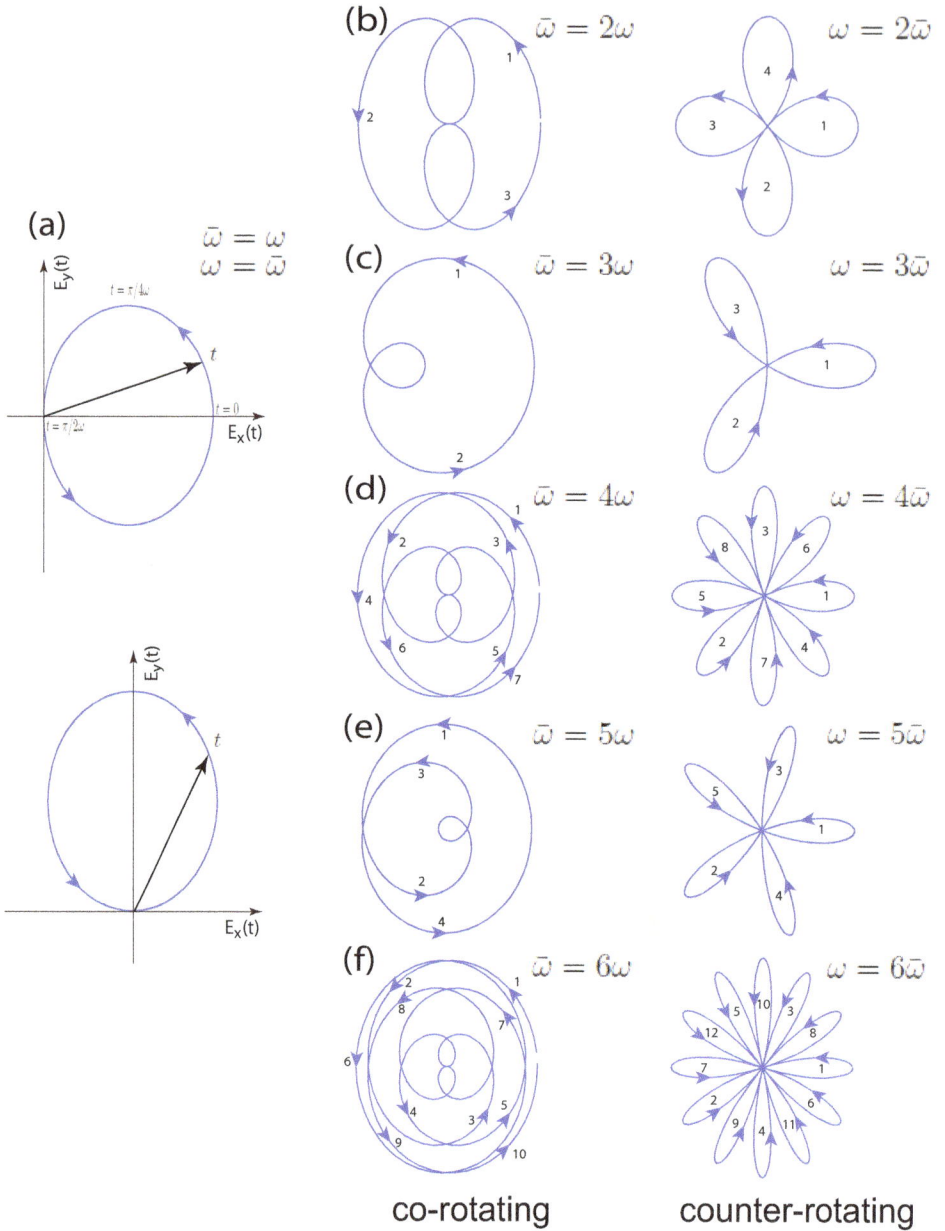

Fig. 2. The electric field vector $\mathbf{E}(t)$ Eq. (5) for the various $(\omega, \bar{\omega})$ combination schemes. Panel (a): For $\omega = \bar{\omega}$, the electric field is a superposition of a circularly polarized pulse and a static field.[37,38] Panels (b-f): For co-rotating combinations $\bar{\omega} = n\omega$ and counter-rotating combinations $\omega = n\bar{\omega}$. Arrows indicate the evolution of the fields with time t.

$\bar{\omega}$, gives Coriolis forces $\bar{\omega} l_z$ and recollisions electron field $E_0 \cos(n\bar{\omega}t)$, Eqs. (13) and (17).

Equations (10-19) confirm that for equal intensity E_0 in co-rotating and counter-rotating pulse combinations, Eq. (5), it is the co-rotating schemes, Eqs. (10), (14), and (18) which always produce large frequency Coriolis forces ωl_z accompanied by lower frequency recollisions forces whereas the counter-rotating schemes, Eqs. (12) and (16) lead to lower frequency Coriolis forces accompanied by much higher frequency recollisions. We illustrate next the net electric vectors in Fig. 2 for the various schemes described above in addition to the counter-rotating combinations $\omega = 4\bar{\omega}$ and $6\bar{\omega}$ and the co-rotating combinations $\bar{\omega} = 4\omega$ and 6ω. We note that for all the counter-rotating combinations $\omega = n\bar{\omega}$, the net electric field includes a zero field amplitude thus maintaining an inversion symmetry. The co-rotating field combinations $\bar{\omega} = n\omega$ show only a reflection symmetry with respect to the x axis. A similar behavior has been found for a $(\omega, 2\omega)$ combination of equal amplitude circular polarizations.[36,49] This corresponds to the $\bar{\omega} = 3\omega$ and $\omega = 3\bar{\omega}$ combinations in Fig. 2. These authors have also concluded that only one quantum orbit contributes to the harmonics in this case. However the HHG spectrum did not show a continuous circularly polarized spectrum.

The case $\omega = \bar{\omega}$ as mentioned above leads to the superposition of a circularly polarized pulse and a static field which in the rotating frame of frequency ω, half the frequency of the incident circularly polarized pulses, leads to a Coriolis force of the same frequency ω as the recollisions force, Eq. (8). It was found that such a combination leads to a circularly polarized MHOHG plateau from which one obtained ~ 100 asec circularly polarized pulses.[40] The intensities required for attosecond pulse generation was $I \geq 10^{14}$ W/cm^2 for both circularly polarized pulse and static fields. Since TeraHertz pulses can approach such intensities and thus generate nearly static fields, numerical simulations of the TDSE for H$_2^+$ showed that attosecond circularly polarized pulses were also generated under similar conditions as the static field scheme.[41] Time-series analysis of the circularly polarized MHOHG spectra confirmed the generation of the harmonics by single recollisions with the parent ion as opposed to the linear polarization excitation spectra.[15]

Equation (7) shows that Coriolis effects can be controlled by the introduction of an external magnetic field $\beta = B/B_0$ (a.u.). The magnetic field furthermore introduces a magnetic confinement potential $\frac{1}{2}\beta^2\rho^2$ in the molecular (x,y) plane (Fig. 2), electron density. Solving exactly the 3D TDSE for the H$_2^+$ molecular ion allowed for the study of the MHOHG spectrum control of electron recollisions with the protons. Results were found of pure even or even and odd harmonic generation for particular field and frequency configurations. Furthermore magnetic fields were found to extend MHOHG plateaus to higher order.[38]

3. Attosecond Magnetic Field Pulse Generation

Intense attosecond magnetic field pulses have been predicted to be produced by intense circularly polarized few cycle or attosecond ultraviolet(UV) laser pulses. Numerical solutions of the TDSE, Eqs. (3-6), yield exact time-dependent functions $\psi(\mathbf{r}, t)$ from which electron accelerations are calculated via the time-dependent Hellmann-Feynman theorem,[15]

$$\ddot{r}(t) = \langle \psi(r,t)| - \partial H/\partial r|\psi(r,t)\rangle - E(t). \tag{20}$$

The MHOHG power spectrum $P_r(\omega)$ is then obtained from the absolute square of the Fourier transform (FT) of Eq. (20),

$$P_r(\omega) = |a_r(\omega)|^2 = |\int \exp(-i\omega t)\ddot{r}(t)dt|^2. \tag{21}$$

We note that the MHOHG spectra can be calculated numerically from the FT of the dipole moment $\langle r(t)\rangle$, velocity $\langle \dot{r}(t)\rangle$, and acceleration $\langle \ddot{r}(t)\rangle$ and must give identical results due to gauge invariance.[50] However it is found that the acceleration form Eq. (20) gives more accurate results due to its emphasis on the recollisions process where the force $-\partial H/\partial r$ is maximum. The temporal profile of the field components $\varepsilon_{x,y}(t)$ for a generated attosecond pulse is obtained further by an inverse FT of the MHOHG amplitudes $a_{x/y}(\omega)$ by,

$$\mathcal{I}_{x/y}(t) = |\varepsilon_{x/y}(t)| = |\int_{\omega_c} \exp(i\omega t)a_{x/y}(\omega)d\omega|. \tag{22}$$

The relative attosecond pulse phase difference φ for circularly polarized pulses is obtained as

$$\varphi(t) = |\arg[\varepsilon_x(t)] - \arg[\varepsilon_y(t)]| = \pi/2. \tag{23}$$

Such new attosecond pulses obtained from circularly polarized MHOHG spectra according to Eqs. (20) and (21), we define now as

$$E(t) = Ef(t)[\hat{e}_x \cos(\omega t) + \hat{e}_y \sin(\omega t)], \tag{24}$$

propagating in the z direction with $\hat{e}_{x/y}$ polarization directions. A smooth n_l cycle $\sin^2(\pi t/n_l\tau)$ pulse envelope $f(t)$ characterizes such an attosecond pulse for maximum electric field amplitude E and intensity $I = c\varepsilon_0 E^2$ and optical cycle $\tau = 2\pi/\omega$. Such a pulse satisfies a total area $\int E(t)dt = 0$.[15] The TDSE at zero static magnetic field ($\beta = 0$), Eq. (3) is solved with the pulse, Eq. (24), providing again the time dependent electron wavefunction $\psi(\mathbf{r}, t)$ for the pulse orientation illustrated in Fig. 1. Such laser-molecule interaction leads to an electronic current $\mathbf{j}(t)$ and corresponding generated magnetic field $\mathbf{B}(r, t)$

$$\mathbf{j}(\mathbf{r}, t) = \frac{i}{2}[\psi(\mathbf{r}, t)\nabla_{\mathbf{r}}\psi^*(\mathbf{r}, t) - \psi^*(\mathbf{r}, t)\nabla_{\mathbf{r}}\psi(\mathbf{r}, t)], \tag{25}$$

$$\mathbf{B}(\mathbf{r}, t) = \frac{\mu_0}{4\pi}\int [\frac{\mathbf{j}(\mathbf{r}', t_r)}{|\mathbf{r} - \mathbf{r}'|^3} + \frac{1}{|\mathbf{r} - \mathbf{r}'|^2 c}\frac{\partial \mathbf{j}(\mathbf{r}', t')}{\partial t}] \times (\mathbf{r} - \mathbf{r}')d^3\mathbf{r}'. \tag{26}$$

Table 1. Maximum local $B_{max}(\mathbf{r}, t)$ and total volume magnetic field $B(t)$ and currents $j(t)$ by circularly polarized attosecond UV pulses with intensity $I = 2 \times 10^{16}$ W/cm^2, wavelength $\lambda = 50$ nm, and duration $3\tau = 500$ asec at different times.

	units	1.0τ	1.25τ	1.5τ	1.625τ	1.75τ	2.0τ
$B_{max}(\mathbf{r}, t)$	T	6.242	12.285	11.971	9.121	6.505	4.167
$B(t)$	T·a_0^3	6.670	17.612	27.797	29.300	27.517	17.320
$j(t)$	fs^{-1}·a_0	5.209	12.568	10.625	10.211	10.294	8.351
	mA·a_0	0.834	2.011	1.700	1.634	1.674	1.336

t_r is the retarded time $t - r/c$ and $\mu_0 = 4\pi \times 10^{-7}$ NA^{-2} (6.692×10^{-4} a.u.). For the static time-independent conditions occurring after the laser pulse, then Eq. (25) reduces to $\mathbf{B}(\mathbf{r}) = \dfrac{\mu_0}{4\pi} \displaystyle\int \dfrac{\mathbf{j}(\mathbf{r}') \times (\mathbf{r} - \mathbf{r}')}{|\mathbf{r} - \mathbf{r}'|^3} d^3\mathbf{r}'$ in accord with the classical Biot-Savart law.[51]

We calculate the maximum local $B_{max}(\mathbf{r}, t)$ and total volume average attosecond magnetic fields and the corresponding currents by integrating $B(t) = |\int \mathbf{B}(\mathbf{r}, t) d\mathbf{r}^3|$ and $j(t) = |\int \mathbf{j}(\mathbf{r}, t) d\mathbf{r}^3|$ over the electron $\mathbf{r}$ space. Table 1 lists values of $B_{max}(\mathbf{r}, t)$, $B(t)$, and $j(t)$ at different moments, illustrated in Fig. 1(b). Both $B(t)$ (in units of T·a_0^3, where a_0 is Bohr radius) and $j(t)$ (fs$^{-1} \cdot a_0$) vary with time, increasing first and then decreasing in phase with the pulse. From Tab. 1 one obtains that the maximum total volume magnetic field is induced at $t = 1.625\tau = 270$ asec with strength $B = 1.172 \times 10^{-4}$ a.u. $= 29.3$ T·a_0^3 (2.93×10^5 Gauss·a_0^3). The maximum local magnetic field $B_{max}(\mathbf{r}, t) = 12.285$ T and the maximum electronic current $j = 0.304$ a.u.$= 12.568$ fs^{-1}·a_0 (2.011 mA·a_0) is produced at time $t = 1.25\tau = 207$ asec, where $E_x = 0$ and $E_y = -E$ (Fig. 1). A time delay $\Delta t = 0.375\tau = 63$ asec occurs between maximum current $j(t = 1.25\tau) = 12.568$ fs^{-1}·a_0 and $B(t = 1.625\tau) = 29.3$ T·a_0^3 (Table 1).

The current electron trajectories are functions of the pulse wavelength λ (frequency ω) and duration $n_l \tau$.[33] We explain this from the classical model,[33] a generalization of the linear polarization model.[14] Assuming the zero initial electron velocities $\dot{x}(t_0) = \dot{y}(t_0) = 0$, where t_0 is the ionization time, the induced time dependent velocities are

$$\begin{aligned}\dot{x}(t) &= -E/\omega\,(\sin\omega t - \sin\omega t_0)\,,\\\dot{y}(t) &= -E/\omega\,(\cos\omega t_0 - \cos\omega t)\,,\end{aligned} \qquad (27)$$

The corresponding displacements are

$$\begin{aligned}x(t) &= -E/\omega^2\,[\cos\omega t_0 - \cos\omega t - (\omega t - \omega t_0)\sin\omega t_0]\,,\\y(t) &= -E/\omega^2\,[\sin\omega t_0 - \sin\omega t + (\omega t - \omega t_0)\cos\omega t_0]\,,\end{aligned} \qquad (28)$$

with $x(t_0) = y(t_0) = 0$ corresponding to recollisions at the center of the molecule ($r = 0$). From Eqs. (27) and (28) it is found that increasing the pulse wavelength λ leads to increase of the maximum induced electron velocity $v = 2E/\omega$ at $\omega t - \omega t_0 =$

$(2n' + 1)\pi$, and the corresponding radii

$$r_{n'} = 2E/\omega^2 \left[1 + (n' + 1/2)^2\pi^2\right]^{1/2},\tag{29}$$

$n' = 0, 1, 2, \cdots$. For a moving point charge the corresponding classical magnetic field can be expressed as

$$\mathbf{B} = \frac{\mu_0}{4\pi} \frac{\mathbf{v} \times \mathbf{r}}{r^3}.\tag{30}$$

From the above equations one then gets the maximum field $B \sim v/r_{n'} \sim \omega$ at times $t_0 + (2n' + 1)\pi/\omega$. Therefore, an increase of λ or lower ω, results in a decrease of the magnetic field due to large radii $r_{n'}$ of the electron, thus reducing the efficiency of the attosecond magnetic field generation. Longer pulse duration have a similar effects. Of note is that we only present the results for the single photon ionization processes. Increasing the pulse frequency results in decrease of the ionization rate in spite of decrease of the radii of the free photoelectron. Consequently, weaker magnetic field pulses are nevertheless produced.

4. Summary

Molecular high-order harmonic generation (MHOHG) is generally an efficient source of coherent radiation by molecules subjected to intense ultrashort ionizing laser pulses. The nonspherical symmetry of molecules eliminates selection rules in spherical atoms. Nevertheless, in both atomic and molecular spectra, recollisions of an ionized electron with its parent ion is the essential process for harmonic generation. Circularly polarized intense laser pulses inhibit in general recollisions with parent ions, so that combinations of different polarization pulses suggest obvious general schemes for recollisions in order to produce circularly polarized harmonics. High order generation of circularly polarized harmonics is thus an essential route for producing circularly polarized attosecond pulses. Schemes involving *co-rotating* and *counter-rotating* intense circularly polarized pulses are analyzed and shown to lead to recollisions in general. Co-rotating pulses result in large Coriolis forces as opposed to counter-rotating combinations of pulses which favour high frequency recollisions. In the limit of near equal frequency co-rotating and counter-rotating pulses, one obtains a scheme involving high frequency pulses in combinations with low frequency Terahertz pulses. Such a combination has been shown to produce copious circularly polarized harmonics for the generation of attosecond pulses.[40,41] The theoretical analysis presented in this paper shows that in principle, one can also use static magnetic fields for controlling Coriolis forces which reduce the efficiency of circularly polairzed MHOHG.Circulalry polarized attosecond pulses provide new tools for creating in matter attosecond coherent electronic currents, which are sources of attosecond magnetic field pulses. These new tools should lead to important advances in the study of magnetic light-matter interaction, in particular in quantifying the magnetic nature of light absorption-emission in ultrafast femto-attosecond nano-optics and spintronics.[43,52]

Acknowledgments

The authors thank RQCHP and Compute Canada for access to massively parallel computer clusters and NSERC, FQRNT for financial support of this research in the ultrafast science programs.

References

1. F. Krausz and M. Ivanov, *Rev. Mod. Phys.* **81**, 163 (2009).
2. Z. Chang and P. Corkum, *J. Opt. Soc. Am. B* **27**, 9 (2010).
3. M. J. J. Vrakking, *Phys. Chem. Chem. Phys.* **16**, 2775 (2014).
4. F. Fillion-Gourdeau, E. Lorin, and A. D. Bandrauk, *Phys. Rev. Lett.* **110**, 013002 (2012); *J. Phys. B* **46**, 175002 (2013).
5. A. D. Bandrauk and M. Y. Ivanov, *Quantum Dynamical Imaging- Numerical and Theoretical Methods.* (Springer, Berlin, 2011).
6. A. D. Bandrauk, J. Manz, and M. J. J. Vrakking, *Chem. Phys.* **366**, 1 (2009).
7. C. Hernández-García, J. A. Pérez-Hernández, T. Popmintchev, M. M. Murnane, H. C. Kapteyn, A. Jaron-Becker, A. Becker, and L. Plaja, *Phys. Rev. Lett.* **111**, 033002 (2013).
8. G. K. Paramonov and A. D. Bandrauk, *AIP Conf. Proc.* **1209**, 7 (2010).
9. T. Renger, V. May, and O. Kühn, *Phys. Rep.* **343**, 137 (2001).
10. A. D. Bandrauk, S. Barmaki, S. Chelkowski, and G. L. Kamta, *Molecular High Order Harmonic Generation*, In eds. K. Yamanouchi, S. L. Chin, P. Agostini, and G. Ferrante, *Progress in Ultrafast Intense Laser Science*, vol.III, pp. 171–205 (Springer, Berlin, 2009).
11. P. Hommelhoff, M. F. Kling, and M. I. Stockman, *Ann. Phys. (Berlin)* **525**, A13 (2013).
12. G. K. Paramonov, O. Kühn, and A. D. Bandrauk, *Phys. Rev. A* **83**, 013418 (2011).
13. T. Zuo, A. D. Bandrauk, and P. B. Corkum, *Chem. Phys. Lett.* **259**, 313 (1996).
14. P. B. Corkum, *Phys. Rev. Lett.* **21**, 1994 (1993).
15. A. D. Bandrauk, S. Chelkowski, and K. J. Yuan, *Int. Rev. At. Mol. Phys.* **2**, 1 (2011).
16. J. L. Krause, K. J. Schafer, and K. C. Kulander *Phys. Rev. Lett.* **68**, 3535 (1992).
17. P. B. Corkum, N. H. Burnett, and F. Brunel, *Phys. Rev. Lett.* **62**, 1259 (1989).
18. A. Kamor, F. Mauger, C. Chandre, and T. Uzer *Phys. Rev. Lett.* **110**, 253002 (2013).
19. X. Wang and J. H. Eberly, *Phys. Rev. Lett.* **103**, 103007 (2009); *J. Chem. Phys.* **137**, 542 (2012).
20. V. D. Mur, S. V. Popruzhenko, and V. S. Popov, *J. Exp. Theor. Phys.* **92**, 777 (2001).
21. C. H. R. Ooi, W. L. Ho, and A. D. Bandrauk, *Phys. Rev. A* **86**, 023410 (2012).
22. I. Barth and O. Smirnova, *Phys. Rev. A* **84**, 063415 (2011).
23. L. V. Keldysh, *Sov. Phys.-JETP* **20**, 1307 (1964).
24. K. J. Yuan, H. Z. Lu, and A. D. Bandrauk, *Chem. Phys. Chem.* **14**, 1496 (2013).
25. X. Zhou, R. Lock, N. Wagner, W. Li, H. C. Kapteyn, and M. M. Murnane *Phys. Rev. Lett.* **102**, 073902 (2009).

26. O. Smirnova, S. Patchkovskii, Y. Mairesse, N. Dudovich, D. Villeneuve, P. Corkum, and M. Y. Ivanov *Phys. Rev. Lett.* **102**, 063601 (2009).
27. A. D. Bandrauk, S. Chelkowski, S. Kawai, and H. Lu *Phys. Rev. Lett.* **101**, 153901 (2008).
28. A. D. Bandrauk, S. Chelkowski, H. Yu, and E. Constant *Phys. Rev. A* **56**, R2537 (1997).
29. A. D. Bandrauk, S. Barmaki, and G. L. Kamta *Phys. Rev. Lett.* **98**, 013001 (2007).
30. M. Odenweller, N. Takemoto, A. Vredenborg, K. Cole, K. Pahl, J. Titze, L. Ph. H. Schmidt, T. Jahnke, R. Dörner, and A. Becker *Phys. Rev. Lett.* **107**, 143004 (2011).
31. X. Xie, A. Scrinzi, M. Wickenhauser, A. Baltuška, I. Barth, and M. Kitzler, *Phys. Rev. Lett.* **101**, 033901 (2008).
32. F. M. Guo, G. Chen, J. G. Chen, S. Y. Li, and Y. J. Yang, *Chin. Phys. B* **22**, 023204 (2012).
33. K. J. Yuan and A. D. Bandrauk, *J. Phys. B* **45**, 074001 (2012).
34. H. Eichmann, A. Egbert, S. Nolte, C. Momma, B. Wellegehausen, W. Becker, S. Long, and J. K. McIver, *Phys. Rev. A* **51**, 3414 (1995).
35. S. Long, W. Becker, and J. K. McIver, *Phys. Rev. A* **52**, 2262 (1995).
36. D. B. Milošević and W. Becker *J. Mod. Opt.* **52**, 233 (2005).
37. T. Zuo and A. D. Bandrauk, *J. Nonlinear Opt. Phys. Mat.* **4**, 533 (1995).
38. A. D. Bandrauk and H. Z. Lu, *Phys. Rev. A* **68**, 043408 (2003).
39. D. B. Milošević and A. F. Starace, *Phys. Rev. A* **60**, 3160 (1999).
40. K. J. Yuan and A. D. Bandrauk, *Phys. Rev. A* **81**, 063412 (2010); *Phys. Rev. A* **83**, 063422 (2011).
41. K. J. Yuan and A. D. Bandrauk, *Phys. Rev. Lett.* **110**, 023003 (2013).
42. K. J. Yuan and A. D. Bandrauk, *Phys. Rev. A* **88**, 013417 (2013).
43. B. Vodungbo, et al, *Opt. Express* **19**, 4346 (2011); *Nat. Commun.* **3**, 999 (2012).
44. G. P. Zhang, W. Hübner, G. Lefkidis, Y. Bai, and T. F. George, *Nat. Phys.* **5**, 499 (2009).
45. A. D. Bandrauk and H. Z. Lu *Phys. Rev. A* **62**, 053406 (2000).
46. B. R. Johnson, J. O. Hirschfelder, and K.-H. Yang *Rev. Mod. Phys.* **55**, 109 (1983).
47. D. Farrelly and T. Uzer *Phys. Rev. Lett.* **74**, 1720 (1995).
48. B. Borca, A. V. Flegel, M. V. Frolov, N. L. Manakov, D. B. Milošević, and A. F. Starace *Phys. Rev. Lett.* **85**, 732 (2000).
49. D. B. Milošević, W. Becker, and R. Kopold, *Phys. Rev. A* **61**, 063403 (2000).
50. A. D. Bandrauk, S. Chelkowski, D. J. Diestler, J. Manz, and K.-J. Yuan *Phys. Rev. A* **79**, 023403 (2009).
51. O. D. Jefimenko, *Electricity and Magnetism: An Introduction to the Theory of Electric and Magnetic Fields* (Electret Scientific Co., Star City, 1989) 2nd ed.; *Am. J. Phys.* **58**, 505 (1990).
52. T. H. Taminiau, S. Karaveli, N. F. van Hulst, and R. Zia *Nat. Commun.* **3**, 979 (2012).

Chapter 11

Many-Electron Response of Gas-Phase Fullerene Materials to Ultraviolet and Soft X-ray Photons

Himadri S. Chakraborty* and Maia Magrakvelidze

Department of Natural Sciences and Center for Innovation and Entrepreneurship, Northwest Missouri State University, 800 University Drive, Maryville, Missouri 64468, USA

This article briefly reviews primary theoretical methods employed from the beginning of the century to predict the response of empty fullerenes, fullerenes endohedrally confining varieties of atoms, and few-layer fullerene nanoonions to the external electromagnetic radiations from ultraviolet to soft x-ray frequencies. Comparisons with available measurements are shown wherever possible and the success of the time-dependent local density approximation (TDLDA) method is predominantly emphasized. For the ionization process, various effects of the coherence, such as, the plasmon resonances, atom-fullerene dynamical hybridization, many-body correlations based on the interchannel coupling in the continuum, and Auger-intercoulombic hybrid multicenter decays of innershell holes are discussed. Finally, a broad outlook is presented that includes attosecond physics with fullerene materials, photoresponse of non-spherical fullerenes as well as fullerenes with off-centered confinements, the bare ion-impact ionization of fullerenes, and explorations beyond spherical systems to cylindrical carbon-nanotubes.

1. Introduction

Oscillations of the correlated electron density in materials stimulated by external radiation are ubiquitous in nature. If a material has boundaries along all or some dimensions, then the charge oscillations quantize, forming standing waves. This quantized oscillation can be described by a paradigmatic many-body quasi-particle, the plasmon, whose frequency for nanometric dimensions of the material falls within the infrared-to-ultraviolet range. Plasmon excitations in a variety of materials, namely, metallic nanoclusters,[1] metallic nanoshells,[2] single-walled carbon nanotubes,[3] thin metallic films,[2] and graphene monolayers[4–6] and bilayers[7,8] have been studied. Closely related collective excitations are lower energy surface plasmons which hold promise as a possible tool for controlling light at subwavelength scales, giving rise to the field of nanophotonics. Carbon fullerenes are known to have a giant dipolar plasmon in the ultraviolet photoionization spectra[9,10] and

---

*himadri@nwmissouri.edu

therefore are of fundamental research interests owing to their eminent stability and symmetry.

Another unique feature of carbon fullerenes is the amount of internal empty space contained within each. Ranging from about a quarter of a nanometer to several nanometers in diameter, this void is, in fact, immense when observed within the context of nanoscience. Such structure intuitively implies that an atom, molecule, cluster, or even a smaller fullerene can be placed into a fullerene, thereby altering the molecular properties of the compound, resulting in the synthesis of a brand-new family of materials - the endohedral fullerenes (endofullerenes),[11] A@C_n in symbol. Endofullerenes are an extremely stable form of nanoscopic molecular matter that can exist even at room temperature. These entrapments have lower-cost sustenance than the traditional field-entrapments, such as the laser cooling or magneto-optical trap processes. Furthermore, they are recently enjoying a rapid improvement in synthesis techniques.[11] Technologically, endofullerenes hold the promise of producing exciting applications. For example: (a) Research is underway to use endohedrally doped fullerenes and bucky onions as seed materials in solid state quantum computations.[12-15] (b) Recent experiments with Ar@C_{60} find evidence of encaged atoms significantly improving the superconducting ability of materials.[16] (c) One proposed biomedical application of endohedral materials is to shield radioactive tracers inside fullerene cages, thereby allowing them to be injected into human bloodstreams to monitor blood flow.[17] Another proposal is using these materials as the carriers during pinpoint medication deliveries to combat diseased human tissue.[18] (d) The ability of fullerenes to sequester metal atoms inside has led researchers to explore their potential as contrast-enhancing agents for magnetic resonance imaging. (e) Very recently, choosing endofullerenes as the acceptor materials in organic photovoltaic devices, a novel route results in higher power consumption efficiency.[19] (f) Finally, the discovery of endofullerenes with trapped noble gases in extraterrestrial environments[20] indicates the astrophysical relevance of their studies.

The encapsulation of a host system in fullerenes, therefore, offers a brilliant natural laboratory to examine the behavior of materials in confinement. Studies of these compounds can not only lead to intriguing effects at the atomic scale or allow us to probe the subtleties of quantum effects in the nanometer region, but also can access the effects of the host-fullerene hybridization. Therefore, understanding the influence of the confining cage on the spectroscopy of the confined materials, and vice versa, are matters of fundamental interest. Moreover, detailed and comprehensive knowledge about the responses of these compounds to external electromagnetic fields becomes especially important as it allows one to assess the potentials and the limitations of their promised applications.

Here is how the article has been organized. In Sec. 2, a succinct account of various theoretical schemes are given with a relatively detailed description of the time-dependent local density approximation (TDLDA) methodology. Section 3 includes some of the recent TDLDA results and their comparisons with measurements and other calculations, wherever possible. Section 4 attempts to identify some possible future research problems.

2. Description of Various Theoretical Methods

There has been a sizable body of theoretical research during last several years which has helped us better understand the details of the photoionization of fullerenes and atomic endofullerenes. Employed theoretical methods can be classified in four groups:

(I) A semiclassical approach using the idea of a dynamical screening of the dielectric fullerene shell has produced a good qualitative description of the low energy photoemissions from the confined atoms.[21] Here the fullerene is modeled as a dielectric shell and classical electrostatic methods are used to calculate the screening factor from the ratio of the confined-to-free atomic absorption rates. The confined absorption rate utilizes the total field on the central atom which is determined on the basis of an iterative model where the primary field on the atom is created jointly by the external photon field and the polarization field of the fullerene. Then the atomic dipole induced by this primary field acts on the fullerene to produce a secondary field on the atom, and iteratively so forth.[22] Some success of this scheme is illustrated in Fig. 3 below.

(II) Important insights of energy dependent oscillations in atomic emissions at higher energies were obtained through calculations in which the confining C_{60} is modeled by static potentials: either a zero-width Direc delta-function-well[23] or a finite-width square-well;[24,25] a nice account can be found in a recent review.[26] However, in either approach above, C_{60} electrons are completely omitted. Moreover, in the square well model, the width and depth of the potential can in principle be arbitrarily chosen as long as they together produce the measured C_{60} electron affinity, giving the potential a semi-empirical quality. A large number of papers were published by Amusia and co-workers by treating the central atom in the random phase approximation (RPA) with exchange (RPAE) method and the host fullerene by the delta function potential; results include: giant endohedral resonances,[27,28] two-shell endohedral atoms[29] and onion-type atoms.[30] Likewise, a number of studies were reported by Dolmatov and Manson[24,26] by using RPAE and Deshmukh, Manson & coworkers[25,31] by relativistic RPA, where both methods use the square-well potential modeling for the confining fullerene. Very recently the R-matrix description of the guest atom has been adopted for the square-well confining model with a good amount of success.[32,33] However, since these static model potentials are oblivious to fullerene electrons, the fundamental limitation of these studies is the omission of the coupling of the electrons of the confined atom to the cage electrons, particularly to their collective motions.

(III) Employing a large-scale one-center expansion type density functional scheme[34] single-electron valence and core photoemissions of several guest atoms have also been performed.[35] Bound and continuum orbitals are evaluated from the Hamiltonian matrix where the electron-electron interaction is treated in a

Kohn-Sham frame. A truncated partial wave approach defines open channels from correct continuum boundary conditions, whereas the remaining partial waves yield closed channels for the description of the wavefunction in the short-range of the multicenter molecular potential. The energy-dependent modulations in the photoelectron intensity ratio of the two highest occupied moleculer orbitals of C_{60} predicted by this method agreed very well with experiments and the TDLDA theory.[36] Another method that fits into this category is also based on a single-center expansion scheme with molecular wave functions denoted by Slater determinants from Hartree-Fock orbitals that utilized a variational approach.[37] While these approaches are particularly superior in the single electron ground and continuum state description by explicitly treating the truncated icosahedral structure of C_{60}, the electronic collective motion was completely omitted in the absence of the many-electron dynamical correlation.

(IV) Finally, a method of particular interest in this article is a density functional approach to describe the ground state electronic structure of the compounds.[10,38] A jellium potential representing sixty C^{4+} ions is constructed as a uniform charge density over a spherical shell with radius R and thickness Δ, augmented by a constant pseudopotential V_0 to insure quantitative accuracy.[39] R is taken to be the known radius of C_{60}, 3.54 Å. The nucleus of the confined atom is placed at the center of the sphere. The Kohn-Sham equations for the system of a total of $240 + N$ electrons (N being the number of electrons of the enclosed atom with $N = 0$ being the empty C_{60}) are then solved to obtain the ground state wave function in the local density approximation (LDA). The parameters V_0 and Δ are determined by requiring both charge neutrality and obtaining the experimental value, 7.54 eV, of the first ionization potential of C_{60}. This procedure yields a value of Δ to be 1.5Å, in excellent agreement with the value inferred from experiment.[36] The successes and limitations of this jellium-based description of C_{60} vis-á-vis experiments and quantum chemical calculations were discussed earlier.[10] However, no method can possibly beat the ease and ability of a jellium-approach to describe photoionization processes of nanometric finite systems by including important electron correlations.

A time dependent LDA, or TDLDA, method[10,38] is employed to calculate the dynamical response of the system to the external dipole field. The perturbation z, the dipole interaction for linearly polarized light, induces a frequency-dependent complex change in the electron density arising from dynamical electron correlations. This can be written, using the LDA susceptibility χ_0, as

$$\delta\rho(\mathbf{r};\omega) = \int \chi_0(\mathbf{r},\mathbf{r}';\omega)\delta V(\mathbf{r}';\omega)d\mathbf{r}', \qquad (1)$$

in which

$$\delta V(\mathbf{r};\omega) = z + \int \frac{\delta\rho(\mathbf{r}';\omega)}{|\mathbf{r} - \mathbf{r}'|}d\mathbf{r}' + \left[\frac{\partial V_{\mathrm{xc}}}{\partial\rho}\right]_{\rho=\rho_0}\delta\rho(\mathbf{r};\omega), \qquad (2)$$

where the second and third term on the right hand side are, respectively, the induced change of the Coulomb and the exchange-correlation potentials. Obviously, besides containing the external perturbation z, δV also includes the dynamical field produced by important electron correlations. The photoionization cross section is then calculated as the sum of independent partial cross sections $\sigma_{n\ell \to k\ell'}$, corresponding to dipole transitions $n\ell \to k\ell'$ with $\ell' = \ell \pm 1$:

$$\sigma_{\mathrm{PI}}(\omega) = \sum_{n\ell} \sigma_{n\ell \to k\ell'} \sim \sum_{n\ell} 2(2\ell+1)|\langle \phi_{k\ell'}|\delta V|\phi_{n\ell}\rangle|^2. \tag{3}$$

The method of calculating photoelectron continuum wavefunctions, $\phi_{k\ell'}$, is detailed in Ref. [10] Clearly, replacing δV in Eq. (3) by z yields the LDA cross section that entirely omits the correlation. Recently, a molecular dynamical density functional package, DMol3, has been used to compute photoionization of a few endofullerene systems.[40] Besides some debatable numerical issues, this approach is useful, but could not yet access level-differential ionization information for applications in photoelectron spectroscopy.

3. Recent Studies Employing TDLDA Methodology

3.1. *Photoionization of Fullerene Plasmons*

For large finite systems such as clusters, one type of plasmon excitation can be classically visualized as a surface resonance where the negative charge density (delocalized electrons) oscillates as an incompressible fluid against the positive background density (ions). The dipole frequency w_s of this surface oscillator is related to the characteristic plasma frequency w_0 by $w_s = w_0/\sqrt{3}$. Using ultraviolet spectroscopy, a giant surface-plasmon was first observed at about 20 eV photon energy for gas phase C_{60}.[9] This measurement coincided with a theoretical prediction[41] that described the carbon core structure in a tight-binding model and the response of C_{60} to external electromagnetic field by RPA.

The emergence of plasmon resonances can be thought of as originating from the formation of collective states under the influence of external electromagnetic field. A nice way to visualize the mechanism is to consider explicitly, in terms of many-body ground $|\Phi_0\rangle$ and excited collective $|\Phi_m\rangle$ states, the complex polarizability α, that results from electron's dipole interactions with the photon,

$$\alpha(\omega) = -\sum_m \left[\frac{|\langle \Phi_m|\zeta|\Phi_0\rangle|^2}{\hbar\omega - \Delta_m + i\delta} - \frac{|\langle \Phi_m|\zeta|\Phi_0\rangle|^2}{\hbar\omega + \Delta_m + i\delta} \right], \tag{4}$$

where $\zeta = \sum_i z_i$ are dipole interactions, $\Delta_m = E_m - E_0$ are many-body excitation energies, and δ is an infinitesimal positive quantity. To a good approximation, $|\Phi_0\rangle$

can be constructed as a linear combination of Slater determinants of single-electron ground states ϕ_i's. Using Eq. 4, the photoabsorption cross section can be expressed as

$$\sigma_{PA}(\omega) = \frac{4\pi\omega}{c}\text{Im}[\alpha(\omega)] = \frac{4\pi\omega}{c}\delta\sum_m\left[\frac{|\langle\Phi_m|\zeta|\Phi_0\rangle|^2}{[\hbar\omega - \Delta_m]^2 + \delta^2} - \frac{|\langle\Phi_m|\zeta|\Phi_0\rangle|^2}{[\hbar\omega + \Delta_m]^2 + \delta^2}\right]. \quad (5)$$

The above expression suggests that the formation of resonance structures in the cross section at the photon energy Δ_m is due to excitations of the ground state Φ_0 to all possible collective excited states Φ_m that the system can support.

Theoretical studies of the dipole photoionization of solid spheres, like metal clusters, predicted one plasmon resonance, indicating effectively one allowed collective excited state and, therefore, a single collective excitation.[42,43] For large enough solid spheres another structure at a higher energy has also been predicted,[1] but it is found to be far weaker than the main resonance and is suspected to originate from a non-local quantum-specific effect.[42] Moving from a solid sphere to a hollow shell, the geometry alters significantly. One direct effect of this change is an alteration in the distribution of ground state single-electron energy levels. This happens due to the following reason: The hollow C_{60} geometry enables the electrons to move inside a spherical shell far from the molecular center (large r), causing considerable weakening of the so called centrifugal barrier potential $\ell(\ell+1)/r^2$. Simultaneously, the narrower radial well of C_{60} compared to the relatively wide well of a solid system considerably separates the $n = 1$ (no radial node) and $n = 2$ (one radial node) levels. As a result, all ℓ-levels for a given n tend to cluster, thereby inducing considerable segregation between π and σ families in the ground state spectrum of C_{60}.[10] In contrast, for a solid sphere a far broader radial potential and associated smaller values of r cause all single electron levels to mix together instead. It is therefore expected that this change in geometry from solid to shell also influences the formation of collective excited states of the respective systems. The hollow geometry allows for two many-body excited states, as opposed to one for the solid geometry. In the semi-classical interpretation of plasmon these two states can be identified with the symmetric (uncompressional sloshing) and antisymmetric (compressional breathing) eigenmodes of vibration of a classical dielectric shell.[44]

In Fig. 1, the total photoionization cross sections calculated in TDLDA for neutral C_{60} and ionic $C_{60}{}^+$ are compared with experiments: measurements for C_{60} were performed at BESSY[9] and on the dipole beamline at BESSY II synchrotron source[45] in Berlin, while that for $C_{60}{}^+$ were carried out on the undulator beamline at Advanced Light Source, Berkley.[46] TDLDA curves are Lorentzian-convoluted with some appropriate width in order to mask Auger-type single-electron decay processes which are undetectable in current experimental environments. Further, in order to compensate for the known limitations[10] of the jellium model, that produces redshifted plasmons, the TDLDA curves are shifted higher in energy in Fig. 1. As seen, the comparison provides a satisfactory qualitative congruence between the TDLDA predictions and measurements of the photoexcitation of two plasmons.

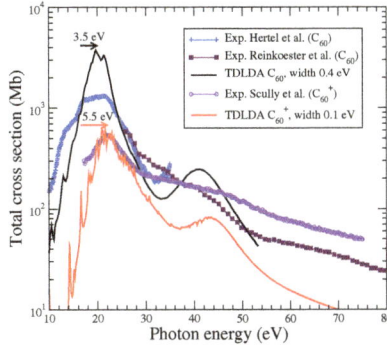

Fig. 1.   TDLDA total photoionization cross sections are compared with measurements for C_{60}[9,45] and $C_{60}{}^+$.[46] Appropriate Lorentzian widths were added to the TDLDA results, which are also suitably blue-shifted to match the experimental giant plasmon resonance.

A critical aspect in applications of plasmons is their excitation frequency and, therefore, controlling the plasmon frequency will constitute a huge step in advancing the technology. Additionally, if multiple plasmons of sufficient strengths and long enough lifetimes can be induced in a system, then that would provide unparalleled versatility in device performance. The simplest prototype of such systems is a bi-fullerene nested nanoonion, such as $C_{60}@C_{240}$. This is because concentric fullerenes having qualitatively similar plasmonic responses can induce a broad range of flexibility in the plasmon generation. This entails a powerful hybridization leading to dramatic variation of plasmon structures when individual charge clouds of neighboring shells strongly overlap. Fullerene onions, which were first synthesized[47] contemporarily with single-walled ordinary fullerenes, are highly stable formations that can exist even at room temperature.

In the nested compound, $C_{60}@C_{240}$, the hybridization between excited collective states of "unperturbed" fullerenes produces excited states of bonding (symmetric) and antibonding (antisymmetric) combinations:

$$|\Phi_m^+\rangle = \sqrt{\alpha}|\Phi_m^{60}\rangle + \sqrt{1-\alpha}|\Phi_m^{240}\rangle \qquad (6)$$

$$|\Phi_m^-\rangle = \sqrt{1-\alpha}|\Phi_m^{60}\rangle - \sqrt{\alpha}|\Phi_m^{240}\rangle \qquad (7)$$

in which α is the mixing parameter. Excitations of the charge cloud of $C_{60}@C_{240}$ to these four collective hybrid states, Eqs. (6) and (7), induce four plasmons. This is seen in the TDLDA cross section presented in Fig. 2(a) in which a four-Lorentzian fit is found to describe the average background shape of the result. Figure 2(b) exhibits the background fits of the onion system along with those for individual fullerenes and it is noted that the more tightly-bound hybrid states producing the lower energy bonding plasmons ($C_{60}+C_{240}$) in a pair. These results suggest a very different characteristics (positions, widths and oscillator strengths) of these cross-mode hybrid plasmons, when compared to those of the "primordial" plasmons of the constituent fullerenes.[48]

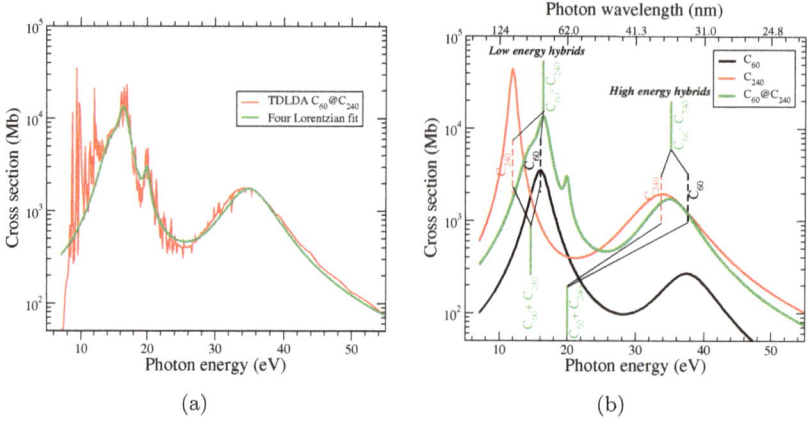

Fig. 2. (a) A four-Lorentzian fit to the TDLDA total photoionization cross section of the $C_{60}@C_{240}$ nanoonion. (b) Lorentzian fits to the calculated photoionization cross sections of isolated C_{60} and C_{240}, and the onion $C_{60}@C_{240}$. The mixing of plasmons from each fullerene form bonding and antibonding plasmon-hybrids in the onion system, as sketched.

3.2. *Giant Photoionization of Fullerene Endohedral Atoms*

Understanding the influence of the confining fullerene cage on the atom inside is a matter of significant fundamental and applied interest. Photoionization investigations offer an excellent probe owing to the weak photon-electron coupling, along with the fact that the photon disappears after the interaction. Semiclassical studies[21] suggested that the photoabsorption cross section of an atom confined in C_{60} is dramatically affected in the neighborhood of the plasmons which occur in the photoionization of the C_{60} shell. This phenomenon is confirmed (at least qualitatively) in *ab initio* quantum mechanical calculations on Ar@C_{60},[49] Mg@C_{60}[50] and Xe@C_{60}[38] using TDLDA where a huge increase of the confined atom's valence cross section at the C_{60} plasmon peak was uncovered. In Fig. 3, the TDLDA enhancement factors for confined Ar 3p and Mg 3s are compared with recently modified semiclassical approach[22] that can distinguish between various central atoms. Besides some detailed structures in the quantum mechanical TDLDA results the qualitative agreement with the semiclassics is reasonable.

The underlying reason for this enhancement is the transfer of oscillator strength from the C_{60} ionization channels to the atomic valence channel through the atom-fullerene dynamical coupling. To scrutinize the mechanism we can resort to the following equation, using the Fano formalism,[51] that represents the leading order correction to the atomic valence photoamplitude owing to its interaction with a number of C_{60} channels $n\ell$.[52]

$$\Delta D(E) = \sum_{n\ell} \int dE' \frac{\langle \psi_{n\ell}(E')| \frac{1}{|\mathbf{r}-\mathbf{r}_{n\ell}|} |\psi(E)\rangle}{E - E'} D_{n\ell}(E'), \qquad (8)$$

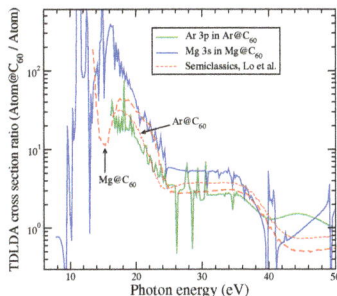

Fig. 3. TDLDA atomic valence photoionization enhancement factors are compared with the semi-classical estimates[22] for Ar@C$_{60}$ and Mg@C$_{60}$.

where the numerator is a measure of the overlaps between ψ and $\psi_{n\ell}$ which are respectively the atomic and C$_{60}$ continuum-channel wavefunctions, and $D_{n\ell}$ represents all the C$_{60}$ independent particle LDA matrix elements. Since the atomic bound wavefunction overlaps somewhat with those of C$_{60}$ due to a weak mixing, this correction term must be non-vanishing. But why is the correction so strikingly huge as seen in Fig. 3? This is due to the following reason. The existence of two plasmon resonances in free C$_{60}$ implies that a coherent mixing of various C$_{60}$ dipole matrix elements forms two regions of constructive interference *via* a mechanism of *phase-coherent* interchannel coupling.[10] Since the valence ionization channels of the caged atoms considered here open approximately at the onset of the low energy (giant) plasmon, it must also participate in the process. Indeed, the correction terms add up coherently, leading to the spectacular enhancement in the low energy plasmon region,[53] as, for instance, seen in Fig. 3.

3.3. *Atom-Fullerene Hybrid Photoionization*

Atom-fullerene hybrid levels with novel ionization behavior[38,54] are interesting, since these levels lead to possibilities of covalent-type bonding of the fullerene with the trapped atom and their influence on the ionization response of the compound to the radiation. In general, spectroscopic examination of such hybrids pave the route to probe wavefunction mixing in other spherical dimer composites, such as, bucky-onions or clusters trapped in fullerene cages.

Only the atomic and fullerene orbitals of the same angular momentum hybridize. This is because the orthogonality property of the spherical harmonics makes the other terms zero. Also, from a perturbation theory viewpoint, to have strong hybridization not only good overlap of the unperturbed wave functions is needed but the binding energies of those levels have to be close, since they respectively guarantee a large numerator and a small denominator of the coupling term. These conditions applied to the known energy levels of the C$_{60}$ π band[10] suggest that the atom's valence and subvalence levels are susceptible to hybridization. For Xe@C$_{60}$

a strong *s-s* hybridization between Xe 5*s* and a C_{60} *s* level of π character was pre-dicted in TDLDA.[54] Also in the case of Zn@C_{60}[55] and Cd@C_{60},[56] TDLDA recently accessed strong *d-d* hybridizations of Zn 3*d* or Cd 4*d* with a C_{60} *d* level.

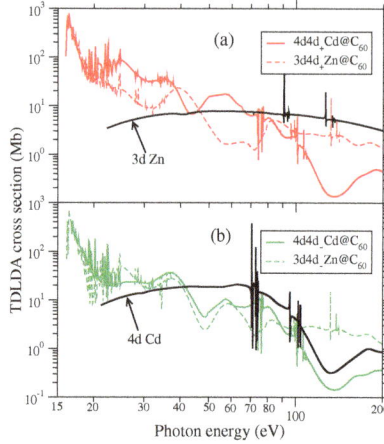

Fig. 4.   TDLDA photoionization cross sections of the symmetric (upper panel) and antisymmetric (lower panel) hybridized *d* states for Cd@C_{60} and Zn@C_{60}. Free Zn 3*d* and free Cd 4*d* results are also shown.

Compared in Fig. 4 are the TDLDA results between Cd@C_{60} with Zn@C_{60} for their symmetric '+' (upper panel) and antisymmetric '−' (lower panel) hybrids. As evident, hybridization produces vastly rich structures when compared with free 4*d* Cd and 3*d* Zn results (shown), which are rather smooth. Comparison between the compounds, on the other hand, reveals a number of significant differences. For each of symmetric and antisymmetric combinations, the C_{60} *d*-orbital mixes with 4*d* Cd and 3*d* Zn by roughly the same amount. This allows approximately similar shapes and magnitudes of plasmon-induced enhancements in the cross sections below 20 eV in Fig. 4. At energies above 20 eV up to 100 eV, strong disagreements between the cross sections of the symmetric pair are noted. The difference is somewhat weaker for the antisymmetric pair but is still significant (note the *log* scale of the cross sections). The mismatch originates from the differences between 4*d* and 3*d* emissions of free Cd and Zn, that constitute roughly half of the hybrid emission, by coherently mixing with the other half, which is fullerene-like. Indeed, as seen, significant differences between TDLDA 4*d* Cd and 3*d* Zn curves, including a more defined shape resonance followed by a Cooper minimum in Cd at 135 eV, corroborates this assumption. Above 100 eV then, the remarkable differences between Cd@C_{60} and Zn@C_{60} predictions in Fig. 4 results from the coherent mixing with 4*d* Cooper minimum in Cd, whereas Zn 3*d* does not have any such minimum.

3.4. *Resonant Hybrid Multicenter Decay Processes*

Intercoulombic decay (ICD), originally predicted by Cederbaum *et al.*[57] and observed initially for Ne clusters,[58] is a unique, naturally abundant nonradiative relaxation pathway of a vacancy in atom A in a cluster or molecule. An outer electron of A fills the vacancy and the released energy, instead of emitting a second electron of A as in standard Auger ionization, transfers to a neighboring atom B *via* Coulomb interactions to ionize B. Over the last decade and a half, a wealth of theoretical[59] and experimental[60] research has gone into studying ICD processes in weakly bound atomic systems. Of particular interest is the resonant ICD (RICD) where the precursor excitation to form an inner-shell vacancy is accomplished by promoting an inner electron to an excited state by an external stimulant, generally electromagnetic radiation. Endofullerene compounds are particularly attractive natural crucibles to study RICD processes. A first attempt to predict ICD in endofullerenes was made by calculating ICD rates for Ne@C_{60}.[61]

ICD of endofullerene molecules can uncover effects not yet known. This is because: (i) endofullerenes being spherical analogues of asymmetric dimers consisting of an atom and a cluster can also induce reverse RICD processes, the decay of cluster innershell excitations through the continuum of the confined atom, of uniquely different character than the forward RICD; and (ii) possibilities of atom-fullerene hybridized final states, predicted to exist abundantly in these systems.[54,55] It has recently been shown by a TDLDA study[62] that for an Ar atom endohedrally sequestered in C_{60}, ICD pathways of photo-generated innershell holes, both in the central atom and the fullerene, can coherently mix with degenerate *intracoulombic* Auger pathways to produce final states with *shared* holes in atom-fullerene hybrid levels. Figure 5(a) presents a schematic of the process which illustrates this novel decay mode that can be called the resonant hybrid Auger-intercoulombic decay or RHA-ICD.

Figure 5(b) displays TDLDA cross sections for the Ar@C_{60} endofullerene hybrid levels (Ar$\pm$ C_{60})$3p$. Features A, B, and C in these curves are resonances that emerge from the decay of Ar $3s \rightarrow 4p, 5p, 6p$ excitations through the continuum of these hybrid levels. The features are similar in shape to the regular $3s$ excitation autoionizing (Auger) resonances in free Ar $3p$ (shown). Remarkably, they are found to be significantly stronger than the Ar-to-C_{60} RICDs (not shown). Another dramatic effect can be noted: The empty C_{60} $3p$ cross section in Fig. 5(b) shows autoionizing resonances corresponding to Auger decays of C_{60} innershell vacancies. But the structures at the corresponding energies in hybrid channels from the decay of C_{60} vacancies are order of magnitude larger than the autoionizing resonances in empty C_{60}; see, for example, the resonances labeled as 1-4 in Fig. 5(b). In essence, Ar and C_{60} innershell vacancies decay significantly more powerfully through the photoionization continua of Ar-C_{60} hybrid levels than through the continua of pure C_{60} levels. It turns out that they emerge from a coherent interference between resonant Auger and ICD channels that produce *divided* vacancies in the final state, vacancies

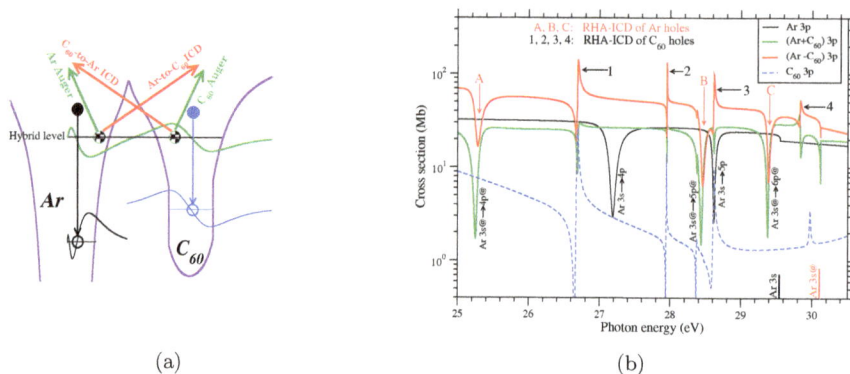

Fig. 5. (a) Schematic of coherent mixing of one-center Auger decays (green) of core holes with corresponding ICDs (red) in the spectra of Ar-C$_{60}$ hybrid electrons. (b) Photoionization cross sections of free Ar 3p and C$_{60}$ 3p levels compared with those of their hybrid pair. Features A-C and 1–4 are identified in the graph.

shared by the confined Ar and the confining C$_{60}$. This hybrid channel offers a unique probe, more powerful than regular ICDs, for multicenter decay processes.[62]

4. Research Outlook

4.1. *Attophysics with Fullerenes and Endofullerenes*

With tremendous advancements in technology for generating attosecond (*as*) isolated pulses and pulse trains, it is now possible to study fundamental phenomena of light-matter interaction with unprecedented precision on *as* timescale.[63,64] In particular, the relative time delay between the photoelectrons from different subshells is expected to probe important aspects of electron correlations which predominantly influence the photoelectron. Pump-probe experiments have been performed to measure the relative delay where extreme ultraviolet (XUV) pulses are used to remove an electron from a particular subshell and subsequently a weak infrared (IR) pulse accesses the temporal information of the emission event. Recently, the relative delay between the 3s and 3p subshells in Ar is measured at three photon energies by interferometric technique.[65,66] Among various theoretical methods employed to investigate 3s-3p relative delay in Ar, TDLDA produced the best agreement with experiments[67] so far along with the multiconfiguration Hartree-Fock approach.[68]

   Therefore, subshell differential time delay studies of empty single-and multi-walled fullerenes as well as endofullerenes can be a major future research area. Effects of the confinement and fullerene cavity, plasmon resonance, Cooper minima, and orbital hybridization on the photoemission time delay can be investigated. Angular distribution profiles of the photoelectron time delay can uncover information on comprehensive details of the process. This can access the delay information near resonances originated from the Auger, ICD and hybrid Auger-ICD (HA-ICD) pathways. The hot topicality of attophysics research can thus stimulate experimental time-domain work in fullerene systems, initiating a novel field of atto-nano science.

4.2. *Photoresponse of Non-Spherical Systems*

Although a spherical or near spherical approximation of a fullerene brings out the robust structural or response properties of the compound, a large number of these compounds, even in the icosahedral fullerene family, are non-spherical. Furthermore, in most endohedral metallofullerenes and lanthanide-fullerenes, the interior atom is located off the center[69] as a result of a charge-transfer process between the atom and the fullerene. For example, the Ce atom was both predicted[70] and shown by the experiment[71] to be triply ionized inside the C_{82} cage, resulting in a significant off-center dislocation of Ce in C_{82}. Theoretical efforts on this track are rather spars. This non-sphericity can induce often dramatic refinements in the plasmonic behavior of the many-electron charge cloud, as well as the nature of atom-fullerene hybridization, as the system responds to the external perturbation. Investigation on these compounds, therefore, can constitute a new and significant direction of fullerene research.

4.3. *Bare Ion-Impact Emissions*

The heavy ion collision technique is an established and powerful tool used to investigate the structural and collisional properties of atoms, molecules, solids and surfaces. For simple ion-atom and ion-molecular collisions, the richness of the method originates in probing the dynamics of ionized electrons in the target-projectile two-center field. However, if the target is large enough to support coherent and collective electronic motions, then the impact spectroscopy will also include plasmon-type effects. If the projectiles are bare ions that are completely stripped of the bound electrons, their interactions with a large target can unambiguously probe the collective plasmonic excitation dynamics driven by the ion field. Moreover, since the ion impact can render the target's non-dipole responses significant, this method can uncover the effects of collective motions in non-dipole ionization channels as well.

Bare ion-impact ionization of various spherical or quasi-spherical fullerenes and their endohedral derivatives can be a promising new direction of research. Energy and angle differential double differential cross sections (DDCS) and single differential cross sections (SDCS) can be calculated to gain detailed insights of collective excitations along the dipole and important non-dipole channels, including the monopole channel. A preliminary, brief study in this direction has recently been conducted in TDLDA for 76 MeV F^{9+} ion colliding with C_{60} with some success.[72] It will also be tremendously exciting to search for signatures of ICD and HA-ICD resonances in such ion-impact processes; a solid drive in this direction comes from exciting recent experiments on ion-impact induced ICDs in dimers.[73] In absence of significant research efforts in this topic, such studies will be valuable in the general field of heavy and highly stripped ion interactions with nanoparticles.

4.4. *Beyond Sphericality: Carbon Nanotubes*

Last but not the least, another important direction for future investigations could be to study tubular systems, especially, carbon nanotubes and their endohedral derivatives. Some interests in the theory-experiment joint study of tubular systems can be found in the photoionization spectroscopy of the image states of multiwalled carbon nanotubes.[74,75] A significant connection exists between this modeling of nanotubes and the fullerene based TDLDA work, since a jellium-like description was also used to model concentric nano-cylinders. Knowledge about physical properties gained through the study of spherical systems can help better understand results for cylindrical systems. Much of the developments in the computational codes for spherical fullerene research can also enable modifications needed for cylindrical nanotubes, if, in a simplified approximation, the cylindrical structure can be separated into a transverse circular symmetry and a longitudinal direction, supporting plane waves for a long enough nanotube. Finally, the authors would like to mention that the TDLDA results presented in this article emerged from an ongoing research program supported by the US National Science Foundation.

References

1. C. Xia, C. Yin, and V. V. Kresin, *Phys. Rev. Lett.* **102**, 156802 (2009).
2. T.-H. Park and P. Nordlander, *Chem. Phys. Lett.* **472**, 228 (2009).
3. C. Kramberger, R. Hambach, C. Giorgetti, M. H. Rümmeli, M. Knupfer, J. Fink, B. Büchner, L. Reining, E. Einarsson, S. Maruyama, F. Sottile, K. Hannewald, V. Olevano, A. G. Marinopoulos, and T. Pichler, *Phys. Rev. Lett.* **100**, 196803 (2008).
4. E.H. Hwang and S. Das Sarma, *Phys. Rev. B* **75**, 205418 (2007).
5. F. Rana, *IEEE Trans. on Nanotechnology* **7**, 91 (2008).
6. M. Jablan, H. Buljan, and M. Soljačić, *Phys. Rev. B* **80**, 245435 (2009).
7. X. F. Wang and T. Chakraborty, *Phys. Rev. B* **75**, 041404(R) (2007).
8. G. Borghi, M. Polini, R. Asgari, and A. H. MacDonald, *Phys. Rev. B* **80**, 241402 (2009).
9. I. V. Hertel, H. Steger, J. de Vries, B. Weisser, C. Menzel, B. Kamke, and W. Kamke, *Phys. Rev. Lett* . **68**, 784 (1992).
10. M. E. Madjet, H. S. Chakraborty, J. M. Rost, and S. T. Manson, *J. Phys. B* **41**, 105101 (2008).
11. A. A. Popov, S. Yang, and L. Dunsch, *Chem. Rev.* **113**, 5989 (2013).
12. C. Ju, D. Suter, and J. Du, *Phys. Rev. A* **75**, 012318 (2007).
13. W. Harneit, *Phys. Rev. A* **65**, 032322 (2002).
14. W. Harneit, C. Boehme, S. Schaefer, K. Huebner, K. Fortiropoulos, and K. Lips, *Phys. Rev. Lett.* **98**, 216601 (2007).
15. C. Ju, D. Suter, and J. Du, *Phys. Lett. A* **375**, 1441 (2011).

16. A. Takeda, Y. Yokoyama, S. Ito, T. Miyazaki, H. Shimotani, K. Yakigaya, T. Kakiuchi, H. Sawa, H. Takagi, K. Kitazawa, and N. Dragoe, *Chem. Commun.* **8**, 912 (2006).
17. J. Zhang, S. Stevenson, and H. C. Dorn, *Acc. Chem. Res.* **46**, 1548 (2013).
18. J. B. Melanko, M. E. Pearce, and A. K. Salem, in *Nanotechnology in Drug Delivery*, edited by M. M. de Villiers, P. Aramwit, G. S. Kwon (Springer, New York, 2009), p. 105.
19. R. B. Ross, C. M. Cardona, D. M. Guldi, S. G. Sankaranarayanan, M. O. Reese, N. Kopidakis, J. Peet, B. Walker, G. C. Bazan, E. V. Keuren, B. C. Holloway, and M. Drees; *Nature Materials* **8**, 208 (2009).
20. L. Becker, R. J. Poreda, and T. E. Bunch, *Proc. Natl. Acad. Sci. USA* **97**, 2979 (2000).
21. S. Lo, A. V. Korol and A. V. Solov'yov, *J. Phys. B* **40**, 3973 (2007).
22. S. Lo, A. V. Korol and A. V. Solov'yov, *Phys. Rev. A* **79**, 063201 (2009).
23. M. Ya. Amusia, A. S Baltenkov, and L. V. Chernycheva, *J. Phys. B* **41**, 165201 (2008) and references therein.
24. V. K. Dolmatov and S. T. Manson, *J. Phys. B* **41**, 165001 (2008) and references therein.
25. K. Govil, A. J. Siji, and P. C. Deshmukh, *J. Phys. B* **42**, 065004 (2009) and references therein.
26. V. K. Dolmatov, Advances in Quantum Chemistry in *Theory of Confined Quantum Systems*, edited by J. R. Sabin and E. Brändas (Academic Press, New York, 2009) Vol. 58, 13.
27. M. Ya. Amusia, A. S. Baltenkov, and L. V. Chernycheva, *JETP Lett.* **87**, 200 (2008).
28. M. Ya. Amusia, A. S. Baltenkov, and L. V. Chernycheva, *JETP Lett.* **89**, 275 (2009).
29. M. Ya. Amusia, L. V. Chernycheva, and E. Z. Liverts, *Phys. Rev. A.* **80**, 032503 (2009).
30. M. Ya. Amusia, L. V. Chernycheva, and E. Z. Liverts, *JETP Lett.* **90**, 350 (2009).
31. J. George, H. R. Varma, P. C. Deshmukh, and S. T. Manson, *J. Phys. B* **45**, 185001 (2012).
32. T. W. Gorczyca, M. F. Hasoglu, and S. T. Manson, *Phys. Rev. A* **86**, 033204 (2012).
33. M. F. Hasoglu, H.-L. Zhou, T. W. Gorczyca, and S. T. Manson, *Phys. Rev. A* **87**, 013409 (2013).
34. M. Venuti, M. Stener, and P. Decleva, *Chem. Phys.* **234**, 135 (1998).
35. M. Stener, G. Fronzoni, D. Toffoli, P. Colavita, S. Furlan, and P. Decleva, *J. Phys. B* **35**, 1421 (2002) and references therein.
36. A. Rüdel, R. Hentges, H. S. Chakraborty, M. E. Madjet, and J. M. Rost, *Phys. Rev. Lett.* **89**, 125503 (2002).
37. J. Jose and R. R. Lucchese, *J. Phys. B* **46**, 215103 (2013).
38. M. E. Madjet, T. Renger, D. E. Hopper, M. A. McCune, H. S. Chakraborty, Jan-M Rost, and S. T. Manson, *Phys. Rev. A* **81**, 013202 (2010).
39. M. J. Puska and R. M. Nieminen, *Phys. Rev. A* **47**, 1181 (1993); M. J. Puska and R. M. Nieminen, *Phys. Rev. A* **49**, 629 (1994).
40. Z. Chen and A. Z. Msezane, *Eur. Phys. J. D* **66**, 184 (2012); *Phys. Rev. A* **86**, 063405 (2012).
41. G. F. Bertsch, A. Bulgac, D. Tománek, and Y. Wang, *Phys. Rev. Lett.* **67**, 2690 (1991).
42. W. Ekardt, *Phys. Rev. B* **31**, 6360 (1985).
43. M. J. Puska, R. M. Nieminen, and M. Manninen, *Phys. Rev. B* **31**, 3486 (1985).

44. Ph. Lambin, A. A. Lucas, and J.-P. Vigneron, *Phys. Rev. B* **46**, 1794 (1992).
45. A. Reinköster, S. Korica, G. Pruemper, J. Viefhaus, K. Godehausen, O. Schwarzkopf, M. Mast, and U. Becker, *J. Phys. B* **37**, 2135 (2004).
46. S. W. J. Scully, E. D. Emmons, M. F. Gharaibeh, R. A. Phaneuf, A. L. D. Kilcoyne, A. S. Schlachter, S. Schippers, A. Mueller, H. S. Chakraborty, M. E. Madjet, and J. M. Rost, *Phys. Rev. Lett.* **94**, 065503 (2005).
47. D. Ugarte, *Nature* **359**, 707 (1992).
48. M. A. McCune, R. De, M. E. Madjet, H. S. Chakraborty, and S. T. Manson, *J. Phys. B Fast Track Commun.* **44**, 241002 (2011).
49. M. E. Madjet, H. S. Chakraborty and S. T. Manson, *Phys. Rev. Lett.* **99**, 243003 (2007).
50. H. S. Chakraborty, M. E. Madjet, J. M. Rost, and S. T. Manson, *Phys. Rev. A* **78**, 013201 (2008).
51. U. Fano, *Phys. Rev.* **124**, 1866 (1961).
52. M. H. Javani, M. R. McCreary, A. B. Patel, M. E. Madjet, H. S. Chakraborty, and S. T. Manson, *Euro. Phys. J. D* **66**, 189 (2012).
53. M. H. Javani, H. S. Chakraborty, and S. T. Manson, *Phys. Rev. A* **89**, 053402 (2014).
54. H. S. Chakraborty, M. E. Madjet, T. Renger, Jan-M. Rost, and S. T. Manson, *Phys. Rev. A* **79**, 061201(R) (2009).
55. J. N. Maser, M. H. Javani, R. De, M. E. Madjet, H. S. Chakraborty, and S. T. Manson, *Phys. Rev. A* **86**, 053201 (2012).
56. M. H. Javani, R. De, M. E. Madjet, S. T. Manson, and H. S. Chakraborty, Preprint arXiv:1405.3567, (2014).
57. L. S. Cederbaum, J. Zobeley, and F. Tarantelli, *Phys. Rev. Lett.* **79**, 4778 (1997).
58. S. Marburger, O. Kugeler, U. Hergenhahn, and T. Möller, *Phys. Rev. Lett.* **90**, 203401 (2003).
59. V. Averbukh, Ph. V. Demekhinb, P. Kolorenc, S. Scheitd, S. D. Stoycheve, A. I. Kuleff, Y.-C. Chiange, K. Gokhberge, S. Kopelkee, N. Sisourate, and L. S. Cederbaume, *J. Electr. Spectr. Relat. Phenom.* **183**, 36 (2011).
60. U. Hergenhahn, *J. Electr. Spectr. Relat. Phenom.* **184**, 78 (2011).
61. V. Averbukh and L. S. Cederbaum, *Phys. Rev. Lett.* **96**, 053401 (2006).
62. M. H. Javani, J. B. Wise, R. De, M. E. Madjet, S. T. Manson, and H. S. Chakraborty, *Phys. Rev. A* **89**, 063420 (2014).
63. E. Goulielmakis, M. Schultze, M. Hofstetter, V. S. Yakovlev, J. Gagnon, M. Uiberacker, A. L. Aquila, E. M. Gullikson, D. T. Attwood, R. Kienberger, F. Krausz, and U. Kleineberg, *Science* **320**, 1614 (2008).
64. F. Krausz and M. Ivanov, *Rev. Mod. Phys.* **81**, 163 (2009).
65. K. Klünder, J. M. Dahlstrm, M. Gisselbrecht, T. Fordell, M. Swoboda, D. Gunot, P. Johnsson, J. Caillat, J. Mauritsson, A. Maquet, R. Taeb, and A. L'Huillier, *Phys. Rev. Lett.* **106**, 143002 (2011).
66. D. Guénot, K. Klünder, C. L. Arnold, D. Kroon, J. M. Dahlström, M. Miranda, T. Fordell, M. Gisselbrecht, P. Johnsson, J. Mauritsson, E. Lindroth, A. Maquet, R. Taïeb, A. L'Huillier, and A. S. Kheifets, *Phys. Rev. A* **85**, 053424 (2012).
67. G. Dixit, H. S. Chakraborty, and M. E. Madjet, *Phys. Rev. Lett.* **111** 203003 (2013).
68. T. Carette, J. M. Dahlström, L. Argenti, and E. Lindroth, *Phys. Rev. A* **87**, 023420 (2013).
69. H. Shinihara, *Rep. Prog. Phys.* **63**, 843 (2000).
70. M. Muthukumar and J. A. Larsson, *J. Phys. Chem. A* **112**, 1071 (2008).
71. A. Müller, S. Schippers, M. Habibi, D. Esteves, J. C. Wang, R. A. Phaneuf, A. L. D. Kilcoyne, A. Aguilar, and L. Dunsch, *Phys. Rev. Lett.* **101**, 133001 (2008).

72. M. E. Madjet, H. S. Chakraborty, and S. T. Manson, *J. Phys. Conf. Ser.* **194**, 102034 (2009).
73. H.-K. Kim, H. Gassert, M. S. Schöffler, J. N. Titze, M. Waitz, J. Voigtsberger, F. Trinter, J. Becht, A. Kalinin, N. Neumann, C. Zhou, L. Ph. H. Schmidt, O. Jagutzki, A. Czasch, H. Merabet, H. Schmidt-Böcking, T. Jahnke, A. Cassimi, and R. Dörner, *Phys. Rev. A* **88**, 042707 (2013).
74. M. Zamkov, H. S. Chakraborty, A. Habib, N. Woody, U. Thumm, and P. Richard, *Phys. Rev. B* **70**, 115419 (2004).
75. M. Zamkov, N. Woody, B. Shan, H. S. Chakraborty, Z. Chang, U. Thumm, and P. Richard, *Phys. Rev. Lett.* **93**, 156803 (2004).

VI

Ultracold Chemistry

Chapter 12

Collisions and Reactions in Ultracold Gases

N. Balakrishnan* and Jisha Hazra

Department of Chemistry, University of Nevada Las Vegas
Las Vegas, NV 89154 USA

Cooling, trapping, and manipulating molecules at temperatures close to absolute zero have become a fertile area of research in recent years, thanks to the highly quantum nature of their interactions and the prospects of controlling the interactions using external electric and magnetic fields. The possibility of observing chemical reactions at the pure quantum level in these systems has led to breakthrough experiments that allow investigation of chemical reactions at the single partial wave level. Here, we highlight recent progress in theoretical investigations of barrierless chemical reactions in ultracold gases and discuss the promise, opportunities, and challenges that these systems present.

1. Introduction

The field of ultracold molecules has seen tremendous growth over the past 10-15 years due to the rapid progress achieved in cooling and trapping of a broad range of atoms and molecules at temperatures between about 100 nK and 1 K.[1-8] In many cases, these methods allow creation/preparation of molecules in a single quantum state, resolved at the spin and hyperfine levels. Interactions of these highly quantized systems can be exquisitely manipulated by external electric or magnetic fields depending on whether the molecules possess a dipole moment or a magnetic moment. Ultracold dipolar gases are of considerable interest in applications ranging from quantum computation and quantum information, simulation of many-body quantum effects, and quantum control of chemical reactions with external fields.[6,8] When prepared in their absolute ground state, these molecules are stable against collisional de-excitation. Inelastic collisions (generally referred to as "bad collisions") release vast amount of energy compared to their initial translational kinetic energy, leading to heating and trap loss. Even in cases where the molecules are prepared in their absolute ground state, collisions with other trapped molecules in the same or different quantum states may lead to inelastic or reactive scattering, as recently demonstrated in elegant experiments performed on fermionic ^{40}K^{87}Rb

*naduvala@unlv.nevada.edu

molecules by the JILA group.[9] Chemical reactions may also occur in collisions of trapped molecules with cold atoms if exothermic reaction channels are energetically allowed. Since many exothermic chemical reactions release energy in the eV scale, the relative kinetic energies of the reaction products are typically a billion times greater than the initial collision energy of molecules prepared at, say, $1\mu K$. Such dramatic difference in initial and final collision energies is not observed in normal chemical reactions. Thus, many high-order partial waves and excited rovibrational levels may be present in the product channels whereas the reactants can be prepared in a single quantum state and reactivity can be observed at the level of a single partial wave.

Among the methods that have been developed to create cold molecules, the buffer gas cooling[10,11] and stark-deceleration schemes[1,12–14] allow cooling of pre-existing molecules. These methods can achieve temperatures in the mK-1K range. The buffer gas cooling method has been successfully applied to molecules such as CaH,[10] and NH[11] while the stark-deceleration scheme has been demonstrated for OH[12,13] and NH,[14] to name a few. The former requires molecules to possess a magnetic moment while the latter requires a dipole moment so that molecules responds to a time-varying electric field. These techniques allow a broader range of molecular species compared to photoassociation[15,16] and Feshbach resonance methods[17,18] which are largely restricted to dimers of alkali metal atoms. However, the methods yield much colder molecules, typically in the μK and sub-μK regimes. The molecules are created in high-lying vibrational levels of the ground electronic state but more recent developments using STIRAP based techniques have allowed their transfer to the absolute rovibrational ground state. More recently, a method based on merged supersonic beams has been proposed that allows continuous scanning of collision energies and it is also applicable to neutral molecules.[19–21] This method has recently been applied to penning ionization studies of H_2, HD and D_2 molecules in collisions with excited He atoms.[21] The different isotopes exhibit different resonance structures in the energy dependence of the ionization cross sections measured in the 1mK to 100 K energy range. The resonances arise from angular momentum barriers which occur at different collision energies for the different isotopes.[21]

Theoretical studies of ultracold collisions have largely focused on inelastic and reactive atom-molecule and molecule-molecule systems with numerous studies reported in the past 15 years.[2–8,22–36] Initial studies were mostly centered on rotational and vibrational quenching of diatomic molecules in collisions with ultracold atoms.[2–5] A series of such calculations performed on H_2, CO, and O_2 in collisions with helium atoms have shown that these processes can become highly efficient at ultracold temperatures as the initial vibrational level of the molecules is increased.[2–5] The first detailed study of ultracold chemical reactivity was reported by Balakrishnan and Dalgarno[29] for the benchmark $F+H_2 \rightarrow HF+H$ reaction. Their calculations using the Stark-Werner (SW)[37] potential energy surface (PES) for the $F+H_2$ system indicated that the reaction occurs with a rate coefficient of about

1.25×10^{-12} cm^3 s^{-1} in the limit of zero temperature. This was a surprising result considering that the reaction has an energy barrier of about 500 K. Thus, the fact that barrier reactions may occur through tunneling at measurable rate coefficient at ultracold temperatures led to similar studies of many other systems, including F+HD/D$_2$,[30–32] Cl+H$_2$/HD,[33] Li+HF,[34,35] O(^{3}P)+H$_2$,[36] etc. Bodo et al.[32] showed significant isotope effect in F+H$_2$/HD/D$_2$ systems and investigated how a smooth change in mass of the hydrogen atom from H to D influences reactivity at ultracold temperatures. Their study, showed that resonances occur for certain values of the hydrogen atom mass, similar to magnetic field induced Feshbach resonances in atom-atom collisions. More recently, Simbotin et al.[38] have reported a systematic study of mass effect in Cl+H$_2$ reaction and found that shape resonances occur for certain mass combinations before the onset of the Wigner threshold regime.

Many studies of barrierless reactions at cold and ultracold temperatures have also been reported in recent years.[39–51] This includes both alkali metal and non-alkali metal atom systems. Soldán et al.[39] and Quéméner et al.[40] investigated Na+Na$_2$ collisions while Cvitaš et al.[42,43] and Quéméner et al.[44] explored Li+Li$_2$ collisions with varying degrees of vibrational excitation of the molecule. Quéméner et al.[45] also examined spin-polarized collisions of K+K$_2$ system. More recently, we have looked into reactions involving non-alkali metal atom systems such as OH+O[46–49] and O(^{1}D)+H$_2$/D$_2$[50,51] with and without vibrational excitation of the molecules. These systems exhibit rate coefficients in the $10^{-11} - 10^{-10}$ cm^3 s^{-1} range, at least about an order of magnitude larger than barrier reactions. However, unlike barrier reactions, rate coefficients of barrierless reactions are less strongly influenced by initial vibrational excitation of the molecule. This is largely due to the statistical nature of such reactions. Also, more recently, methods based on multi-channel quantum defect theory (MQDT) have been applied to estimate overall reaction rate coefficients of barrierless reactions.[52,53] These methods have shown that, in some cases, ultracold reactions attain the universal regime, implying that, once the short-range region is reached the reaction occurs with 100% probability. A more rigorous description of ultracold chemical reactions within the MQDT framework has been recently formulated by Hazra et al.[54,55] In this approach, full quantum close-coupling (CC) calculations were performed at short-range and the MQDT formalism was employed at long-range. The method, demonstrated for the D+H$_2$ →HD+H collision with full rovibrational quantum state resolution of the HD product, was found to yield comparable results as full close-coupling calculations but at significantly less computational effort. Long-range theories based on the Langevin model have also been applied for the characterization of ultracold reactions though with mixed success.

Since cooling, trapping, manipulation of ultracold molecules involve external fields (electric, magnetic, optical or some combination of these), collisional studies of cold and ultracold molecules also need to include the effect of external fields. This, brings in significant additional complexity as splitting of energy levels due to

Stark or Zeeman effect needs to be taken into account. A number of such calculations have been reported for both atom-molecule and molecule-molecule collisions in recent years but due to the sheer complexity of such calculations, they have been largely restricted to rotational level or spin changing collisions in the lowest vibrational level. Application to helium collisions with NH, OH, O_2 and other molecules as well as self collisions of NH have been performed by a number of authors.[56–63] Krems and Dalgarno[59] have provided detailed formulation of atom-molecule and molecule-molecule collisions in external fields, taking open-shell character of both the atom and molecule collision partners. This approach makes use of the uncoupled formalism for the various angular momenta. However, Tscherbul and Dalgarno[64] have shown that one may still be able to use the total angular momentum representation in collisions in external fields though the total angular momentum is not a good quantum number (it is not conserved) in this case. However, this approach was shown to be computationally more efficient than the uncoupled formalism despite yielding some unphysical (ghost) eigenvalues. By far, most quantum close-coupling studies of molecular collisions in external fields have been applied to non-reactive collisions involving spin and hyperfine changes but not chemical reaction. The theory of chemical reactions in external field was first formulated by Tscherbul and Krems[65] and applied to the Li+HF reaction. However, due to the very large basis set required for such calculations, they have not been replicated for other systems, except for NH+NH collisions within the rigid rotor model by Janssen et al.[63] In recent years, methods based on long-range models and MQDT formalisms have become the preferred approach to describe chemical reactions in the presence of external electric and or magnetic field though they do not yield product quantum state resolved reaction rate coefficients. Quéméner and Bohn[66] have recently examined the effect of relative orientation of electric and magnetic fields in ultracold reactions within the MQDT formalism.

Here we provide an overview of recent studies of atom-molecule chemical reactions and inelastic molecule-molecule collisions with full-rovibrational quantum state resolution of the reaction products but in the absence of external field. For the chemical reaction studies, we focus on more recent work on barrierless reactions. This chapter is organized as follows: Section 2 provides an overview of the theoretical approach employed in calculations of barrierless chemical reactions. Results of reactive scattering calculations are presented in Section 3. In section 4, we highlight rovibrational energy transfer in ultracold molecule-molecule collisions taking H_2+H_2 system as an illustrated example. We discuss highly efficient energy transfer processes involving rotational and vibrational energy exchange in this system. A brief account of recent progress in MQDT-based approaches for ultracold reactions is given in Section 5. Conclusions and outlook are presented in section 6.

2. Methodology for Reactive Scattering

The theory of reactive scattering in atom-diatom systems is well-developed and has been described in great detail in many prior publications.[67-71] Only a brief account relevant to the present context is provided here. The formalism has been developed both in mass-scaled Jacobi coordinates $(R_\tau, r_\tau, \gamma_\tau$ where R_τ is the atom-molecule center of mass distance for a given atom-diatom arrangement channel τ, r_τ is the diatom distance, and γ_τ is the angle between R_τ and r_τ) and hyperspherical coordinates. While Jacobi coordinates are convenient and easy to use for non-reactive calculations, they are not ideal for reactive scattering studies as different sets of Jacobi coordinates need to be employed for the different atom-diatom fragments in a triatomic A+BC system. Therefore, hyperspherical coordinates are widely employed for describing atom-diatom exchange reactions. The hyperspherical coordinates involve a radial coordinate, $\rho = \sqrt{R_\tau^2 + r_\tau^2}$, denoted as hyper radius which is independent of the arrangement channel, and two hyper angles. Many different choices and conventions are adopted in the literature for the definition of the two hyper angles. In the Delves coordinate (DC)[67] system, they are chosen as $(\Theta_\tau, \gamma_\tau)$ where $\Theta_\tau = \arctan(r_\tau/R_\tau)$. In the adiabatically-adjusting principle axis hyperspherical (APH) coordinates of Pack and Parker,[69] hyper angles are (θ, ϕ) where θ is given in terms of the Jacobi coordinates and ϕ is a kinematic angle that varies continuously in its range. Note that the APH coordinates are independent of τ and they allow an evenhanded description of all three arrangement channels in an A+BC system compared to the DCs. This is important in the strong interaction region where all three atoms are in close proximity. In our approach, we adopt the formalism of Pack and Parker,[69] in which the APH coordinates (ρ, θ, ϕ) are used in the strong interaction region (inner chemically important region) and the DC[67] system is used in the outer region. Since the DCs are more closely related to the Jacobi coordinates, it is easier to match numerically computed solutions at large hyper radius to known asymptotic solutions in Jacobi coordinates. This requires mapping of the wave function matrix (or its log-derivative) at some intermediate distance outside the region of strong interaction from the APH to DC system and continued propagation in the DCs to an asymptotically large value of ρ.

The Hamiltonian for a triatomic system in APH coordinate systems is written as

$$H = -\frac{\hbar^2}{2\mu\rho^5}\frac{\partial}{\partial\rho}\rho^5\frac{\partial}{\partial\rho} + \frac{\hat{\Lambda}^2}{2\mu\rho^2} + V(\rho,\theta,\phi), \tag{1}$$

where μ is the three body reduced mass, $\hat{\Lambda}$ is the grand angular momentum operator,[72] and $V(\rho, \theta, \phi)$ is the adiabatic potential energy surface of the triatomic system. The eigenfunctions of $\hat{\Lambda}^2$ are known as the hyperspherical harmonics $Y_{\lambda JM}^{nm}(\omega)$.[72] In APH coordinates (in the inner region), the total nuclear wave function is expanded as[72]

$$\Psi_i^{JMpq}(\omega,\rho) = 4\sqrt{2}\sum_t \rho^{-5/2}\Gamma_t^{Jpq}(\rho)\Phi_t^{JMpq}(\omega;\rho_\xi), \tag{2}$$

where the superscripts p and q on Ψ denote, respectively, the inversion parity and particle exchange symmetry for identical particles within the triatomic system, and M denotes the projection of J on the space-fixed (SF) axis. The radial expansion coefficients, Γ^{Jpq}, explicitly depend on ρ while the five-dimensional (5D) surface functions, Φ^{JMpq}, depend on the two hyperangles θ and ϕ and three Euler angles, α, β and γ. These five angles are collectively denoted as ω. Due to presence of strong coupling between the radial (ρ) and the angular (θ, ϕ) parts within the PES, $V(\rho,\theta,\phi)$, one needs to apply a "sector adiabatic" approach for the solution of the Schrödinger equation in hyperspherical coordinates. In this approach, the entire range of hyperradius is partitioned into a large number of sectors and the 5D surface functions are evaluated at the center of each sector, ρ_ξ, assuming that they do not change within a sector. The surface functions and corresponding adiabatic energies, ε^{Jpq}, are obtained by solving the equation

$$\left[\frac{\hat{\Lambda}^2}{2\mu\rho_\xi^2} + \frac{15\hbar^2}{8\mu\rho_\xi^2} + V(\rho_\xi,\theta,\phi) - \varepsilon_t^{Jpq}\right]\Phi_t^{JMpq}(\omega;\rho_\xi) = 0 \tag{3}$$

at each sector. The surface functions, Φ^{JMp}, are further expanded in terms of primitive orthonormal basis sets in θ and ϕ according to

$$\Phi_t^{JMp} = \frac{1}{\sqrt{2\pi}}\sum_{l=\mu}^{l_{max}}\sum_{m=-m_{max}}^{m_{max}}\sum_{\Omega=-\Omega_{max}}^{\Omega_{max}} \tag{4}$$
$$\times\, d_{\mu,\nu}^l(\theta)e^{im\phi}\tilde{D}_{\Omega M}^J(\alpha,\beta,\gamma)$$

where $\tilde{D}_{\Omega M}^J$ are normalized Wigner rotational D functions, Ω is the projection of J on the body-fixed (BF) axis, and $d_{\mu,\nu}^l$[72] is expressed in terms of Jacobi polynomials $P_{l-\mu}^{(\mu-\nu,\mu+\nu)}(\cos\theta)$. The parameters l_{max} and m_{max} control the size of the basis sets in θ and ϕ. A hybrid discrete variable representation (DVR) in θ and a finite basis representation (FBR) in ϕ is used to solve the eigenvalue problem involving the surface function hamiltonian. An implicitly Restarted Lanczos Method and Sylvester algorithm are used for the diagonalization of the DVR Hamiltionian which includes tensor products of kinetic energy operators. Additionally, using a Sequential Diagonalization Truncation (SDT) technique the hamiltonian matrix is kept to a reasonable size.

Outside the region of strong interaction the coupled-channel equations resulting from the Schrödinger equation are solved using the DCs. In this case, the total wave function is expanded in a complete set of ρ_ξ dependent vibrational wavefunctions $\Upsilon_t^J(\theta_\tau;\rho_\xi)$, coupled angular functions $\mathcal{Y}_t^J$, and a radial expansion coefficient Γ_t^{JM}

to yield,

$$\Psi_i^{JM} = \frac{2}{\rho^{5/2} \sin 2\theta_\tau} \sum_t \Gamma_{ti}^J(\rho) \Upsilon_t^J(\theta_\tau; \rho_\xi) \mathcal{Y}_t^{JM}(\hat{S}, \hat{s}) \tag{5}$$

where i and t are the collective initial and intermediate vibrational and rotational quantum numbers, $\{v_\tau, j_\tau, \ell_\tau\}$. The "sector adiabatic" method is also employed here to calculate the surface functions $\Upsilon_n^J(\theta_\tau; \rho_\xi)$ and the corresponding adiabatic energies. These are computed using a one-dimensional Numerov propagator[73] at each sector. The Hamiltonian in the DC has the similar form as in APH except that the expression for $\hat{\Lambda}^2$ has a different form[73] and the variables of the three body PES are also different.

Solution of the Schrödinger equation is obtained by propagation of radial coupled channel equations from a small value of ρ in the classically forbidden region to a large asymptotic value of ρ. Here, we propagate the R-matrix, which is defined as $R(\rho) = \Gamma(\rho) \left[\frac{d\Gamma(\rho)}{d\rho} \right]^{-1}$ for each collision energy using the log-derivative method of Johnson.[75] At the boundary of each sector, the R-matrix is transformed from a sector centered at ρ_ξ to another sector centered at $\rho_{\xi+1}$ via an overlap matrix.[72,73] Since two sets of hyperspherical coordinates are employed, an intermediate transformation of the R-matrix is required from the last sector of the APH to the first sector of Delves by an unitary (orthogonal) transformation via

$$\mathbf{R}_{\text{Delves}}^J = \mathbf{U}^J \mathbf{R}_{\text{APH}}^J \tilde{\mathbf{U}}^J. \tag{6}$$

Explicit expression for $\mathbf{U}^J$ is given in Eq.(140) of Pack and Parker.[69] At $\rho = \rho_{max}$, the scattering boundary conditions are applied to the R-matrix to evaluate the reactance matrix $\mathbf{K}^J$ via $\mathbf{K}^J = (\mathbf{R}\mathcal{F} - \mathcal{B})^{-1}(\mathbf{R}\mathcal{E} - \mathcal{A})$ and the scattering matrix $\mathbf{S}^J$ from $\mathbf{K}^J$ using the transformation $\mathbf{S}^J = [I + i\mathbf{K}^J][I - i\mathbf{K}^J]^{-1}$. The scattering boundary conditions are applied in Jacobi coordinates and it requires a projection of channel wavefunctions from Delves to Jacobi coordinates. The details of the projection and explicit forms of the four matrices, $\mathcal{F}, \mathcal{B}, \mathcal{E}$ and $\mathcal{A}$, are given in Eqs. (116) to (121) of Pack and Parker.[69] Finally, for a given collision energy E_c and total angular momentum quantum number J, the elastic scattering probability $P_{i,\text{el}}^J(E_c)$, for an initial quantum state i can be obtained as

$$P_{i,\text{el}}^J(E_c) = P_{i \leftarrow i}^J(E_c) = |S_{i \leftarrow i}^J(E_c)|^2. \tag{7}$$

Similarly, by summing over all final states f, initial state-selected inelastic $P_{i,\text{in}}^J(E_c)$ and reactive scattering probabilities $P_{i,\text{re}}^J(E_c)$ are obtained via

$$P_{i,\text{in/re}}^J(E_c) = \sum_{f^{\text{in/re}}} P_{f \leftarrow i}^J(E_c) = \sum_{f^{\text{in/re}}} |S_{f \leftarrow i}^J(E_c)|^2. \tag{8}$$

Corresponding elastic, inelastic and reactive cross sections for initial rotational level $j = 0$ are evaluated according to

$$\sigma_{i,\text{el}}^{J}(E_c) = \frac{\pi}{k^2} \left| 1 - S_{i \leftarrow i}^{J}(E_c) \right|^2 \tag{9}$$

$$\sigma_{i,\text{in/re}}^{J}(E_c) = \frac{\pi}{k^2} \sum_{f^{\text{in/re}}} \left| S_{f \leftarrow i}^{J}(E_c) \right|^2 \tag{10}$$

where $k^2 = 2\mu_{A,BC} E_c / \hbar^2$ is the wave vector in the incident channel with the reduced mass $\mu_{A,BC} = \frac{m_A(m_B + m_C)}{m_A + m_B + m_C}$.

3. Results

Here we discuss properties of two benchmark chemical reactions at cold and ultra-cold temperatures. While these reactions have been extensively studied at higher temperatures only recently have their dynamics been explored in the ultracold regime.[46–51] The goal of these studies is to gain insights into barrierless chemical reactions in the sub-Kelvin regime and also to explore the effect of vibrational excitation of the reagent molecule on the reactivity at vanishingly small kinetic energies.

3.1. O+OH reaction

The hydroxyl radical is an important species in combustion chemistry, earth's atmosphere, and astrophysical environments. In the earth's lower atmosphere it acts as a powerful oxidant and it is the primary radiating species in the mesosphere where it is produced in vibrational levels as high as $v = 10$ in $H+O_3 \rightarrow OH+O_2$ reaction. In astrophysics it has been implicated as a key species in oxygen chemistry. Due to its sizable dipole-moment and open-shell character, the OH molecule has also become an important target for cooling and trapping experiments. Indeed, Stuhl et al.[74] have recently reported stark-deceleration followed by evaporative cooling of OH down to 50 mK, paving groundwork to potentially study chemical reactions involving a key chemical species at cold and ultracold temperatures.

Using the method outlined in section 2, we have carried out detailed investigation of the $OH(v)+O \rightarrow H+O_2$ reaction at temperatures ranging from $1\mu K$ to 100 K for vibrational levels $v = 0-3$.[50] Fig. 1 compares non-thermal rate coefficients obtained as the product of the cross section and relative velocity for the $v = 3$ reaction against non-reactive vibrational quenching and elastic scattering. It is seen that the rate coefficients for the reactive and vibrational quenching processes become finite in the limit of zero collision energy in accordance with the Wigner threshold law. The reactive and quenching rate coefficients are comparable though a slight preference is seen for reactive scattering at low collision energies. A comparison of rate coefficients for initial vibrational levels $v = 0 - 3$ as a function of the temperature is provided in Fig. 2.[49] While the rate coefficients increase with vibrational levels for $v = 1 - 3$,

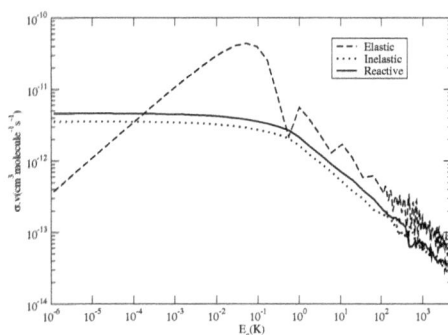

Fig. 1. Elastic (dashed curve), inelastic (dotted curves), and reactive (black solid curve) rate coefficients for O+OH($v = 3, j = 0$) collisions as a function of the collision energy in K for zero total angular momentum ($J = 0$).[50] Reproduced with permission.

they are all lower than that for $v = 0$ at low temperatures. The enhanced reactivity of $v = 0$ was attributed to the presence of a quasibound level of the O$\cdots$OH($v = 0$) system that lies close to the energy of the O+OH($v = 0$) threshold. Quasibound or virtual states that lie close to a channel threshold can significantly influence collisional outcomes at vanishingly small energies. As the vibrational level of OH is raised, the energy of the quasibound level is displaced from the channel threshold and its effect becomes less dramatic.

3.2. *O(^1D)+H$_2$ reaction*

The O(^1D)+H$_2$ system has long been considered as a benchmark reaction for barrierless chemical reactions. It has been the topic of numerous theoretical and experimental investigations. However, its dynamics has not been explored at temperatures lower than 100 K. Our recent studies[49] explored the reaction dynamics of this system in the cold and ultracold limit with and without vibrational excitation of the H$_2$ molecule. It was found that vibrational excitation of H$_2$ has almost no effect on reactivity. This was attributed to the statistical nature of the reaction once the deeply bound intermediate H$_2$O complex is formed. Fig. 3 shows a comparison of the rate coefficients for the reaction for vibrational levels $v = 0 - 2$.

Fig. 4 shows rotational distribution of the product OH molecules resulting from O(^1D)+H$_2$($v = 0, j = 0$) reaction in vibrational levels $v = 0 - 4$ at collision energies of 10^{-10} eV ($\sim 1.16\mu$K) and 0.1 eV (~ 1160 K). It is seen that the distribution is

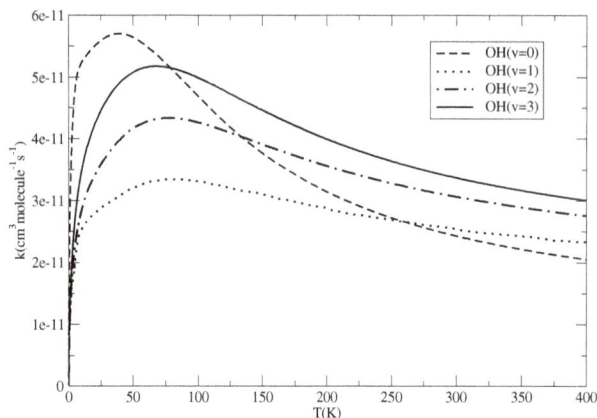

Fig. 2.   Reactive rate coefficients for OH($v = 0$) (dashed curve), OH($v = 1$) (dotted curve), OH($v = 2$) (dashed-dot curve), and OH($v = 3$) (solid curve) as a function of the temperature.[50] Reproduced with permission.

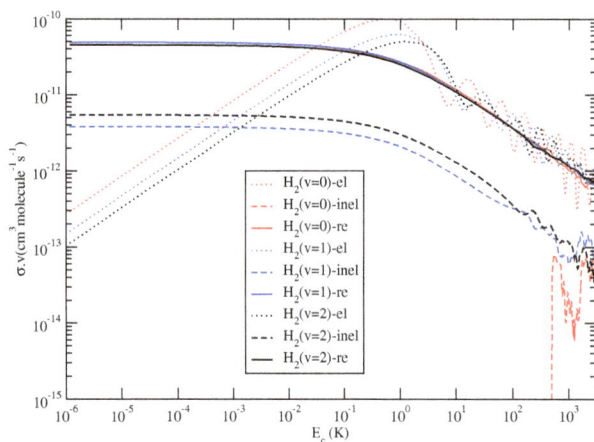

Fig. 3.   Rate coefficients for O(^1D)+H$_2$ collisions as functions of the incident kinetic energy in K for zero total angular momentum ($J = 0$). Red curve ($v = 0$); blue curve ($v = 1$); black curve ($v = 2$). Elastic rate coefficients are labeled by dotted lines, inelastic rate coefficients are labeled by dashed lines whereas reactive rate coefficients are labeled by solid lines.[49] Reproduced with permission.

dominated by a few high lying rotational levels in the ultracold regime whereas a range of rotational levels are comparably populated at the higher collision energy. This shows high specificity of the reaction at low collision energies.

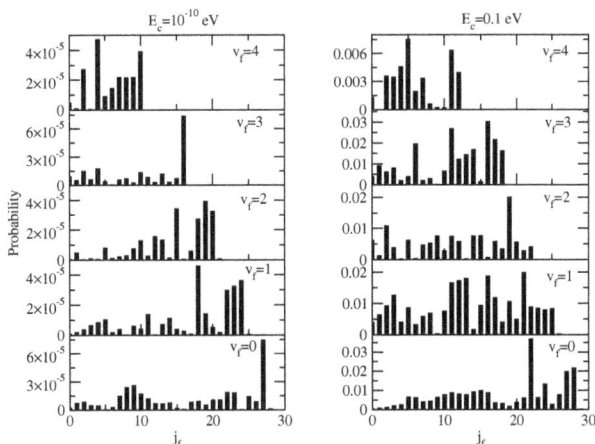

Fig. 4. Rotational distribution of the product OH molecule in $O(^1D)+H_2(v = 0, j = 0) \rightarrow OH(v_f, j_f)+H$ reaction at two different collision energies for zero total angular momentum $(J = 0)$. Left panel is for a collision energy of 10^{-10} eV and the right panel corresponds to a collision energy of 0.1 eV.[49] Reproduced with permission.

We have investigated the effect of hydrogen isotope in this reaction by considering reactivity of $O(^1D)+D_2$ for vibrational levels $v = 0 - 2$. Like H_2, the reactivity was found to be largely insensitive to the initial vibrational levels of D_2. Though not shown here, our results for $v = 0$ are in excellent agreement with available experimental results for temperatures in the 220-450 K range.[51] Unfortunately, there are no experimental results to compare with the theoretical predictions at lower temperatures. In Fig. 5 we show the kinetic isotope effect (KIE) defined as the ratio k_D/k_H, the ratio of the rate coefficients for the D_2 and H_2 reactions as a function of the temperature. The ratio remains fairly constant at about 0.75 for temperatures above 200 K but shows strong sensitivity to temperature below 100 K.

4. Quasiresonant Energy Transfer in Molecule–Molecule Collisions

While atom-molecule collisions have been extensively investigated at cold and ultracold temperatures, similar studies of molecule-molecule collisions, including rotational and vibrational levels of both molecules, have not been reported for many systems. In fact, the only system for which such studies have been reported at cold and ultracold temperatures is the H_2+H_2 system,[24-28] due to its low mass, closed-shell electronic structure, and the availability of an accurate potential energy surface that includes all six internal degrees of freedom. Unlike atom-diatom collisions, molecule-molecule collisions can involve exchange of rotational and/or vibrational energies between the two molecules that involve conservation of their rotational angular momentum and nearly the conservation of internal energy. For example, if the exchange of rotational energy involves two different vibrational levels of the

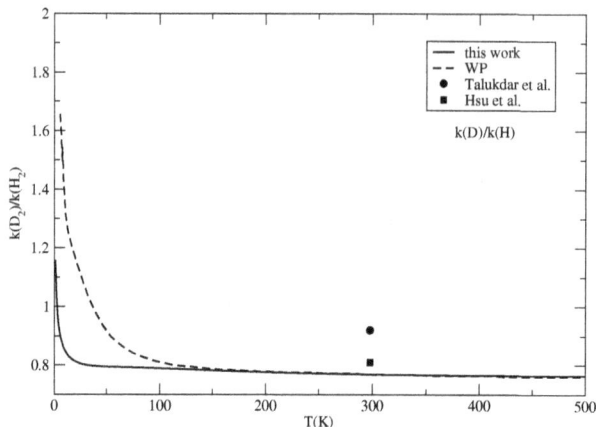

Fig. 5. The kinetic isotope effect for vibrational level $v = 0$ from the present calculation (solid black curve) is compared with the WP calculations of Sun *et al.*[76] (dashed red curve), experimental results of Talukdar *et al.*[77] (blue dot) and Hsu *et al.*[78] (green square). Reproduced with permission.

molecules, due to the slightly different centrifugal distortion of the vibrational levels, the transition involves a small energy gap and it becomes a quasi-resonant inelastic process. An example of such a quasi-resonant rotation-rotation (QRRR) transfer is $H_2(v = 1, j = 0) + H_2(v = 0, j = 2) \rightarrow H_2(v = 1, j = 2) + H_2(v = 0, j = 0)$ collision.[24,26] In this case a non-rotating vibrationally excited para-H_2 molecule in $v = 1$ picks up two quanta of rotational energy from a vibrationally ground state para-H_2 in $j = 2$ rotational level. The process has an energy defect of 24.45 K. Though many other final states are possible in this collision even in the limit of zero collision energy, the QRRR process dominates over all other energy transfer mechanism by more than two orders of magnitude. This is illustrated in Fig. 6 in which cross sections for the QRRR process from a full close-coupling calculation is compared with the total inelastic quenching cross section. It is seen that the total inelastic cross section is almost entirely composed of the QRRR process with the next leading inelastic process that involves pure rotational quenching contributing less than 1%.

Similar QRRR process has also been observed in ortho-H_2 collisions,[27] such as $H_2(v = 1, j = 1) + H_2(v = 0, j = 3) \rightarrow H_2(v = 1, j = 3) + H_2(v = 0, j = 1)$, where the vibrationally excited molecule picks up two quanta of rotation from the vibrationally cold molecule. This process has an energy defect of about 45 K and occurs with similar efficiency as that of the para-H_2 case. A different quasi-resonant mechanism occurs in collisions of an ortho-H_2 with a para-H_2 molecule. In this

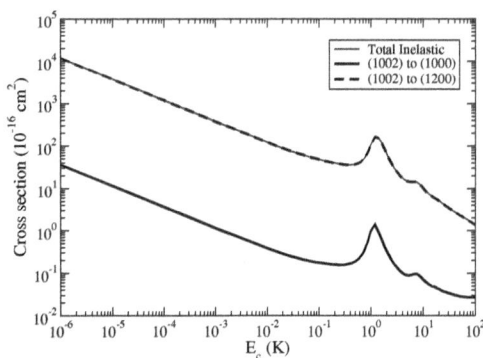

Fig. 6.   Comparison of total inelastic cross sectons for $H_2(v = 1, j = 0)+H_2(v = 0, j = 2)$ collisions with state-to-state cross sections corresponding to the near-resonant (1200) and the next leading (1000) final states as a function of collision energy.[26] Reproduced with permission.

case, since ortho-para transitions are not allowed, rotational energy exchange is not permitted and the quasi resonant mechanism involves a vibrational exchange,[28] as in $H_2(v = 1, j = 0)+H_2(v = 0, j = 1) \rightarrow H_2(v = 0, j = 0)+H_2(v = 1, j = 1)$. Though the energy defect for this process is only about 8.5 K, it is much less efficient than the QRRR process. This is because, pure vibrational energy transfer is driven by the isotropic part of the interaction potential and is generally much less efficient than pure rotational transfer. The latter is controlled by the angular anisotropy of the interaction potential. As the quasi-resonant processes become more efficient and specific at low energies, cold and ultracold molecules offer interesting opportunities for studying novel energy transfer processes in molecular collisions.

5. Long-Range Theories for Cold and Ultracold Collisions

While it is desirable to perform explicit close-coupling (CC) calculations of molecular collisions in the cold and ultracold regimes, such calculations are computationally intractable for most molecular systems when all quantum numbers and external field effects are included. As discussed in the introduction, the MQDT based methods have become more widespread in characterizing cold collisions. The MQDT formalism was originally developed by Seaton and Fano[79–81] to understand the spectra of Rydberg atoms. Since then, it has been successfully extended to more general contexts[82,83] and applied to both resonant and non-resonant scattering in a variety of atomic collision processes,[84–89] ion-atom collisions,[90,91] atom-molecule systems,[92,93] and molecule-molecule reactive scattering.[52,53,94,95] For ultracold chemical reactions within the MQDT framework two complementary approaches are used. In one case, inclusion of the short-range interaction is circumvented with suitable parametrized boundary conditions.[52] This introduces a scattering phase shift from which the total reaction probability can be estimated. This approach essentially assumes a 100%

absorption at short-range and the Schrödinger equation is solved in the long-range potential. In the second approach, the reaction rate coefficients are analytically estimated from the quantum threshold (QT) laws that are applicable in the presence of an electric field.[96] While these approaches are reasonable for extracting the total reaction rate coefficient they do not yield product resolved rate coefficients or discriminate among products if more than one product channel is involved in the reaction.

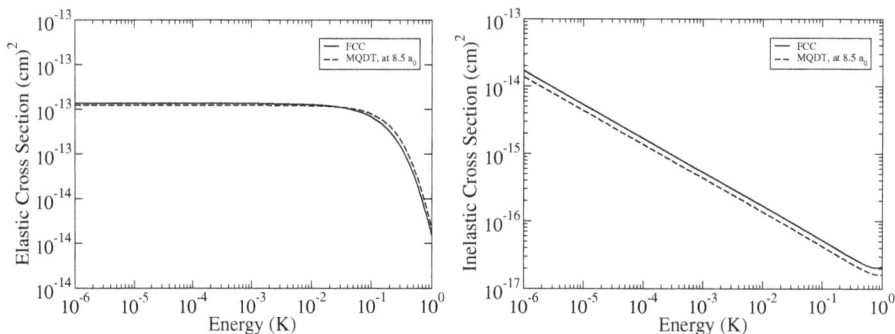

Fig. 7.   Elastic (left) and inelastic (right) cross sections for the scattering of two para-H_2 molecules, $H_2(v = 0, j = 2)+H_2(v = 0, j = 0)$. The inelastic process corresponds to quenching of the rotationally excited molecule. The black solid curves represent the FCC calculation, while the dashed curves denote the MQDT results at a matching distance of 8.5 a_0.

We have recently proposed an alternative approach in which full close-coupling (FCC) calculation is performed at short-range and the MQDT formalism is applied in the long-range.[54,55] This hybrid CC-MQDT approach was shown to yield comparable results as full close-coupling calculations for the quasi-resonant transitions in H_2-H_2 discussed above and chemical reaction in $D+H_2(v, j)$ collisions for vibrational levels $v = 0 - 7$. Since the CC calculations are restricted to a relatively short radial distance in the hybrid approach, the method is computationally less demanding than full CC calculations, yet capable of yielding the same information as the latter. A comparison between the FCC calculation and the hybrid CC-MQDT approach is presented in Fig. 7 for scattering of two para-para H_2 molecules. This is a pure rotational de-excitation process where a rotationally excited para-$H_2(v = 0, j = 2)$ molecule hits another rotationless para-$H_2(v = 0, j = 0)$ molecule and yields two ground state para H_2 molecules. The main inelastic transition involves $(v_1, j_1, v_2, j_2) = (0, 0, 0, 2) \rightarrow (0, 0, 0, 0)$. To account for the dominant s-wave scattering at ultralow energies, only cross sections for total angular momentum $J = 2$ are evaluated for both elastic and inelastic collisions. The solid black curves denote the FCC result for both elastic (left panel) and inelastic (right panel) collisions obtained at an asymptotic matching distance of $R_\infty = 100$ a_0. The

MQDT results (dashed curves) are derived from a short-range K-matrix obtained by matching the log-derivative matrix of the CC wave functions with MQDT reference functions at 8.5 a_0. The details of the matching procedure and evaluation of the scattering cross sections from the short-range K-matrix are given in Hazra et al.[54] It is clear from the figure that the MQDT calculations nearly reproduce the results of FCC calculation for a matching distance of 8.5 a_0 for collisional energies ranging from 1μ K to 1K. We hope to extend the formalism for open shell systems and collisions in external fields in the near future.

6. Conclusions and outlook

As the field of cold and ultracold molecules continues to evolve a detailed understanding of their collisions and interactions is required for harnessing their use in many overlapping areas such as quantum information, quantum computation, many-body quantum simulations, quantum control of chemical reactions, etc. We have discussed recent progress in accurate description of inelastic collisions and reactions of cold and ultracold molecules taking examples of chemically relevant systems that are well studied at higher temperatures. We believe that the continued developments in both theoretical and experimental fronts will unravel many more fascinating aspects of cold and ultracold molecules that are yet to be understood.

Acknowledgments

This work was supported in part by NSF grant PHY-1205838 (N.B.) and ARO MURI grant No. W911NF-12-1-0476. We thank John Bohn, Brandon Ruzic, Phillip Stancil, Robert Forrey, and Brian Kendrick for productive collaboration on many aspects of the work described here.

References

1. H. L. Bethlem and G. Meijer, Production and application of translationally cold molecules, *Int. Rev. Phys. Chem.* **22**, 73–128 (2003).
2. R. V. Krems, Molecules near absolute zero and external field control of atomic and molecular dynamics, *Int. Rev. Phys. Chem.* **24**, 99–118 (2005).
3. P. F. Weck and N. Balakrishnan, Importance of long-range interactions in chemical reactions at cold and ultracold temperatures, *Int. Rev. Phys. Chem.* **25**, 283–311 (2006).
4. R. V. Krems, Cold controlled chemistry, *Phys. Chem. Chem. Phys.* **10**, 4079–4092 (2008).
5. G. Quéméner, N. Balakrishnan, and A. Dalgarno, *in Cold Molecules: Theory, Experiment, Applications, R.Krems, W. C. Stwalley, and B. Friedrich (Eds.)*, CRC Press (2009), pp.69-124.
6. L. D. Carr, D. DeMille, R. V. Krems, and J. Ye, Cold and ultracold molecules: science, technology and applications, *New. J. Phys.* **11**, 055049 (2009).

7. S. Y. T. van de Meerakker, H. L. Bethlem, N. Vanhaecke, and G. Meijer, Manipulation and Control of Molecular Beams, *Chem. Rev.* **112**, 4828–4878 (2012).

8. G. Quéméner and P. S. Julienne, Ultracold Molecules under Control, *Chem. Rev.* **112**, 4949–5011 (2012).

9. S. Ospelkaus, K.-K. Ni, D. Wang, M. H. G. de Miranda, B. Neyenhuis, G. Quéméner, P. S. Julienne, J. L. Bohn, D. S. Jin, and J. Ye, Quantum-State Controlled Chemical Reactions of Ultracold Potassium-Rubidium Molecules, *Science* **327**, 853–857 (2010).

10. J. D. Weinstein, R. deCarvalho, T. Guillet, B. Friedrich, and J. M. Doyle, Magnetic trapping of calcium monohydride molecules at millikelvin temperatures, *Nature* **395**, 148–150 (1998).

11. W. C. Campbell, E. Tsikata, H.-I. Lu, L. D. van Buuren, and J. M. Doyle, Magnetic trapping and Zeeman relaxation of NH($X\ ^3\Sigma^-$), *Phys. Rev. Lett.* **98**, 213001 (2007).

12. S. Y. T. van de Meerakker, P. H. M. Smeets, N. Vanhaecke, R. T. Jongma, and G. Meijer, Deceleration and electrostatic trapping of OH radicals, *Phys. Rev. Lett.* **94**, 023004 (2005).

13. B. C. Sawyer, B. L. Lev, E. R. Hudson, B. K. Stuhl, M. Lara, J. L. Bohn, and J. Ye, Magnetoelectrostatic trapping of ground state OH molecules, *Phys. Rev. Lett.* **98**, 253002 (2007).

14. S. Hoekstra, M. Metsälä, P. C. Zieger, L. Scharfenberg, J. J. Gilijamse, G. Meijer, and S. Y. T. van de Meerakker, Electrostatic trapping of metastable NH molecules, *Phys. Rev. A* **76**, 063408 (2007).

15. J. T. Bahns, W. Stwalley, and P. L. Gould, Formation of cold (T $\leq$ 1K) molecules, *Adv. At. Mol. Opt. Phys.* **42**, 171–224 (2000).

16. F. Masnou–Seeuws and P. Pillet, Formation of ultracold molecules (T $<$ 200 μK) via photoassociation in a gas of laser-cooled atoms, *Adv. At. Mol. Opt. Phys.* **47**, 53–127 (2001).

17. E. A. Donley, N. R. Claussen, S. T. Thompson, and C. E. Wieman, Atom-molecule coherence in a Bose-Einstein condensate, *Nature* **417**, 529–533 (2002).

18. C. Chin, R. Grimm, P. Julienne, and E. Tiesinga, Feshbach resonances in ultracold gases, *Rev. Mod. Phys.* **82**, 1225 (2010).

19. E. Narevicius and M. G. Raizen, Toward cold chemistry with magnetically decelerated supersonic beams, *Chem. Rev.* **112**, 4879–4889 (2012).

20. Y. Shagam and E. Narevicius, Sub-Kelvin collision temperatures in merged neutral beams by correlation in phase-space, *J. Phys. Chem. C* **117**, 22454–22461 (2013).

21. E. Lavert-Ofir, Y. Shagam, A. B. Henson, S. Gersten, J. Kłos, P. S. Żuchowski, J. Narevicius, and E. Narevicius, Observation of the isotope effect in sub-kelvin reaction, *Nature Chemistry* **6**, 332–335 (2014).

22. N. Balakrishnan, V. Kharchenko, R. C. Forrey, and A. Dalgarno, Complex scattering lengths in multi-channel atom-molecule collisions, *Chem. Phys. Lett.* **280**, 5–9 (1997).

23. N. Balakrishnan, R. C. Forrey, and A. Dalgarno, Quenching of H_2 vibrations in ultra-cold ^{3}He and ^{4}He collisions, *Phys. Rev. Lett.* **80**, 3224–3227 (1998).

24. G. Quéméner, N. Balakrishnan, and R. V. Krems, Vibrational energy transfer in ultracold molecule-molecule collisions, *Phys. Rev. A* **77**, 030704(R) (2008).

25. G. Quéméner and N. Balakrishnan, Quantum calculations of H_2-H_2 collisions: from ultracold to thermal energies, *J. Chem. Phys.* **130**, 114303 (2009).

26. N. Balakrishnan, G. Quéméner, R.C. Forrey, R. J. Hinde and P. C. Stancil, Full-dimensional quantum dynamics calculations of H_2-H_2 collisions, *J. Chem. Phys.* **134**, 014301 (2011).

27. S. Fonseca dos Santos, N. Balakrishnan, S. Lepp, Quéméner, R. C. Forrey, R. J. Hinde, and P. C. Stancil, Quantum dynamics of rovibrational transitions in H_2-H_2

collisions: Internal energy and rotational angular momentum conservation effects, *J. Chem. Phys.* **134**, 214303 (2011).

28. S. Fonseca dos Santos, N. Balakrishnan, R. C. Forrey, and P. C. Stancil, Vibration-vibration and vibration-translation energy transfer in H_2-H_2 collisions: a critical test of experiment with full-dimensional quantum dynamics, *J. Chem. Phys.* **138**, 104302 (2013).

29. N. Balakrishnan and A. Dalgarno, Chemistry at ultracold temperatures, *Chem. Phys. Lett.* **341**, 652–656 (2001).

30. E. Bodo, F. A. Gianturco, and A. Dalgarno, F + D_2 reaction at ultracold temperatures, *J. Chem. Phys.* **116**, 9222–9227 (2002).

31. N. Balakrishnan and A. Dalgarno, On the isotope effect in F + HD reaction at ultracold temperatures, *J. Phys. Chem. A* **107**, 7101–7105 (2003).

32. E. Bodo, F. A. Gianturco, N. Balakrishnan, and A. Dalgarno, Chemical reactions in the limit of zero kinetic energy: virtual states and Ramsauer minima in F + H_2 → HF + H, *J. Phys. B: At. Mol. Opt. Phys.* **37**, 3641–3648 (2004).

33. N. Balakrishnan, On the role of van der Waals interaction in chemical reactions at low temperatures, *J. Chem. Phys.* **121**, 5563–5566 (2004).

34. P. F. Weck and N. Balakrishnan, Quantum dynamics of the Li+HF → H+LiF reaction at ultralow temperatures, *J. Chem. Phys.* **122**, 154309 (2005).

35. P. F. Weck and N. Balakrishnan, Heavy atom tunneling in chemical reactions: Study of H+LiF collisions, *J. Chem. Phys.* **122**, 234310 (2005).

36. P. F. Weck and N. Balakrishnan, Reactivity enhancement of ultracold $O(^3P)$+H_2 collisions by van der Waals interactions, *J. Chem. Phys.* **123**, 144308 (2005).

37. K. Stark and H.-J. Werner, An accurate multireference configuration interaction calculation of the potential energy surface for the F+H_2 → HF+H reaction, *J. Chem. Phys.* **104**, 6515–6530 (1996).

38. I. Simbotin, S. Ghosal, and R. Côté, Threshold resonance effects in reactive processes, *Phys. Rev. A* **89**, 040701(R) (2014).

39. P. Soldán, M. T. Cvitaš, J. M. Hutson, P. Honvault, and J.-M. Launay, Quantum dynamics of ultracold Na + Na_2 collisions, *Phys. Rev. Lett.* **89**, 153201 (2002).

40. G. Quéméner, P. Honvault, and J.-M. Launay, Sensitivity of the dynamics of Na + Na_2 collisions on the three-body interaction at ultralow energies, *Eur. Phys. J. D* **30**, 201–207 (2004).

41. M. T. Cvitaš, P. Soldán, J. M. Hutson, P. Honvault, and J.-M. Launay, Ultracold collisions involving heteronuclear alkali metal dimers, *Phys. Rev. Lett.* **94**, 200402 (2005).

42. M. T. Cvitaš, P. Soldán, J. M. Hutson, P. Honvault, and J.-M. Launay, Ultracold Li + Li_2 collisions: Bosonic and fermionic cases, *Phys. Rev. Lett.* **94**, 033201 (2005).

43. M. T. Cvitaš, P. Soldán, J. M. Hutson, P. Honvault, and J.-M. Launay, Interactions and dynamics in Li+Li_2 ultracold collisions, *J. Chem. Phys.* **127**, 074302 (2007).

44. G. Quéméner, J.-M. Launay, and P. Honvault, Ultracold collisions between Li atoms and Li_2 diatoms in high vibrational states, *Phys. Rev. A* **75**, 050701(R) (2007).

45. G. Quéméner, P. Honvault, J.-M. Launay, P. Soldán, D. E. Potter, and J. M. Hutson, Ultracold quantum dynamics: Spin-polarized K+K_2 collisions with three identical bosons or fermions, *Phys. Rev. A* **71**, 032722 (2005).

46. G. Quéméner, N. Balakrishnan, and B. K. Kendrick, Quantum dynamics of the O+OH→H+O_2 reaction at low temperatures, *J. Chem. Phys.* **129**, 224309 (2008).

47. G. Quéméner, N. Balakrishnan, and B. K. Kendrick, Formation of molecular oxygen in ultracold O+OH collisions, *Phys. Rev. A* **79**, 022703 (2009).

48. J. C. Juanes-Marcos, G. Quéméner, B. K. Kendrick, and N. Balakrishnan, Ultracold collisions and reactions of vibrationally excited OH radicals with oxygen atoms, *Phys. Chem. Chem. Phys.* **13**, 19067–19076 (2011).

49. G. B. Pradhan, N. Balakrishnan and B. K. Kendrick, Ultracold collisions of $O(^1D)$ and H_2: The effects of H_2 vibrational excitation on the production of vibrationally and rotationally excited OH, *J. Chem. Phys.* **138**, 164310 (2013).

50. G. B. Pradhan, J. C. Juanes-Marcos, N. Balakrishnan and B. K. Kendrick, Chemical reaction versus vibrational quenching in low energy collisions of vibrationally excited OH with O, *J. Chem. Phys.* **139**, 194305 (2013)

51. G. B. Pradhan, N. Balakrishnan and B. K. Kendrick, Quantum dynamics of $O(^1D)+D_2$ reaction: isotope and vibrational excitation effects, *J. Phys. B: At. Mol. Opt. Phys.* **47**, 135202 (2014).

52. Z. Idziaszek and P. S. Julienne, Universal Rate Constants for Reactive Collisions of Ultracold Molecules, *Phys. Rev. Lett.* **104**, 113202 (2010).

53. B. Gao, Universal Model for Exoergic Bimolecular Reactions and Inelastic Processes, *Phys. Rev. Lett.* **105**, 263203 (2010).

54. J. Hazra, B. Ruzic, N. Balakrishnan, and J. L. Bohn, Multichannel quantum defect theory for ro-vibrational transitions in ultracold molecule-molecule collisions, *Phys. Rev. A* **90**, 032711 (2014).

55. J. Hazra, B. Ruzic, J. L. Bohn, and N. Balakrishnan, Quantum Defect Theory for cold chemistry with product-quantum-state resolution, *Phys. Rev. A* **90**, 062703 (2014).

56. A. Volpi and J. L. Bohn, Magnetic-field effects in ultracold molecular collisions, *Phys. Rev. A* **65**, 052712 (2002).

57. A. V. Avdeenkov and J. L. Bohn, Collisional dynamics of ultracold OH molecules in an electrostatic field, *Phys. Rev. A* **66**, 052718 (2002).

58. A. V. Avdeenkov and J. L. Bohn, Linking Ultracold Polar Molecules, *Phys. Rev. Lett.* **90**, 043006 (2003).

59. R. V. Krems and A. Dalgarno, Quantum mechanical theory of atom-molecule and molecular collisions in a magnetic filed: spin depolarization, *J. Chem. Phys.* **120**, 2296–2307 (2004).

60. C. Ticknor and J. L. Bohn, Influence of magnetic fields on cold collisions of polar molecules, *Phys. Rev. A* **71**, 022709 (2005).

61. G. Quéméner, N. Balakrishnan, and A. Dalgarno, *in Cold Molecules: Theory, Experiment, Applications*, R. V. Krems, W. C. Stwalley, and B. Friedrich (Eds.), CRC Press (2009), pp.125-166.

62. M. L. González-Martínez and J. M. Hutson, Ultracold atom-molecule collisions and bound states in magnetic fields: Tuning zero-energy Feshbach resonances in He-NH($X^3\Sigma^-$), *Phys. Rev. A* **75**, 022702 (2007).

63. L. M. C. Janssen, A. van der Avoird, and G. C. Groenenboom, Quantum reactive scattering of Ultracold NH($X^3\Sigma^-$) radicals in a magnetic trap, *Phys. Rev. Lett.* **110**, 063201 (2013).

64. T. V. Tscherbul and A. Dalgarno, Quantum theory of molecular collisions in a magnetic field: Efficient calculations based on the total angular momentum representation, *J. Chem. Phys.* **133**, 184104 (2010).

65. T. V. Tscherbul and R. V. Krems, Quantum theory of chemical reactions in the presence of electromagnetic fields, *J. Chem. Phys.* **129**, 034112 (2008).

66. G. Quéméner and J. L. Bohn, Ultracold molecular collisions in combined electric and magnetic fields, *Phys. Rev. A* **88**, 012706 (2013).

67. L. M. Delves, Tertiary and general-order collisions, *Nucl. Phys.* **9**, 391–399 (1959); Tertiary and general-order collisions (II), **20**, 275–308 (1960).

68. W. H. Miller, Coupled equations and the minimum principle for collisions of an atom and a diatomic molecule, including rearrangements, *J. Chem. Phys.* **50**, 407 (1969).
69. R. T Pack and G. A. Parker, Quantum reactive scattering in three dimensions using hyperspherical (APH) coordinates. Theory, *J. Chem. Phys.* **87**, 3888–3921 (1987), and references therein.
70. B. Lepetit, J. M. Launay and M. Le. Dourneuf, Quantum study of electronically non-adiabatic collinear reactions. I. Hyperspherical description of the electronuclear dynamics, *Chem. Phys.* **106**, 103–110 (1986).
71. J. M. Launay and M. Le. Dourneuf, Hyperspherical close-coupling calculation of integral cross sections for the reaction $H+H_2 \rightarrow H_2+H$, *Chem. Phys. Letts.* **163**, 178–188 (1989).
72. B. K. Kendrick, R. T. Pack, R. B. Walker, and E. F. Hayes, Hyperspherical surface functions for nonzero total angular momentum. I. Eckart singularities, *J. Chem. Phys.* **110**, 6673–6693 (1999) and references therein.
73. G. A. Parker, R. B. Walker, B. K. Kendrick and R. T. Pack, Accurate quantum calculations on three-body collisions in recombination and collision-induced dissociation. I. Converged probabilities for the $H+Ne_2$ system, *J. Chem. Phys.* **117**, 6083–6102 (2002).
74. B. K. Stuhl, M. T. Hummon, M. Yeo, G. Quéméner, J. L. Bohn, and J. Ye, Evaporative cooling of the dipolar hydroxyl radical, Nature **492**, 396–400 (2012).
75. B. R. Johnson, The multichannel log-derivative method for scattering calculations *J. Comput. Phys.* **13**, 445–449 (1973).
76. Sun Z, Lin S Y and Zheng Y, Adiabatic and non-adiabatic quantum dynamics calculation of $O(^1D) + D_2 \rightarrow OD + D$ reaction, *J. Chem. Phys.* **135**, 234301 (2011).
77. R. K. Talukdar and A. R. Ravishankara, Rate coefficients for $O(^1D) + H_2$, D_2, HD reactions and H atom yield in $O(^1D) + HD$ reaction *Chem. Phys. Lett.* **253**, 177–183 (1996).
78. Y.-T. Hsu, J.-H. Wang and K. Liu, Reaction dynamics of $O(^1D)+ H_2$, D_2, and HD: Direct evidence for the elusive abstraction pathway and the estimation of its branching *J. Chem. Phys.* **107**, 2351–2356 (1997).
79. M. J. Seaton, Quantum defect theory I. General formulation, *Proc. Phys. Soc.* **88**, 801–814 (1966).
80. M. J. Seaton, Quantum defect theory, *Rep. Prog. Phys.* **46**, 167–257 (1983).
81. U. Fano and A. R. P. Rau, *Atomic Collisions and Spectra* (Academic Press, Orlando, FL, 1986).
82. F. H. Mies, A multichannel quantum defect analysis of diatomic predissociation and inelastic atomic scattering, *J. Chem. Phys.* **80**, 2514–2525 (1984).
83. C. H. Greene, A. R. P. Rau, and U. Fano, General form of the quantum-defect theory. II, *Phys. Rev. A* **26**, 2441–2459 (1984).
84. J. P. Burke, C. H. Greene, and J. L. Bohn, Multichannel cold collisions: simple dependences on energy and magnetic field, *Phys. Rev. Lett.* **81**, 3355–3358 (1998).
85. F. H. Mies and M. Raoult, Analysis of threshold effects in ultracold atomic collisions, *Phys. Rev. A* **62**, 012708 (2000).
86. M. Raoult and F. H. Mies, Feshbach resonance in atomic binary collisions in the Wigner threshold law regime, Phys. Rev. A **70**, 012710 (2004).
87. B. Gao, E. Tiesinga, C. J. Williams, and P. S. Julienne, Multichannel quantum-defect theory for slow atomic collisions, *Phys. Rev. A* **72**, 042719 (2005).
88. T. M. Hanna, E. Tiesinga, and P. S. Julienne, Prediction of Feshbach resonances from three input parameters, *Phys. Rev. A* **79**, 040701 (2009).

89. B. P. Ruzic, C. H. Greene, and J. L. Bohn, Quantum defect theory for high-partial-wave cold collisions, *Phys. Rev. A* **87**, 032706 (2013).

90. Z. Idziaszek, T. Calarco, P. S. Julienne, and A. Simoni, Quantum theory of ultracold atom-ion collisions, *Phys. Rev. A* **79**, 010702 (2009).

91. B. Gao, Universal Properties in Ultracold Ion-Atom Interactions, *Phys. Rev. Lett.* **104**, 213201 (2010).

92. J. F. E. Croft, A. O. G. Wallis, J. M. Hutson, and P. S. Julienne, Multichannel quantum defect theory for cold molecular collisions, *Phys. Rev. A* **84**, 042703 (2011).

93. J. F. E. Croft, J. M. Hutson, and P. S. Julienne, Optimized multichannel quantum defect theory for cold molecular collisions , *Phys. Rev. A* **86**, 022711 (2012).

94. Z. Idziaszek, G. Quéméner, J. L. Bohn, and P. S. Julienne, Simple quantum model of ultracold polar molecule collisions, *Phys. Rev. A* 82, 020703(R) (2010).

95. G. R. Wang, T. Xie, Y. Huang, W. Zhang, and S. L. Cong, Quantum defect theory for the van der Waals plus dipole-dipole interaction, *Phys. Rev. A* **86**, 062704 (2012).

96. G. Quéméner and John L. Bohn, Strong dependence of ultracold chemical rates on electric dipole moments, *Phys. Rev. A* **81** 022702 (2010).

Index

www.ingramcontent.com/pod-product-compliance
Lightning Source LLC
Chambersburg PA
CBHW081512190326
41458CB00015B/5352